T0140100

Intelligent Systems Reference Library

Volume 79

Series editors

Janusz Kacprzyk, Polish Academy of Sciences, Warsaw, Poland
e-mail: kacprzyk@ibspan.waw.pl

Lakhmi C. Jain, University of Canberra, Canberra, Australia, and
University of South Australia, Adelaide, Australia
e-mail: Lakhmi.Jain@unisa.edu.au

About this Series

The aim of this series is to publish a Reference Library, including novel advances and developments in all aspects of Intelligent Systems in an easily accessible and well structured form. The series includes reference works, handbooks, compendia, textbooks, well-structured monographs, dictionaries, and encyclopedias. It contains well integrated knowledge and current information in the field of Intelligent Systems. The series covers the theory, applications, and design methods of Intelligent Systems. Virtually all disciplines such as engineering, computer science, avionics, business, e-commerce, environment, healthcare, physics and life science are included.

More information about this series at http://www.springer.com/series/8578

Animesh Adhikari · Jhimli Adhikari

Advances in Knowledge Discovery in Databases

 Springer

Animesh Adhikari
Department of Computer Science
Parvatibai Chowgule College
Margao, Goa
India

Jhimli Adhikari
Department of Computer Science
Narayan Zantye College
Bicholim, Goa
India

ISSN 1868-4394   ISSN 1868-4408   (electronic)
Intelligent Systems Reference Library
ISBN 978-3-319-36606-7   ISBN 978-3-319-13212-9   (eBook)
DOI 10.1007/978-3-319-13212-9

Springer International Publishing AG Switzerland is part of Springer Science+Business Media (www.springer.com)

Preface

Data mining and knowledge discovery is a fast growing field of study. Numerous applications of data mining are reported on various domains. Due to significant advancements in collection and storage technologies, data mining techniques remain important tools in knowledge discovery and decision-making processes. People have now realized the importance of data, and with the help of data mining techniques, data become more meaningful than before.

This book presents research results on transactional data, time-stamped data, and multiple databases. Also, data mining applications are found on different other data such as uncertain data, social network data, sensor data, biological data, high-dimensional data, imbalance data, big data, privacy preserving data, text data, heterogeneous data, noisy data, outliers, and graph data.

Data analysis using market basket data dominated at the early stages, and research results are now also being reported. Mining time-stamped data seems to be a natural activity as most of the data generated are related to time. Mining multiple databases is relatively a recent topic of data mining. Various research results started coming since the year 2003. We present here our research work in the field of knowledge discovery in databases (KDD) done during the last ten years.

The authors of this book extend their gratitude to Professor Lakhmi Jain for recognizing our work, and Dr. Thomas Ditzinger for accepting our work on behalf of Springer. We thank Ms. Gajalakshmi for coordinating the production process. We thank Ms. Mathangi for overseeing the entire production process.

India

Animesh Adhikari
Jhimli Adhikari

Contents

Authors' Biography

Animesh Adhikari is an Associate Professor in the Department of Computer Science, Parvatibai Chowgule College, Goa, India. He received the Master of Technology and Ph.D. degrees, both in Computer Science, from Indian Statistical Institute and Goa University, respectively. His areas of interest include data mining, decision support systems, database systems, and artificial intelligence. He has published two research monographs, one book review, twelve international journal papers and six international conferences papers.

Jhimli Adhikari is an Assistant Professor in the Department of Computer Science, Narayan Zantye College, Bicholim, Goa, India. She received her Master of Computer Application and Ph.D. degrees from Jadavpur University and Goa University, respectively. Her areas of interest include data mining and artificial intelligence. She has published one research monograph, five international journal papers, and two international conferences papers.

Chapter 1
Introduction

Knowledge discovery in databases remains an active area of research since its inception in the late 1980s. Numerous applications of knowledge discovery in databases are reported in various domains. One of the popular domains is market basket data and the early research results are based on it. On the other hand, the time-stamped data are being generated continuously as time component is related to virtually all data. As a result, it remains an active area of research over time. But, research reports on mining multiple related databases started coming in early 2000s. The domain of multiple related databases is expanding since many applications generate multiple databases. This book presents some advances in the above three areas. This chapter ends with a note of future trends of data mining.

1.1 Background

Knowledge Discovery in Databases (KDD) is an automatic, exploratory analysis and modeling of large data repositories. KDD is the organized process of identifying valid, novel, useful, and understandable patterns from large and complex data sets (Maimon and Rokach 2010; Han et al. 2011). The term KDD was first coined by Gregory Piatetsky-Shapiro at the International Joint Conference on Artificial Intelligence, 1989 in Detroit, USA (Zhang and Wu 2011). A set of databases or a large set of data is a source of knowledge for an organization; data mining and knowledge discovery becomes a natural activity. With the advancement of technologies, data collection channels become more cost effective in collecting a huge amount of data. People have realized that a large source of data could be a source of meaningful knowledge, and that can be used in the decision making processes. A simple SQL query or on-line analytic processing (OLAP) might not be sufficient for many complex data analyses. Thus, knowledge discovery in databases become an urgent necessity of modern society.

Research of data mining started in late 1980s. Piatetsky-Shapiro and Frawley (1991) edited a book on early collection of research papers on KDD. Later research results are edited by Fayyad et al. (1996) in their book *Advances in Knowledge*

© Springer International Publishing Switzerland 2015
A. Adhikari and J. Adhikari, *Advances in Knowledge Discovery in Databases*,
Intelligent Systems Reference Library 79, DOI 10.1007/978-3-319-13212-9_1

Discovery and Data Mining. Since then many intelligent and interesting algorithms are proposed for mining a database (Agrawal and Srikant 1994; Han et al. 2000; Savasere et al. 1995). Many popular text books are now available in the market (see at the link http://www.kdnuggets.com/publications/books.html).

Applications of data mining are found in different areas of computing, and it has now become a frontier area of research of computing. Numerous real-world applications of data mining techniques in a variety of areas, including human performance, geospatial, bioinformatics, on- and off-line customer transaction activity, security-related computer audits, network traffic, text and image, and manufacturing quality [see Part III of Ye (2003)]. A recent coverage of data mining techniques and applications is presented by Maimon and Rokach (2010).

1.2 The Proposed Book

This book presents advances in data mining in the following areas:

1. Market basket database
2. Time-stamped databases
3. Multiple related databases

Data analysis using a single database is presented in Chaps. 2–5. Analysis using time-stamped databases is done in Chaps. 7, 9, 13 and 14. The remaining chapters deal with multiple databases, where some chapters dealing time-stamped databases also deal with multiple databases. A brief note of each of the above topics is give below.

1.2.1 Mining Market Basket Database

Frequent pattern mining is a fundamental operation behind several common data mining tasks. It leads to mining associations and correlations between itemsets. More than 15 years have passed since Agrawal et al. (1993) introduced support-confidence framework for mining association rules in a database. Since then, there has been an orchestrated effort focused on a variety of ways of making the data mining in large databases as efficient as possible. With this regard, many interesting data mining algorithms (Agrawal and Srikant 1994; Coenen et al. 2004; Han et al. 2000; Toivonen 1996; Wu et al. 2004) have been proposed.

Measuring association among variables is a fundamental task being at the heart of data mining. Tan et al. (2003) have described several key properties of twenty one interestingness measures proposed in statistics, machine learning and data mining literature. Hershberger and Fisher (2005) discuss measures such as Phi, Pearson's contingency coefficient, Cramer's V coefficient for making association analysis of two variables. These measures of association are used with chi-square and other contingency table analyses.

Pattern recognition in data is a well known classical problem that falls under the ambit of data analysis (Adhikari et al. 2014; Bishop 2013; Theodoridis 2011). Agrawal and Srikant (1995) introduced a problem of mining sequential patterns from a database of customer sales transactions and presented three algorithms for solving this problem.

Although market basket analysis originated from transactions made during shopping, it is important to realize that there are many other areas in which it can be applied. These include: analysis of credit card purchases, analysis of telephone calling patterns, identification of fraudulent medical insurance claims, and analysis of telecom service purchases.

1.2.2 Mining Time-Stamped Databases

Many business processes are required to collect a huge volume of data over a long period of time. Data mining methods assume implicitly that the domain under consideration is stable over time, and thus provide a rather static view on knowledge hidden in the data available so far. But many real databases are dynamic, and hence grow over time. Moreover, many real world databases contain time information. For instance, data about the sold items including its production date, data about the sales includes selling date, data about the warranty case together with the date of the claim, and the expiry date of warranty and the payment. Time stamp seems to be a natural attribute to objects described in a database. Knowledge discovery in a time-stamped database is an interesting and well known research issue (Bettini et al. 2000; Mitsa 2010; Hsu et al. 2008).

Data mining and knowledge discovery processes often involve analyzing data by ignoring time. Most of the cases temporal data are treated as an unordered collection of events. Analyzing data involving temporal dimension has received attention in the recent time (Adhikari and Rao 2010; Leonard and Wolfe 2005). Data analyses based on time dimension might offer significant knowledge to an organization. Time component is present almost in every data. Thus, time-based data analyses are present everywhere.

Time-stamped data mining can be used in forecasting (Matsubara et al. 2012) and numerous time related patterns. Some applications, in direction of time related patterns, are discussed here.

Periodic frequent patterns are special kind of frequent patterns that occur periodically (regularly) within a dataset (Tanbeer et al. 2009). In this approach, the time of occurrence of each transaction is taken into account for periodic frequent pattern mining. Temporal association rule can be defined as a pair (R, T), where R is an association rule and T is a temporal feature, such as a period or a calendar. There are three interesting measures regarding the discovery of association rules viz., support, confidence (Agrawal et al. 1993) and informativeness (Smyth and Goodman 1992).

The conditional probability of the consequent occurring given the antecedent is referred to as confidence of the rule. Informativeness is a measure that computes the usefulness of a rule in terms of the information it provides. An event sequence is a set of events that are ordered in time, whereas an event episode is a set of events that are time-stamped. Event sequences are investigated in order to predict events or to determine correlations of events (Agrawal and Srikant 1995). There is a distinction between event sequences and event episodes (Manilla et al. 1997). Since an event episode is a set of events that are time-stamped, the distance between the atomic events matters. Under the assumption that every event has an assigned value, event data can be further refined into (a) time-synchronous event data, in which an accurate time-stamp is important, (b) ordinal event data, where the ordering of the events according to time plays an important role, (c) aggregateable event data, which can be summarized for a particular interval, and (d) hierarchical event data, where the grouping is defined based on a hierarchical structure in the meta data. Sequential pattern mining has received a particular attention in the last decade (Agrawal and Srikant 1995; Zaki 2001; Wang and Han 2004). A sequence is a time-ordered list of objects. These can be time-stamped at regular or irregular time intervals. The objective is to extract patterns from a set of sequences of instantaneous events that satisfy some user-specified constraints. These constraints can vary from just a support threshold that defines frequency of a set of gaps, windows (Zaki 2000; Srikant and Agrawal 1996), or regular expression constraints (Garofalakis et al. 1999) in view of focusing more into the mining process. Another approach to discover temporal patterns in sequential data is the frequent episode discovery framework (Manilla et al. 1997). In the sequential patterns mining framework a collection of sequences are given, and the task is to discover ordered sequences of items that occur in sufficiently many of those sequences. In the frequent episodes mining framework the data are given as a single long sequence, and the task is to unearth temporal patterns, called episode, and it may occur often along that sequence.

1.2.3 Mining Multiple Related Databases

We are at the age of big data. There are various reasons for encountering a big data. Two main reasons for accumulating big data are (i) the invention of cost effective data collection channels and storage devices, and (ii) enormous data are being generated from a variety of applications. There are other reasons also. When a large organization transacts from the different branches, the collection of such related databases could be very large. Mining and analysis of multiple data sources offers many opportunities. But, it poses a number of challenges (Adhikari et al. 2010a). An obvious challenge is the size of the databases. A single computer might take an unreasonable amount of time to process the database. Sometimes it may not be possible to mine the amalgamated database.

Zhang et al. (2003) proposed local pattern analysis to mine multiple databases. We have proposed an extended model of local pattern analysis (Adhikari and Rao 2008). A few specialized as well as generalized techniques are introduced by us (Adhikari et al. 2010b). A coding-based technique is proposed to store pattern-base in building multi-database mining applications (Adhikari et al. 2010a).

Numerous applications on multi-databases are reported in the recent time. A multi-branch company is often interested in high-frequency rules because they are supported by most of its branches for corporate profitability. Wu and Zhang (2003) have proposed a weighting model for synthesizing high-frequent association rules from different data sources. Kum et al. (2006) have proposed ApproxMAP algorithm, to mine approximate sequential patterns, called consensus patterns, from large sequence databases in two steps. First, sequences are organized into similarity groups, called clusters. Then, consensus patterns are mined directly from each cluster through multiple alignments. Hu and Zhong (2006) have presented a conceptual model with dynamic multi-level workflows corresponding to a mining-grid centric multi-layer grid architecture, for multi-aspect analysis in building an e-business portal on the Wisdom Web. The authors have showed that this integrated model would help to dynamically organize status-based business processes that govern enterprise application integration. To reduce the search cost in the data from all databases, we need to identify which databases are most likely relevant to a data mining application. For this purpose, Wu et al. (2005) have proposed an algorithm for selecting relevant databases. Ratio rules are aimed at capturing the quantitative association knowledge. Yan et al. (2006) have extended this framework to mining ratio rules from distributed and dynamic data sources. Authors have proposed an integrated method to mining ratio rules from distributed and changing data sources, by first mining the ratio rules from each data source separately through a novel, robust and adaptive one-pass algorithm, and then integrating the rules of each data source in a simple probabilistic model. Biological databases contain a wide variety of data types, often with rich relational structure. Consequently multi-relational data mining techniques frequently are applied to biological data. Page and Craven (2003) have presented several applications of multi-relational data mining to biological data, taking care to cover a broad range of multi-relational data mining techniques. The field of bioinformatics is expanding rapidly. In this field large multiple as well as complex relational tables are dealt with frequently. Wang et al. (2005) present various techniques in biological data mining and data management. The book also includes preprocessing tasks such as data cleaning and data integration as applied to biological data. Principal component analysis (PCA) is frequently used for constructing the reduced representation of the data. The method often reduces the dimensionality of the original data by a large factor and constructs features that capture the maximally varying directions in the data. Kargupta et al. (2000) have proposed a technique of computing the collective principal component analysis from heterogeneous sites. Some more applications are introduced here in this book.

1.3 Experimental Settings

Several experiments are conducted for validations of the algorithms presented in Chaps. 2, 3, 7, 9, 10, 12, 14–16; and the experimental results justify the proposed approach. All these experiments have been implemented on a 1.6 GHz Pentium processor with 256 MB of memory using visual C++ (version 6.0) software.

The experiments in Chaps. 4–6, 8 and 11 are carried out using a 2.8 GHz Pentium D dual processor with 512 MB of memory using 6.0 version of visual C++ programming language.

For mining calendar-based periodic patterns on different databases, the experiments are done on a 2.4 GHz, core i3 processor with 4 GB of memory, running Windows 7 HB, using Visual C++, version 6.0 software (Chap. 13).

1.4 Future Directions

Data mining has come to prominence over the last two decades as a discipline in its own right which offers benefits with respect to many domains, both commercial and academic. Broadly data mining can be viewed as an application domain, as opposed to a technology (Coenen 2011). As data collection technologies and storage devices become cost effective and diverse in nature, more data are being captured in short span of time, and we enter in the age of "Big Data". Big Data is a new term used to identify the datasets that due to their large size and complexity, we can not manage them with our current methodologies or data mining software tools. The Big Data challenge is becoming one of the most exciting opportunities for the next years (Fan and Bifet 2012).

There are many challenges in dealing with big data such as absolute volume, growth, data management/integration, rights management, and privacy. One obvious challenge is the size of the data. It's not enough to have enough persistent storage space to keep it, we also need backup space, space to process it, caches and so on—it all adds up. Another is the growth and speed of that growth of the data volume. One can plan for any size, but it's not easy to adjust the plans if unexpected growth rates come along.

Another challenge that we must look into is the integration between company's internal data holdings, external public data (like the World Wide Web, or commercial third-party sources—like Twitter—which play an important role in the modern news ecosystem), and customer data (customers would like to see their own internal, proprietary data be inter-operable with company's data). We need to respect the rights associated with each data set, as we deal with company's own data, third party data and customers' data. Also, we must be very careful regarding privacy of data (see the link: Big data analytics at Thomson Reuters. Interview with Jochen L. Leidner 2013).

References

Adhikari A, Rao PR (2008) Synthesizing heavy association rules from different real data sources. Pattern Recogn Lett 29(1):59–71

Adhikari J, Rao PR (2010) Measuring influence of an item in a database over time. Pattern Recogn Lett 31(1):179–187

Adhikari A, Ramachandrarao P, Pedrycz W (2010a) Developing multi-database mining applications. Springer, Berlin

Adhikari A, Ramachandrarao P, Prasad B, Jhimli Adhikari J (2010b) Mining multiple large data sources. Int Arab J Inf Technol 7(3):241–249

Adhikari A, Adhikari J, Pedrycz W (2014) Data analysis and pattern recognition in multiple databases. Springer, Berlin

Agrawal R, Srikant R (1994) Fast algorithms for mining association rules. In: Proceedings of international conference on very large data bases, pp 487–499

Agrawal R, Srikant R (1995) Mining sequential patterns. In: Proceedings of the 11th international conference on data engineering, pp 3–14

Agrawal R, Imielinski T, Swami A (1993) Mining association rules between sets of items in large databases. In: Proceedings of ACM SIGMOD conference management of data, pp 207–216

Bettini C, Jajodia S, Wang XS (2000) Time granularities in databases, data mining, and temporal reasoning. Springer, Berlin

Big data analytics at Thomson Reuters. Interview with Jochen L. Leidner (2013) http://www.odbms.org/blog/2013/11/big-data-analytics-at-thomson-reuters-interview-with-jochen-l-leidner/

Bishop CM (2013) Pattern recognition and machine learning. Springer, Berlin

Coenen F (2011) Data mining: past, present and future. Knowl Eng Rev 26(1):25–29

Coenen F, Goulbourne G, Leng PH (2004) Tree structures for mining association rules. Data Min Knowl Disc 8(1):25–51

Fan W, Bifet A (2012) Mining big data: current status, and forecast to the future. SIGKDD Explor 14(2):1–5

Fayyad UM, Piatetsky-Shapiro G, Smyth P, Uthurusamy (eds) (1996) Advances in knowledge discovery and data mining. AAAI/MIT Press, Cambridge

Garofalakis MN, Rastogi R, Shim K (1999) SPIRIT: sequential pattern mining with regular expression constraints. In: Proceedings of VLDB, pp 223–234

Han J, Pei J, Yiwen Y (2000) Mining frequent patterns without candidate generation. In: Proceedings of ACM SIGMOD conference on management of data, pp 1–12

Han J, Kamber M, Pei J (2011) Data mining: concepts and techniques, 3rd edn. Morgan Kaufmann, Los Altos

Hershberger SL, Fisher DG (2005) Measures of association, encyclopedia of statistics in behavioral science. Wiley, New York

Hsu W, Lee ML, Wang J (2008) Temporal and spatio-temporal data mining. IGI Publishing, Hershey

Hu J, Zhong N (2006) Organizing multiple data sources for developing intelligent e-business portals. Data Min Knowl Disc 12(2–3):127–150

Kargupta H, Huang W, Krishnamurthy S, Park B, Wang S (2000) Collective PCA from distributed and heterogeneous data. In: Proceedings of the fourth European conference on principles and practice of knowledge discovery in databases, pp 452–457

Kum H-C, Chang HC, Wang W (2006) Sequential pattern mining in multi-databases via multiple alignment. Data Min Knowl Disc 12(2–3):151–180

Leonard M, Wolfe B (2005) Mining transactional and time series data. In: Proceedings of SUGI 30, paper 080-30

Maimon O, Rokach L (2010) Data mining and knowledge discovery handbook. Springer, Berlin

Manilla H, Toivonen H, Verkamo L (1997) Discovery of frequent episodes in event sequences. Data Min Knowl Disc 1(3):259–289

Matsubara Y, Sakurai Y, Faloutsos C, Iwata T, Yoshikawa M (2012) Fast mining and forecasting of complex time-stamped events. In: Proceedings of KDD, pp 271–279

Mitsa T (2010) Temporal data mining. CRC Press, Boca Raton

Page D, Craven M (2003) Biological applications of multi-relational data mining. SIGKDD Explor 5(1):69–79

Piatetsky-Shapiro G, Frawley WJ (eds) (1991) Knowledge discovery in databases. AAAI/MIT Press, Cambridge

Savasere A, Omiecinski E, Navathe S (1995) An efficient algorithm for mining association rules in large databases. In: Proceedings of the 21st international conference on very large data bases, pp 432–443

Smyth P, Goodman M (1992) An information theoretic approach to rules induction from databases. IEEE Trans Knowl Data Eng 4(4):301–316

Srikant R, Agrawal R (1996) Mining sequential patterns: generalizations and performance improvements. In: Proceedings of EDBT, pp 3–17

Tan P-N, Kumar V, Srivastava J (2003) Selecting the right interestingness measure for association patterns. In: Proceedings of SIGKDD conference, pp 32–41

Tanbeer SK, Ahmed CF, Jeong BS, Lee YK (2009) Discovering periodic-frequent patterns in transactional databases. Pacific Asia knowledge discovery in databases, pp 242–253

Theodoridis S (2011) Pattern recognition. Elsevier, Amsterdam

Toivonen H (1996) Sampling large databases for association rules. In: Proceedings of the 22-th international conference on very large data bases, pp 134–145

Wang J, Han J (2004) BIDE: efficient mining of frequent closed sequences. In: Proceedings of the 20th international conference on data engineering, pp 79–90

Wang JT, Zaki MJ, Toivonen HT, Shasha DE (2005) Data mining in bioinformatics. Springer, Berlin

Wu X, Zhang S (2003) Synthesizing high-frequency rules from different data sources. IEEE Trans Knowl Data Eng 14(2):353–367

Wu X, Zhang C, Zhang S (2004) Efficient mining of both positive and negative association rules. ACM Trans Inf Syst 22(3):381–405

Wu X, Zhang C, Zhang S (2005) Database classification for multi-database mining. Inf Syst 30 (1):71–88

Yan J, Liu N, Yang Q, Zhang B, Cheng Q, Chen Z (2006) Mining adaptive ratio rules from distributed data sources. Data Min Knowl Disc 12(2–3):249–273

Ye N (2003) The handbook of data mining. CRC Press, Boca Raton

Zaki MJ (2000) Sequence mining in categorical domains: incorporating constraints. In: Proceedings of CIKM, pp 422–429

Zaki MJ (2001) SPADE: an efficient algorithm for mining frequent sequences. Mach Learn J 42(1/2):31–60

Zhang S, Wu X (2011) Fundamentals of association rules in data mining and knowledge discovery. Wiley Interdisc Rew Data Min Knowl Disc 1(2):97–116

Zhang S, Wu X, Zhang C (2003) Multi-database mining. IEEE Comput Intell Bull 2(1):5–13

Chapter 2
Synthesizing Conditional Patterns in a Database

Though frequent itemsets and association rules express interesting association among items of frequently occurring itemsets in a database, there may exist other types of interesting associations among the items. A critical analysis of frequent itemsets would provide more insight about a database. In this paper, we introduce the notion of conditional pattern in a database. Conditional patterns are interesting and useful for solving many problems. We propose an algorithm for mining conditional patterns in a database. Experiments are conducted on three real datasets. The results of the experiments show that conditional patterns store significant nuggets of knowledge about a database.

2.1 Introduction

Association analysis of items (Agrawal et al. 1993; Antonie and Zaïane 2004), and selecting right interestingness measures (Hilderman and Hamilton 1999; Tan et al. 2002) are two significant tasks at the heart of many data mining problems. An association analysis is generally associated with interesting patterns in a database, and the interestingness of a pattern is expressed by using some measures. A pattern would become interesting if the values of interestingness measures satisfy some conditions. Positive association rules (Agrawal et al. 1993) and negative association rules (Antonie and Zaïane 2004) are examples of two patterns that are synthesized from the itemset patterns in a database. Positive association rules are expressed by a forward implication $X \rightarrow Y$, where X and Y are itemsets in the database. X and Y are called the antecedent and consequent of the association rule respectively. The meaning attached to this type of association rules is that if all the items in X are purchased by a customer then it is likely that all the items in Y are purchased by the same customer at the same time. On the other hand, negative association rules are expressed by one of the following three forward implications: $X \rightarrow \neg Y$, $\neg X \rightarrow Y$, and $\neg X \rightarrow \neg Y$, where X and Y are itemsets in the given database. Let us consider the negative association rules of the form $X \rightarrow \neg Y$. The meaning attached to the negative association rules of the form $X \rightarrow \neg Y$ is that if all the items in X are

© Springer International Publishing Switzerland 2015
A. Adhikari and J. Adhikari, *Advances in Knowledge Discovery in Databases*,
Intelligent Systems Reference Library 79, DOI 10.1007/978-3-319-13212-9_2

purchased by a customer then it is unlikely that all the items in Y are purchased by the same customer at the same time. Though association rules express interesting association among items in frequent itemsets, they might not be sufficient for all kinds of association analysis of items in a given database.

The importance of an itemset could be judged by its support (Agrawal et al. 1993). *Support* (*supp*) of an itemset X in database D is the fraction of transactions in D containing X. Itemset X is *frequent* in D if $supp(X, D) \geq \alpha$, where α is user defined *minimum support level*. Itemset $X = \{x_1, x_2, \ldots, x_m\}$ corresponds to Boolean expression $x_1 \wedge x_2 \wedge \cdots \wedge x_m$. Thus, if the itemset $\{x_1, x_2, \ldots, x_m\}$ contains in a transaction then the Boolean expression $x_1 \wedge x_2 \wedge \cdots \wedge x_m$ is true for that transaction. On the other hand, if the itemset $\{x_1, x_2, \ldots, x_m\}$ does not contain in a transaction then the Boolean expression $x_1 \wedge x_2 \wedge \cdots \wedge x_m$ is false for that transaction. In general, let E be a Boolean expression on items in D. Then, $supp(E, D)$ is the fraction of transactions in D that satisfy E.

Frequent itemset mining has received significant attention in KDD community. Several implementations of mining frequent itemsets (FIMI 2004) have been reported. Frequent itemsets are important patterns in a database, since they determine the major characteristics of a database. Wu et al. (2005) have proposed a solution of inverse frequent itemset mining. Authors argued that one could efficiently generate a synthetic market basket database from the frequent itemsets and their supports. Let X and Y be two itemsets in database D. The characteristics of database D are revealed more by the pair $(X, supp(X, D))$ than that of $(Y, supp(Y, D))$, if $supp(X, D) > supp(Y, D)$. Thus, it is important to study frequent itemsets more than infrequent itemsets. Negative association rules are generated from infrequent itemsets. Thus, their applications in different problem domains are limited. The goal of this chapter is to study some kind of association among items which is not immediately available from frequent itemsets and association rules.

If X is frequent in D then every non-null subset of X is also frequent in D. Consider the following example.

Example 2.1 Let $D = \{\{a, b\}, \{a, b, c, d\}, \{a, b, c, h\}, \{a, b, g\}, \{a, b, h\}, \{a, c\}, \{a, c, d\}, \{b\}, \{b, c, d, h\}, \{b, d, g\}\}$. The frequent itemsets in D at minimum support level 0.2 are given as follows: $\{a\}(0.7), \{b\}(0.8), \{c\}(0.5), \{d\}(0.4), \{g\}(0.2), \{h\}(0.3), \{a, b\}(0.5), \{a, c\}(0.4), \{a, d\}(0.2), \{a, h\}(0.2), \{b, c\}(0.3), \{b, d\}(0.3), \{b, g\}(0.2), \{b, h\}(0.3), \{c, d\}(0.3), \{c, h\}(0.2), \{a, b, c\}(0.2), \{a, b, h\}(0.2), \{a, c, d\}(0.2), \{b, c, d\}(0.2), \{b, c, h\}(0.2)$. $X(\eta)$ denotes frequent itemset X with support η. Suppose we wish to study association among items in $\{a, b, c\}$. A frequent itemset mining algorithm could mine the following details about items in $\{a, b, c\}$.

Table 2.1 provides the information on how frequently a non-null subset of $\{a, b, c\}$ occurs in D. Such information might not be sufficient for all types of queries and analyses of items in $\{a, b, c\}$.

A positive association rule finds positive association between two disjoint non-null itemsets. Positive association rules in D are synthesized from frequent itemsets

Table 2.1 Frequent itemset $\{a, b, c\}$ and its non-null subsets at $\alpha = 0.2$

Itemset	$\{a\}$	$\{b\}$	$\{c\}$	$\{a, b\}$	$\{a, c\}$	$\{b, c\}$	$\{a, b, c\}$
Support	0.7	0.8	0.5	0.5	0.4	0.3	0.2

Table 2.2 Association rules generated from $\{a, b, c\}$ at $\alpha = 0.2$ and $\beta = 0.5$

Association rule	Support	Confidence
$\{a, c\} \rightarrow \{b\}$	0.2	0.66667
$\{b, c\} \rightarrow \{a\}$	0.2	0.66667

in D. A positive association rule $r: X \rightarrow Y$ in D is characterized by its support and confidence measures (Agrawal et al. 1993). *Support* of association rule $r: X \rightarrow Y$ in D is the fraction of transactions in D containing both X and Y. *Confidence* (*conf*) of association rule r in D is the fraction of transactions in D containing Y among the transactions containing X. An association rule r in D is interesting if $supp(r, D) \geq \alpha$, and $conf(r, D) \geq \beta$, where β is the *minimum confidence level*. The parameters α and β are user-defined inputs to an association rule mining algorithm. We synthesize association rules from $\{a, b, c\}$ of Example 2.1 as follows (Example 2.2).

Example 2.2 We continue here the discussion of Example 2.1. The interesting association rules generated from $\{a, b, c\}$ are given in Table 2.2.

The chapter is organized as follows. In Sect. 2.6, we introduce conditional pattern in a database. We discuss properties of conditional patterns in Sect. 2.3. In Sect. 2.4, we propose an algorithm for extracting conditional patterns in a database. The results of the experiments are given in Sect. 2.5. Also, we present an application of conditional patterns in this section. We discuss related work in Sect. 2.6.

2.2 Conditional Pattern

With reference to Examples 2.1 and 2.2, the study of items in $\{a, b, c\}$ might be incomplete if we know only the supports and the association rules with respect to non-null subsets of $\{a, b, c\}$. Thus, the information provided in Tables 2.1 and 2.2 might not be sufficient for all types of queries and analyses related to items in $\{a, b, c\}$. In fact, there are some queries related to items in $\{a, b, c\}$ whose answers are not immediately available from Tables 2.1 and 2.2. A few examples of such queries are given below.

- Find the support that a transaction contains item a but not items b and c, with respect to $\{a, b, c\}$.
- Find the support that a transaction contains items a and b but not item c, with respect to $\{a, b, c\}$.

The above queries correspond to a specific type of pattern in a database. Some of these of patterns could have significant supports, since $\{a, b, c\}$ is a frequent itemset. In general, if we wish to study the association among the items in Y with negation of items in $X - Y$, then such analysis is not immediately available from frequent itemsets and positive association rules, for itemsets X and Y in a database such that $Y \subseteq X$. Such association analyses could be interesting, since the corresponding Boolean expressions could have high supports. Therefore, we need to mine such patterns for effective analyses of items in frequent itemsets.

Let $\langle Y, X \rangle$ be a pattern that a transaction in a database contains all the items of Y, but not items of $X - Y$, for itemsets X and Y in the database such that $Y \subseteq X$. Let $supp\langle Y, X, D \rangle$ be the support that a transaction in database D contains all the items of Y, but not items of $X - Y$, for itemsets X and Y in D such that $Y \subseteq X$. A pattern of type $\langle Y, X \rangle$ is called a *conditional pattern* (Adhikari and Rao 2008). A conditional pattern $\langle Y, X \rangle$ has two components: *pattern itemset* (Y) and *reference itemset* (X). Thus, a conditional pattern $\langle Y, X \rangle$ is associated with two values: $supp\langle Y, X, D \rangle$ and $supp(X, D)$. $supp\langle Y, X, D \rangle$ and $supp(X, D)$ are called *conditional support* (*csupp*) and *reference support* (*rsupp*) of conditional pattern $\langle Y, X \rangle$ in D, respectively. The conditional support and reference support of conditional pattern $\langle Y, X \rangle$ in D are denoted by $csupp\langle Y, X, D \rangle$ and $rsupp\langle Y, X, D \rangle$, respectively. In other words, $supp\langle Y, X, D \rangle$ and $supp(X, D)$ are denoted by $csupp\langle Y, X, D \rangle$ and $rsupp\langle Y, X, D \rangle$, respectively. A conditional pattern $\langle Y, X \rangle$ in D is *interesting* if $csupp\langle Y, X, D \rangle \geq \delta$ and $rsupp\langle Y, X, D \rangle \geq \alpha$, where δ is the *minimum conditional support level*. The parameters α and δ are user defined inputs to a conditional pattern mining algorithm.

Figure 2.1 give more insight about above two queries.

The shaded region in Fig. 2.1a is a set of transactions in D such that each transaction contains item a but not items b and c, with respect to $\{a, b, c\}$. The shaded region in Fig. 2.1b is a set of transactions in D such that each transaction contains items a and b but not item c, with respect to $\{a, b, c\}$. Thus, we get following formulas.

$$supp\langle \{a\}, \{a, b, c\}, D \rangle = supp(\{a\}, D) - supp(\{a, b\}, D) \\ - supp(\{a, c\}, D) + supp(\{a, b, c\}, D) \tag{2.1}$$

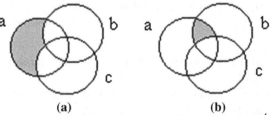

(a)   **(b)**

Fig. 2.1 *Shaded regions* in (**a**) and (**b**) correspond to conditional supports of $\langle \{a\}, \{a, b, c\} \rangle$ and $\langle \{a, b\}, \{a, b, c\} \rangle$ in D, respectively

Table 2.3 Conditional patterns with respect to {a, b, c} in D

Conditional pattern	csupp	Conditional pattern	csupp
$\langle \{a\}, X \rangle$	0	$\langle \{a, c\}, X \rangle$	0.2
$\langle \{b\}, X \rangle$	0.2	$\langle \{b, c\}, X \rangle$	0.1
$\langle \{c\}, X \rangle$	0	$\langle X, X \rangle$	1.0
$\langle \{a, b\}, X \rangle$	0.3		

Table 2.4 Non-trivial conditional patterns with respect to {a, b, c} at $\delta = 0.2$ and $\alpha = 0.2$

Conditional pattern	csupp	rsupp	Conditional pattern	csupp	rsupp
$\langle \{b\}, \{a, b, c\} \rangle$	0.2	0.2	$\langle \{a, c\}, \{a, b, c\} \rangle$	0.2	0.2
$\langle \{a, b\}, \{a, b, c\} \rangle$	0.3	0.2			

$$supp\langle \{a, b\}, \{a, b, c\}, D \rangle = supp(\{a, b\}, D) - supp(\{a, b, c\}, D) \qquad (2.2)$$

A conditional pattern $\langle Y, X \rangle$ in a database is *trivial* if $Y = X$. A trivial conditional pattern is known when the corresponding frequent itemset gets extracted from the database. Thus, trivial conditional patterns get mined during mining of frequent itemsets. In the following example, we identify conditional patterns in D with respect to {a, b, c} of Example 2.1.

Example 2.3 Consider the frequent itemsets in D of Example 2.1. The conditional patterns with respect to $X = \{a, b, c\}$ in D are given in Table 2.3.

We define *size* of an itemset X as the number of items in X, denoted by $|X|$. Based on the sizes of pattern itemset and reference itemset, we could categorize conditional patterns in a database. The conditional patterns $\langle \{a\}, X \rangle, \langle \{b\}, X \rangle$ and $\langle \{c\}, X \rangle$ belong to the same category. But, the conditional patterns $\langle \{a\}, X \rangle$ and $\langle \{a, b\}, X \rangle$ are of different categories. In general, two conditional patterns $\langle X, Y \rangle$ and $\langle P, Q \rangle$ in D are of the *same category*, if $|X| = |P|$ and $|Y| = |Q|$, for X, Y, P, Q are itemsets in D. All the conditional patterns mined with respect to a frequent itemset are not interesting. The interesting conditional patterns with respect to {a, b, c} in D are given in Table 2.4.

We observe that $csupp\langle Y, X, D \rangle \leq supp(Y, D)$, for $Y \subset X$. Nonetheless, $csupp\langle Y, X, D \rangle$ could be high, if Y is frequent in D. Thus, it is necessary to study such patterns in a database for effective analyses of items in frequent itemsets. The problem could be stated as follows.

We are given a database D of customer transactions. Extract interesting non-trivial conditional patterns from D.

2.3 Properties of Conditional Patterns

In this section, we present some interesting properties of conditional patterns in a database. Before presenting the properties, we introduce some notations. Let $X = \{x_1, x_2, \ldots, x_m\}$ and $Y = \{y_1, y_2, \ldots, y_p\}$. Then, $supp(X \cup Y, D)$ and $supp(X \cap Y, D)$ refer to $supp((x_1 \wedge x_2 \wedge \cdots \wedge x_m) \vee (y_1 \wedge y_2 \wedge \cdots \wedge y_p), D)$ and $supp((x_1 \wedge x_2 \wedge \cdots \wedge x_m) \wedge (y_1 \wedge y_2 \wedge \cdots \wedge y_p), D)$ respectively.

Lemma 2.1 *Let E be a Boolean expression that a transaction contains at least one item of itemset X in database D. Then,*

$$supp(E, D) = \sum_{Y \subseteq X, Y \neq \phi} supp\langle Y, X, D \rangle \qquad (2.3)$$

Proof We re-state theorem of total probability (Feller 1968) in terms of supports as follows: For any m Boolean expressions $X_1, X_2, \ldots, X_m$ in database D we have,

$$supp\left(\cup_{i=1}^{m} X_i, D\right) = \sum_{i=1}^{m} supp(X_i, D)$$

$$- \sum_{i<j;\, i,\, j=1}^{m} supp(X_i \cap X_j, D) + \cdots + (-1)^{m-1} supp\left(\cap_{i=1}^{m} X_i, D\right).$$

The events $\langle Y, X \rangle$ and $\langle Z, X \rangle$ are mutually exclusive, for $Y \neq Z, Y \subseteq X$ and $Z \subseteq X$. Thus, $supp(\langle Y, X \rangle \cap \langle Z, X \rangle, D) = 0$, for $Y \neq Z, Y \subseteq X$ and $Z \subseteq X$. □

Let $X = \{a, b, c\}$. With reference to Examples 2.1 and 2.3, $supp(a \vee b \vee c, D) = 1$ and $supp(a \vee b \vee c, D) = supp\langle\{a\}, X, D\rangle + supp\langle\{b\}, X, D\rangle + supp\langle\{c\}, X, D\rangle + supp\langle\{a, b\}, X, D\rangle + supp\langle\{a, c\}, X, D\rangle + supp\langle\{b, c\}, X, D\rangle + supp\langle X, X, D\rangle$. Thus, it validates Lemma 2.1.

Lemma 2.2 $supp(X, D) \leq \sum_{Y \subseteq X, Y \neq \phi} supp\langle Y, X, D \rangle$, *for any two itemsets X and Y in database D such that $Y \subseteq X$.*

Proof Let $X = \{x_1, x_2, \ldots, x_m\}$. Then, X corresponds to Boolean expression $x_1 \wedge x_2 \wedge \cdots \wedge x_m$ in D. Let E be a Boolean expression that a transaction contains at least one item of itemset X in D. Then, $supp(E, D) = \sum_{Y \subseteq X, Y \neq \phi} supp\langle Y, X, D \rangle$, (Lemma 2.1) $= supp\langle X, X, D \rangle + Q$, where $Q \geq 0$. Then, $supp(E, D) = supp(X, D) + Q$, since $supp(X, D) = supp\langle X, X, D \rangle$. The lemma follows. □

With reference to Examples 2.1 and 2.3, let $X = \{a, b, c\}$. $supp(X, D) = 0.2$. Now, $supp\langle\{a\}, X, D\rangle + supp\langle\{b\}, X, D\rangle + supp\langle\{c\}, X, D\rangle + supp\langle\{a, b\}, X,$

$D\rangle + supp\langle\{a, c\}, X, D\rangle + supp\langle\{b, c\}, X, D\rangle + supp\langle X, X, D\rangle = 1.0 \geq 0.2$. Thus, it validates Lemma 2.2.

Lemma 2.3 *The conditional supports of* $\langle X, Y\rangle$ *and* $\langle X, Z\rangle$ *in a database may not be equal, for any three itemsets X, Y and Z in the database such that* $X \subseteq Y$ *and* $X \subseteq Z$.

Proof The itemsets $Y - X$ and $Z - X$ may not be the same. Thus, the lemma follows. $\square$

A conditional pattern is so named due to Lemma 2.3. Using Example 2.1, we get $supp\langle\{a, b\}, \{a, b, h\}, D\rangle = 0.3$ and $supp\langle\{a, b\}, \{a, b, d\}, D\rangle = 0.4$. We observe that $supp\langle\{a, b\}, \{a, b, h\}, D\rangle \neq supp\langle\{a, b\}, \{a, b, d\}, D\rangle$.

Lemma 2.4 *There is no fixed ordered relationship between conditional supports of* $\langle Y, X\rangle$ *and* $\langle Z, X\rangle$ *in a database, for any three itemsets X, Y and Z in the database such that* $Z \subseteq Y \subseteq X$.

Proof Let X, Y and Z be three itemsets X, Y and Z in database D such that $supp\langle Y, X, D\rangle \leq supp\langle Z, X, D\rangle$, for some $Z \subseteq Y \subseteq X$. Also, there may exist another three itemsets P, Q and R in database D such that $supp\langle R, P, D\rangle \leq supp\langle Q, P, D\rangle$, for some $R \subseteq Q \subseteq P$. The proof is based on a counter example. With reference to database D of Example 2.1, let $X = \{a, b, c\}$, $Y = \{a, b\}$ and $Z = \{a\}$. Then, $supp\langle Z, X, D\rangle = supp\langle\{a\}, \{a, b, c\}, D\rangle = 0$, and $supp\langle Y, X, D\rangle = supp\langle\{a, b\}, \{a, b, c\}, D\rangle = 0.3$. Thus, $supp\langle Z, X, D\rangle \leq supp\langle Y, X, D\rangle$, for $Z \subseteq Y \subseteq X$. Let $A = \{b, d, h\}$, $B = \{b, d\}$ and $C = \{b\}$. Then, $supp\langle C, A, D\rangle = supp\langle\{b\}, \{b, d, h\}, D\rangle = 0.3$, and $supp\langle B, A, D\rangle = supp\langle\{b, d\}, \{b, d, h\}\rangle = 0.2$. Thus, $supp\langle B, A, D\rangle \leq supp\langle C, A\rangle$, for $C \subseteq B \subseteq A$. $\square$

We could synthesize a set of frequent itemsets from a set of association rules. In particular, let $r_1: X \rightarrow Y$ and $r_2: X \rightarrow Z$ be two positive association rules in D, where X, Y and Z are three frequent itemsets in D. The set of frequent itemsets is synthesized from $\{r_1, r_2\}$ is $\{X, XY, XZ\}$. In a similar way, we could synthesize a set of frequent itemsets from a set of conditional patterns. In particular, let $cp_1: \langle x_1 \wedge x_2, x_1 \wedge x_2 \wedge x_3\rangle$, and $cp_2: \langle x_1 \wedge x_3, x_1 \wedge x_2 \wedge x_3\rangle$ are two conditional patterns in D, where x_i is an item in D, for $i = 1, 2, 3$. The set of frequent itemsets is synthesized from $\{cp_1, cp_2\}$ is $\{\{x_1, x_2\}, \{x_1, x_3\}, \{x_1, x_2, x_3\}\}$.

Example 2.4 Let us consider Table 2.2. The set of frequent itemsets synthesized from the set of positive association rules is given as follows: $\{\{a, c\}(0.4), \{b, c\} (0.3), \{a, b, c\}(0.2)\}$. Let us consider Table 2.4. The set of frequent itemsets synthesized from the set of conditional patterns is given as follows: $\{\{a, b\}(0.5), \{a, c\}(0.4), \{a, b, c\}(0.2)\}$.

From Example 2.4, we could conclude that the association rules and conditional patterns in a database may not represent the same information about a database. This because of the fact that the amount of information conveyed by association rules in a database is dependent on β, at a given α. Also, the information conveyed by the conditional patterns in a database is dependent on δ, at a given α. Thus, we have the following definition.

Definition 2.1 A set of association rules A and a set of conditional patterns C in a database convey the same information about a given database if the set of frequent itemsets synthesized from A is the same as the set of frequent itemsets synthesized from C.

Lemma 2.5 *The set association rules in a database at $\beta = \alpha$ and the set of conditional patterns in the database at $\delta = 0$ represent the same information about the database at a give α.*

Proof Let S be a set of frequent itemsets in database D. Also, let CLOSURE $(S) = \{s: (s \in S), \text{ or } (s \neq \phi \text{ and } s \subseteq p \in S)\}$. Let SFIS($D$, i) be the set of frequent itemsets of size i, for $i = 1, 2, \dots$. The set of frequent itemsets synthesized from association rules in D at $\beta = \alpha$ is equal to CLOSURE $(\cup_{i \geq 2} SFIS(D, i))$. Also, the set of frequent itemsets synthesized from the conditional patterns in D at $\delta = 0$ is equal to CLOSURE $(\cup_{i \geq 2} SFIS(D, i))$. $\qquad\qquad\square$

With reference to Example 2.1, the frequent itemsets in D at $\alpha = 0.4$ are given as follows: $\{a, b\}(0.5)$, $\{a, c\}(0.4)$. The association rules in D at $\beta = 0.4$ are given in Table 2.5.

The set frequent itemsets synthesized from the above association rules is equal to $\{\{a\}(0.7), \{b\}(0.8), \{c\}(0.5), \{a, b\}(0.5), \{a, c\}(0.4)\}$. The conditional patterns in D at $\delta = 0.4$ are given in Table 2.6.

The set of frequent itemsets synthesized from the above conditional patterns is equal to $\{\{a\}(0.7), \{b\}(0.8), \{c\}(0.5), \{a, b\}(0.5), \{a, c\}(0.4)\}$. Thus, the set of frequent itemsets synthesized from the above association rules and the set of

Table 2.5 Association rules in D at $\alpha = 0.4$ and $\beta = 0.4$

Association rule (r)	supp(r, D)	conf(r, D)
$\{a\} \rightarrow \{b\}$	0.5	0.71
$\{b\} \rightarrow \{a\}$	0.5	0.63
$\{a\} \rightarrow \{c\}$	0.4	0.57
$\{c\} \rightarrow \{a\}$	0.4	0.80

Table 2.6 Conditional patterns in D at $\alpha = 0.4$ and $\delta = 0$

Conditional pattern	csupp	rsupp	Conditional pattern	csupp	rsupp
$\langle \{a\}, \{a, b\} \rangle$	0.2	0.5	$\langle \{a\}, \{a, c\} \rangle$	0.3	0.4
$\langle \{b\}, \{a, b\} \rangle$	0.3	0.5	$\langle \{c\}, \{a, c\} \rangle$	0.1	0.4

frequent itemsets synthesized from the above conditional patterns are the same at $\beta = \alpha$ and $\delta = 0$. Thus, it validates Lemma 2.5.

Lemma 2.6 *Let the conditional pattern* $\langle Y, X \rangle$ *in database D is interesting at conditional support level δ and support level α. Then itemset Y is frequent at level $\alpha + \delta$.*

Proof $supp\langle Y, X, D \rangle \geq \delta$ and $supp(X, D) \geq \alpha$, since $\langle Y, X \rangle$ is interesting in D at conditional support level δ and support level α. The patterns X and $\langle Y, X \rangle$ in D can not occur in a transaction simultaneously. $supp(X, D) \geq \alpha$ implies $supp(Y, D) \geq \alpha$, since $Y \subseteq X$. Also, $supp\langle Y, X, D \rangle \geq \delta$ and thus, $supp(Y, D) \geq (\alpha + \delta)$. ☐

With reference to Example 2.3, $\langle \{b\}, \{a, b, c\} \rangle$ is interesting conditional pattern in D at $\delta = 0.2$ and $\alpha = 0.2$. With reference to Example 2.1, supp({b}, D) = $0.8 \geq 0.2 + 0.2 = 0.4$. Thus, it validates Lemma 2.6.

Lemma 2.7 *Let* $X_1, X_2, ..., X_m$ *be itemsets in database D such* $X_i \subseteq X_{i+1}$, *for i = 1, 2, ..., m − 1. Then,* $supp\langle Y, X_i, D \rangle \geq supp\langle Y, X_{i+1}, D \rangle$, *for* $Y \subseteq X_i$ *at every i = 1, 2, ..., m − 1.*

Proof Let $Y = \{a_1, a_2, ..., a_p\}$. Let $Z = X_{k+1} - X_k$, for $i = k$. Also let, $X_k = \{b_1, b_2, ..., b_q\}$, and $Z = \{c_1, c_2, ..., c_r\}$. Consider the following two Boolean expressions: $E_1 = a_1 \wedge a_2 \wedge \cdots \wedge a_p \wedge \neg b_1 \wedge \neg b_2 \wedge \cdots \wedge \neg b_q$ and $E_2 = a_1 \wedge a_2 \wedge \cdots \wedge a_p \wedge \neg b_1 \wedge \neg b_2 \wedge \cdots \wedge \neg b_q \wedge \neg c_1 \wedge \neg c_2 \wedge \cdots \wedge \neg c_r$. The Boolean expressions E_1 and E_2 correspond to conditional patterns $\langle Y, X_k \rangle$ and $\langle Y, X_{k+1} \rangle$, respectively. The expression E_2 is more restrictive than the expression E_1. Thus, $supp(E_1, D) \geq supp(E_2, D)$. ☐

With reference to database D of Example 2.1, let $Y = b$, $X_1 = \{a, b\}$ and $X_2 = \{a, b, c\}$. We have $supp\langle Y, X_1, D \rangle = 0.3$ and $supp\langle Y, X_2, D \rangle = 0.2$. We observe that $supp\langle Y, X_1, D \rangle \geq supp\langle Y, X_2, D \rangle$.

2.4 Mining Conditional Patterns

For mining conditional patterns in a database, we need to find their conditional supports. We calculate $supp\langle Y, X, D \rangle$ in terms of supports of relevant frequent itemsets, for $Y \subseteq X$. Let $X = Y \cup Z$, where $Z = \{a_1, a_2, ..., a_p\}$. The following theorem is useful for synthesizing conditional supports using relevant frequent itemsets in D.

Lemma 2.8 *Let X, Y and Z are itemsets in database D such that* $X = Y \cup Z$, *where* $Z = \{a_1, a_2, ..., a_p\}$. *Then,*

$$supp\langle Y, X, D\rangle = supp(Y, D) - \sum_{i=1}^{p} supp(Y \cap \{a_i\}, D) + \sum_{i<j;\ i,j=1}^{p} supp(Y \cap \{a_i, a_j\}, D)$$

$$- \sum_{i<j<k;\ i,j,k=1}^{p} supp(Y \cap \{a_i, a_j, a_k\}, D) + \cdots + (-1)^p \qquad (2.4)$$

$$\times supp(Y \cap \{a_1, a_2, \ldots, a_p\}, D)$$

Proof We shall prove the result using method of induction on p. For $p = 1$, $X = Y \cap \{a_1\}$. Then, $supp\langle Y, X, D\rangle = supp(Y, D) - supp(Y \cap \{a_1\}, D)$. Thus, the result is true for $p = 1$. Let us assume that the result is true for $p = m$. □

We shall prove that the result is true for $p = m + 1$. Let $Z = \{a_1, a_2, \ldots, a_{m+1}\}$. Due to the addition of item a_{m+1}, many supports are required to be added to or, subtracted from the expression of $supp\langle Y, X, D\rangle$, for $p = m$. For example, $supp(Y \cap \{a_{m+1}\}, D)$ is required to be subtracted, $supp(Y \cap \{a_i, a_{m+1}\}, D)$ is required to be added, for $1 \le i \le m$, and so on. Finally, the term $(-1)^{m+1} \times supp(Y \cap \{a_1, a_2, \ldots, a_{m+1}\}, D)$ is required to be added. Thus, the expression of $supp\langle Y, X, D\rangle$ at $p = m + 1$, is given as follows.

$$supp\langle Y, X, D\rangle = supp(Y, D) - \sum_{i=1}^{m+1} supp(Y \cap \{a_i\}, D) + \sum_{i<j;\ i,j=1}^{m+1} supp(Y \cap \{a_i, a_j\}, D)$$

$$- \sum_{i<j<k;\ i,j,k=1}^{m+1} supp(Y \cap \{a_i, a_j, a_k\}, D) + \cdots + (-1)^{m+1}$$

$$\times supp(Y \cap \{a_1, a_2, \ldots, a_{m+1}\}, D).$$

Formulas (2.1) and (2.2) validate above theorem. We shall use this formula in the proposed algorithm to compute conditional support of a conditional pattern.

Lemma 2.9 *The maximum number of non-trivial conditional patterns is equal to* $\sum_{X \in SFIS(D);\ |X| \ge 2} 2^{|X|-2}$, *where SFIS(D) is the set of frequent itemsets in database D.*

Proof The number of subsets of X is equal to $2^{|X|-2}$ such that $Y \ne \phi$, for $Y \subset X$. Each such subset of X corresponds to a non-trivial conditional pattern with reference to X. Thus, the lemma follows. □

The interestingness of a conditional pattern is judged by its conditional support and reference support. By combining both the measures one could define many interestingness measures of a conditional pattern. An appealing measure of interestingness of a conditional pattern $\langle Y, X\rangle$ in database D could be $csupp\langle Y, X, D\rangle + rsupp\langle Y, X, D\rangle$.

2.4.1 Algorithm Design

For mining conditional patterns in a database, we make use of an existing frequent itemset mining algorithm (Agrawal and Srikant 1994; Han et al. 2000; Savasere et al. 1995). There are two approaches of mining conditional patterns in a database.

In the first approach, we could synthesize conditional patterns from current frequent itemset extracted during the mining process. As soon as a frequent itemset is found during the mining process, we could call an algorithm of finding conditional patterns that generates conditional patterns from the current frequent itemset. When a frequent itemset is extracted, then all the non-null subsets of the frequent itemest have already been extracted. Thus, we could synthesize all the conditional patterns from the current frequent itemset extracted from the database. In the second approach, we could synthesize conditional patterns from the frequent itemsets in the given database after mining of all frequent itemsets. Thus, all the frequent itemsets are processed at the end of mining task. These two approaches seem to be the same so far as the computational complexity is concerned. In this chapter, we have followed the second approach of synthesizing conditional patterns. During the process of mining frequent itemsets, the frequent itemsets of smaller size get extracted before the frequent itemsets of larger size. The frequent itemsets are stored in array *SFIS* and get sorted based on their size automatically. During the processing of current frequent itemset, all the non-null subsets are available before the current itemset in *SFIS*.

Before presenting proposed algorithm of synthesizing the conditional patterns, we first state how we have designed the synthesizing algorithm. The frequent itemsets of size one can not generate conditional patterns. Thus, the algorithm skips processing frequent itemsets of size one. There are $2^{|X|-1}$ non-null subsets of an itemset X. Each non-null subset of X may correspond to an interesting conditional pattern, for $|X| \geq 2$. The subset X of X corresponds to a trivial conditional pattern. Thus, we need to process $2^{|X|-2}$ subsets of X.

One could view a conditional pattern as an object having following attributes: *pattern*, *reference*, *csupp*, and *rsupp*. We use an array *CP* to store conditional patterns in a database. The y attribute of ith conditional pattern is accessed by notation $CP(i) \cdot y$. Also, a frequent itemset could be viewed as an object described by a set of attributes. A frequent itemset could be described by the following attributes: itemset and supp. Let N be the number of frequent itemsets in the given database D. The variables i and j are used to index the frequent itemset being processed and the conditional pattern being synthesized, respectively. An algorithm for synthesizing interesting non-trivial conditional patterns is presented below.

Algorithm 2.1. Synthesize interesting non-trivial conditional patterns in a database.
procedure *conditional-pattern-synthesis* (*N, SFIS*)
Input:
N: number of frequent itemsets in the given database
SFIS: array of frequent itemsets in the given database
Output:
Interesting non-trivial conditional patterns
01:   let $i = 1$;
02:   let $j = 1$;
03:   **while** ($|SFIS(i)| = 1$) **do**
04:      increase i by 1;
05:   **end while**
06:   **while** ($i \leq N$) **do**
07:      $CP(j).rsupp = SFIS(i).supp$; $CP(j).reference = SFIS(i).itemset$;
08:      let $sum = 0$;
09:      **for** $k = 1$ to ($2^{|SFIS(i).itemset| - 1}$) **do**
10:         let $tempItemset = k$-th subset of $SFIS(i).itemset$;
11:         **if** ($SFIS(i).itemset = tempItemset$) **then goto** line 24; **end if**
12:         let $kk = 1$;
13:         **while** ($kk \leq i$) **do**
14:            **if** ($SFIS(kk).itemset = tempItemset$) **then**
15:               $sum = sum + (-1)^{|SFIS(kk).itemset| - |tempItemset|} \times SFIS(kk).supp$;
16:               **goto** line 21;
17:            **end if**
18:            increase kk by 1;
19:         **end while**
20:      **end for**
21:      **if** ($sum \geq \delta$) **then**
22:         $CP(i).csupp = sum$; $CP(i).pattern = tempItemset$;
23:         increase j by 1;
24:      **end if**
25:      increase i by 1;
26:   **end while**
27:   sort conditional patterns on ($csupp + rsupp$) in non-increasing order;
28:   **for** $k = 1$ to j **do**
29:      display k-th conditional pattern;
30:   **end for**
31:   **end procedure**

In this section, we explain and justify the statements of the above algorithm. The important parts of the algorithm are explained as follows: The frequent itemsets of size one generate trivial conditional patterns. Thus, we have skipped processing frequent itemsets of size one using lines 3–5. We synthesize conditional patterns using lines 6–26. There are $2^{|X|-1}$ non-null subsets for an itemset X. Each subset is considered using a *for*-loop in lines 9–20. The algorithm synthesizes conditional patterns with reference to a frequent itemset X, for $|X| \geq 2$. The algorithm bypasses processing itemset Y, if $Y = X$. When we synthesize conditional patterns with reference to a frequent itemset, we have already finished synthesizing its subsets. All the non-null subsets appear on or before the frequent itemset in *SFIS*. Thus, if a frequent itemset X located at position i, then we search for a subset of X from index 1 to i in *SFIS*, since *SFIS* is sorted non-decreasing order on length of an itemset. Thus, it justifies the condition of *while*-loop at line 13. Formula (2.4) expresses $supp\langle Y, X, D\rangle$ in terms of $supp(Y \cap Z, D)$, for all $Z \subseteq X - Y$. The co-efficient of $supp(Y \cap Z, D)$ is $(-1)^{|Z|}$ in the expression of $supp\langle Y, X, D\rangle$. Thus, $supp\langle Y, X, D\rangle = \sum_{Z \subseteq X-Y} (-1)^{|Z|} \times supp\langle Y, X, D\rangle$. This formula has been applied at line 15 to calculate $supp\langle Y, X, D\rangle$. A conditional pattern is interesting if the conditional support is greater than or equal to δ, provided the reference support of the itemset is greater than or equal to α. We need not check the reference support, since we deal with the frequent itemsets. In line 21, we check whether the currently synthesized conditional pattern is interesting. The details of a synthesized conditional pattern are stored using lines 7 and 22. At line 27, we sort all interesting conditional patterns in the given database. Finally, we display interesting conditional patterns using lines 28–30.

Lemma 2.10 *Algorithm conditional-pattern-synthesis executes in $O(N^2 \times 2^P)$ time, where N is the number of frequent itemsets in the database.*

Proof Lines 3–5 take $O(N)$ time. The *while*-loop at line 6 repeats maximum N times. Let the average size of the frequent itemsets of size greater than 1 be p. Thus, the *for*-loop at line 9 repeats 2^{p-1} times. The *while*-loop at line 13 repeats maximum N times. Thus, the time complexity of lines 6–26 is equal to $O(N^2 \times 2^P)$. The time complexity of line 27 is equal to $O(N \times 2^P \times \log(N \times 2^P))$, since the number of conditional patterns is equal to $O(N \times 2^P)$. The time complexity of lines 28–30 is equal to $O(N \times 2^P)$. Therefore, the time complexity of the algorithm is maximum $\{O(N^2 \times 2^P), O(N \times 2^P \times \log(N \times 2^P))\}$.   □

2.5 Experiments

We have carried out several experiments to study the effectiveness of our approach. We present experimental results using three real databases. Database *retail* (Frequent itemset mining dataset repository 2004) is obtained from an anonymous Belgian retail supermarket store. Databases *BMS-Web-Wiew-1* and *BMS-Web-Wiew-2* can be found from KDD CUP 2000 (Frequent itemset mining dataset repository 2004). We present some characteristics of these databases in Table 2.7.

Let *NT*, *AFI*, *ALT*, and *NI* denote the number of transactions, the average frequency of an item, the average length of a transaction, and the number of items in the corresponding database respectively. Top five interesting conditional patterns of available categories are shown in Table 2.8. We have implemented apriori algorithm for the purpose of mining conditional patterns in the given databases. The

Table 2.7 Database characteristics

Database	NT	ALT	AFI	NI
retail	88,162	11.305755	99.673800	10,000
BMS-Web-Wiew-1	1,49,639	2.000000	155.711759	1,922
BMS-Web-Wiew-2	3,58,278	2.000000	7165.560000	100

Table 2.8 Top 5 conditional patterns of each category available in *retail* at $\alpha = 0.05$ and $\delta = 0.03$

Conditional pattern	csupp	rsupp
⟨{39}, {1, 39}⟩	0.520451	0.066332
⟨{39}, {8, 39}⟩	0.524421	0.062362
⟨{39}, {0, 39}⟩	0.526871	0.059912
⟨{39}, {2, 39}⟩	0.525612	0.061171
⟨{39}, {3, 39}⟩	0.525714	0.061069
⟨{39}, {39, 41, 48}⟩	0.210317	0.083551
⟨{39}, {32, 39, 48}⟩	0.221603	0.061274
⟨{39}, {38, 39, 48}⟩	0.208106	0.069213
⟨{48}, {39, 41, 48}⟩	0.139482	0.083551
⟨{32}, {32, 39, 48}⟩	0.049432	0.061274
⟨{39,48}, {32, 39, 48}⟩	0.269277	0.061274
⟨{39, 48}, {39, 41, 48}⟩	0.247000	0.083551
⟨{39, 48}, {38, 39, 48}⟩	0.261337	0.069213
⟨{38, 39}, {38, 39, 48}⟩	0.048127	0.069213
⟨{32, 39}, {32, 39, 48}⟩	0.034629	0.061274

conditional patterns in a database are ranked based on the sum of conditional support and reference support.

In both *BMS-Web-Wiew-1* and *BMS-Web-Wiew-2*, only one category of conditional patterns is available, since the maximum length of a transaction in each of these two databases is 2.

We have also conducted experiments for finding time needed to mine conditional patterns in different databases. The execution time for finding conditional patterns in a database increases as the size, i.e., the number of transactions contained in a database increases. We observe this phenomenon in Figs. 2.2 and 2.3. We have also conducted experiments to find time needed to synthesize conditional patterns in a database. The time (only) for synthesizing conditional patterns in each of the above databases is equal to 0 ms at the respective values of α and δ shown in Tables 2.8, 2.9 and 2.10.

We have also conducted experiments for finding the number of conditional patterns in a database at a given α. The number of conditional patterns in a database decreases as α increases. We observe this phenomenon in Figs. 2.4 and 2.5.

We have also conducted experiments for finding execution time needed for mining conditional patterns in a database at a given α. The execution time needed

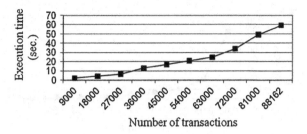

Fig. 2.2 Execution time versus the number of transactions in *retail*

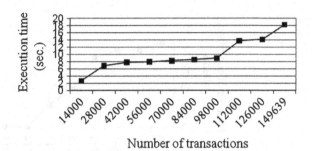

Fig. 2.3 Execution time versus the number of transactions in *BMS-Web-Wiew-1*

Table 2.9 Top 5 conditional patterns of each category available in *BMS-Web-Wiew-1* at $\alpha = 0.01$ and $\delta = 0.009$

Conditional pattern	csupp	rsupp
$\langle\{5\}, \{1, 5\}\rangle$	0.235453	0.013740
$\langle\{5\}, \{3, 7\}\rangle$	0.236135	0.013058
$\langle\{5\}, \{5, 7\}\rangle$	0.235293	0.013900
$\langle\{5\}, \{5, 9\}\rangle$	0.236335	0.012858
$\langle\{7\}, \{7, 9\}\rangle$	0.203563	0.011568

Table 2.10 Top 5 conditional patterns of each category available in *BMS-Web-Wiew-2* at $\alpha = 0.009$ and $\delta = 0.007$

Conditional pattern	csupp	rsupp
$\langle\{7\}, \{1, 7\}\rangle$	0.174072	0.022943
$\langle\{7\}, \{6, 7\}\rangle$	0.185401	0.011614
$\langle\{7\}, \{7, 9\}\rangle$	0.175810	0.021204
$\langle\{7\}, \{0, 7\}\rangle$	0.185702	0.011312
$\langle\{7\}, \{2, 7\}\rangle$	0.185747	0.011268

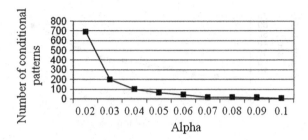

Fig. 2.4 Number of conditional patterns versus α for *retail*

Fig. 2.5 Number of conditional patterns versus α for *BMS-Web-Wiew-1*

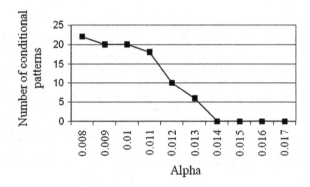

for mining conditional patterns in a database decreases as α increases. We observe this phenomenon in Figs. 2.6 and 2.7.

Also, we have conducted experiments to study the relationship between the size of a database and the number of conditional patterns in it. The experiments are conducted on databases *retail* and *BMS-Web-Wiew-1*. The results of the experiments are shown in Figs. 2.8 and 2.9. From the graphs in Figs. 2.8 and 2.9, we could conclude that there is no universal relationship between the size of a database and the number of conditional patterns in it.

Fig. 2.6 Execution time versus α for *retail*

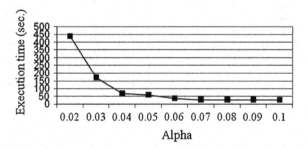

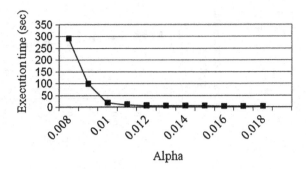

Fig. 2.7 Execution time versus α for *BMS-Web-Wiew-1*

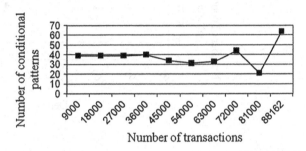

Fig. 2.8 Number of conditional patterns versus the number of transactions in *retail*

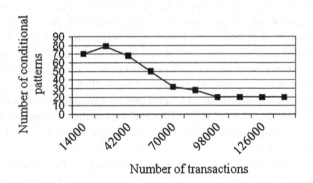

Fig. 2.9 Number of conditional patterns versus the number of transactions in *BMS-Web-Wiew-1*

Also, we have conducted experiments to study the relationship between the number of conditional patterns and conditional support. The experiments have been conducted on databases *retail* and *BMS-Web-Wiew-1*. The number of conditional patterns in a database decreases as δ increases. We observe this phenomenon in Figs. 2.10 and 2.11.

Fig. 2.10 Number of
conditional patterns versus δ
for *retail*

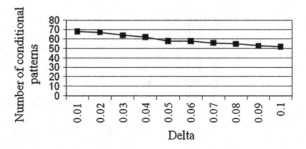

Fig. 2.11 Number of
conditional patterns versus
δ for *BMS-Web-Wiew-1*

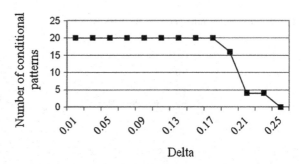

2.5.1 An Application

Adhikari and Rao (2007) have proposed a technique for mining arbitrary Boolean expressions induced by frequent itemsets using conditional patterns in a database. The pattern itemset of a conditional pattern with respect to itemset $X = \{a_1, a_2, ..., a_m\}$ is of the form $\{b_1, b_2, ..., b_m\}$, where $b_i = a_i$, or $\neg a_i$, for $i = 1, 2, ..., m$. Let $\Psi(X)$ be the set of all such pattern itemsets with respect to X. Then $\Psi(X)$ could be called as the *generator* of Boolean expressions induced by X. $\Psi(X)$ contains $2^m - 1$ pattern itemsets. A pattern itemset of the corresponding conditional pattern is also called a *minterm*, or *standard product*. Every Boolean expression of items of X could be constructed using pattern itemsets in $\Psi(X)$. In particular, let $X = \{a, b, c\}$. Then, $\Psi(X) = \{\{a, b, c\}, \{a, b, \neg c\}, \{a, \neg b, c\}, \{a, \neg b, \neg c\}, \{\neg a, b, c\}, \{\neg a, b, \neg c\}, \{\neg a, \neg b, c\}\}$. Boolean expression $\neg b \wedge c$ could be expressed by the pattern itemsets as follows: $(a \wedge \neg b \wedge c) \vee (\neg a \wedge \neg b \wedge c)$. Every Boolean expression could be expressed by pattern itemsets in the corresponding generator. A Boolean expression expressed as a sum of pattern itemsets is said to be in *canonical form*. Each pattern itemset corresponds to a set of transactions in D. In the following, we show how each pattern itemset with respect to $\{a, b, c\}$ corresponds to a set of transactions in D.

The shaded region in Fig. 2.12a contains the set of transactions containing the items a, b and c with respect to $\{a, b, c\}$. Thus, it corresponds to the pattern itemset of $\langle\{a, b, c\}, \{a, b, c\}\rangle$. The shaded region in Fig. 2.12b contains the set of transactions containing the items a, and b, but not the item c, with respect to

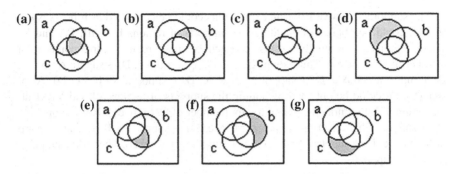

Fig. 2.12 Generator of {a, b, c}. **a** a ∧ b ∧ c, **b** a ∧ b ∧ ¬c, **c** a ∧ ¬b ∧ c, **d** a ∧ ¬b ∧ ¬c, **e** ¬a ∧ b ∧ c, **f** ¬a ∧ b ∧ ¬c, **g** ¬a ∧ ¬b ∧ c

{a, b, c}. Thus, it corresponds to the pattern itemset of ⟨{a, b}, {a, b, c}⟩. The shaded region in Fig. 2.12c contains the set of transactions containing the items a and c, but not the item b, with respect to {a, b, c}. Thus, it corresponds to the pattern itemset of ⟨{a, c}, {a, b, c}⟩. The shaded region in Fig. 2.12d contains the set of transactions containing the item a, but not the items b and c, with respect to {a, b, c}. Thus, it corresponds to the pattern itemset of ⟨{a}, {a, b, c}⟩. The shaded region in Fig. 2.12e contains the set of transactions containing the items b and c, but not the item a, with respect to {a, b, c}. Thus, it corresponds to the pattern itemset of ⟨{b, c}, {a, b, c}⟩. The shaded region in Fig. 2.12f contains the set of transactions containing the item b, but not the items a and c, with respect to {a, b, c}. Thus, it corresponds to the pattern itemset of ⟨{b}, {a, b, c}⟩. Finally, the shaded region in Fig. 2.12g contains the set of transactions containing the item c, but not the items a and b, with respect to {a, b, c}. Thus, it corresponds to the pattern itemset of ⟨{c}, {a, b, c}⟩.

Let X = {a, b, c}. Consider the Boolean expressions E_1({a, b, c}) = c ∨ (a ∧ ¬b) and E_2({a, b, c}) = ¬a ∧ ¬c given in Fig. 2.13.

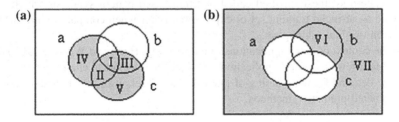

Fig. 2.13 Boolean expressions E_1({a, b, c}) = c ∨ (a ∧ ¬b) and E_2({a, b, c}) = ¬a ∧ ¬c represent the shaded areas of (**a**) and (**b**) respectively

The supports of above Boolean expressions could be computed as follows. supp (E_1, D) could be obtained by adding the supports of regions I, II, III, IV, and V. These regions are mutually exclusive. Each of these regions corresponds to a member of $\Psi(\{a, b, c\})$. Thus, $supp(E_1, D) = supp(a \wedge b \wedge c, D) + supp(a \wedge \neg b \wedge c, D) + supp(\neg a \wedge b \wedge c, D) + supp(a \wedge \neg b \wedge \neg c, D) + supp(\neg a \wedge \neg b \wedge c, D)$. Also, $supp(E_2, D)$ could be obtained by adding the supports of regions VI and VII. But, the region VII does not correspond to any member of $\Psi(\{a, b, c\})$. Now, $supp(\neg E_2, D) = supp(a \wedge b \wedge c, D) + supp(a \wedge \neg b \wedge c, D) + supp(\neg a \wedge b \wedge c, D) + supp(a \wedge \neg b \wedge \neg c, D) + supp(\neg a \wedge \neg b \wedge c, D) + supp(a \wedge b \wedge \neg c, D)$. Therefore, $supp(E_2, D) = 1 - supp(\neg E_2, D)$.

2.6 Related Work

Agrawal et al. (1993) introduce association rule and support-confidence framework and an algorithm to mine frequent itemsets. The algorithm is sometimes called AIS after the authors' initials. Since then, many algorithms have been reported to generate association rules in a database. Association rule mining finds interesting association between two itemsets in a database. Agrawal and Srikant (1994) introduce apriori algorithm that uses breadth-first search strategy to count the support of itemsets. The algorithm uses an improved candidate generation function, which exploits the downward closure property of support and makes it more efficient than AIS. Han et al. (2000) describe the data mining method FP-growth that uses an extended prefix-tree structure to store the databases in a compressed form. FP-growth adopts a divide-and-conquer approach to decompose both the mining tasks and databases. It uses a pattern fragment growth method to avoid the costly process of candidate generation and testing. Savasere et al. (1995) have introduced partition algorithm. The database is scanned only twice. For the first scan, the database is partitioned and in each partition support is counted. Then the counts are merged to generate potential frequent itemsets. In the second scan, the potential frequent itemsets are counted to find the actual frequent itemsets.

In the context of pattern synthesis, Viswanath et al. (2006) have proposed a novel pattern synthesis method called partition based pattern synthesis which can generate an artificial training set of exponential order when compared with that of the given original training set.

In the context of other applications of data mining, Hong and Weiss (2001) have examined a few successful application areas and their technical challenges to show how the demand for data mining of massive data warehouses has fuelled advances in automated predictive methods.

2.7 Conclusion and Future Work

Frequent itemsets could be considered as the basic ingredient of a database. Thus, we could analyze the characteristics of a database in more detail by mining various patterns with respect to frequent itemsets. This chapter introduces conditional patterns in a database and proposes an algorithm to mine them. Thus, we could reveal more characteristics of a database using conditional patterns. Also, we have observed that conditional patterns store significant nuggets of knowledge about a database that are not immediately available from frequent itemsets and association rules. In Sect. 2.5.1, we have presented an application of conditional patterns in a database. In future also, we shall search for more applications of conditional patterns in a database.

References

Adhikari A, Rao PR (2007) A framework for synthesizing arbitrary Boolean expressions induced by frequent itemsets. In: Proceedings of 3rd Indian international conference on artificial intelligence, pp 5–23

Adhikari A, Rao PR (2008) Mining conditional patterns in a database. Pattern Recogn Lett 29 (10):1515–1523

Agrawal R, Srikant R (1994) Fast algorithms for mining association rules. In: Proceedings of 20th very large databases (VLDB) conference, pp 487–499

Agrawal R, Imielinski T, Swami A (1993) Mining association rules between sets of items in large databases. In: Proceedings of ACM SIGMOD conference management of data, pp 207–216

Antonie M-L, Zaïane OR (2004) Mining positive and negative association rules: an approach for confined rules. In: Proceedings of PKDD, pp 27–38

Feller W (1968) An introduction to probability theory and its applications, vol 1, 3rd edn. Wiley, New York

FIMI (2004) http://fimi.cs.helsinki.fi/src/

Frequent itemset mining dataset repository (2004) http://fimi.cs.helsinki.fi/data

Han J, Pei J, Yiwen Y (2000) Mining frequent patterns without candidate generation. In: Proceedings of ACM SIGMOD conference management of data, pp 1–12

Hilderman RJ, Hamilton HJ (1999) Knowledge discovery and interestingness measures: a survey. In: Technical report CS-99-04, Department of Computer Science, University of Regina

Hong SJ, Weiss SM (2001) Advances in predictive models for data mining. Pattern Recogn Lett 22(1):55–61

Savasere A, Omiecinski E, Navathe S (1995) An efficient algorithm for mining association rules in large databases. In: Proceedings of the 21st international conference on very large data bases, pp 432–443

Tan P-N, Kumar V, Srivastava J (2002) Selecting the right interestingness measure for association patterns. In: Proceedings of SIGKDD conference, pp 32–41

Viswanath P, Murty MN, Bhatnagar S (2006) Partition based pattern synthesis technique with efficient algorithms for nearest neighbor classification. Pattern Recogn Lett 27(14):1714–1724

Wu X, Wu Y, Wang Y. Li Y (2005) Privacy-aware market basket data set generation: a feasible approach for inverse frequent set mining. In: Proceedings of SIAM international conference on data mining, pp 103–114

Chapter 3
Synthesizing Arbitrary Boolean Expressions Induced by Frequent Itemsets

Frequent itemsets determine major characteristics of a transactional database. An arbitrary Boolean expression can be thought as a generalized form of a query. It offers important knowledge to an organization. It is important to mine arbitrary Boolean expressions induced by frequent itemsets. In this chapter, we have introduced the concept of generator of an itemset, and showed that every Boolean function can be synthesized by its generator. The concept of conditional pattern has been introduced in Chap. 2. We discussed a simple and elegant framework for synthesizing generator of an itemset and designed an algorithm for this purpose. Experimental results are provided on four different databases.

3.1 Introduction

An itemset could be thought as a basic type of pattern in a transactional database. Itemset patterns influence heavily KDD research in the following ways: Many interesting algorithms have been reported on mining itemset patterns in a database (FIMI 2004; Muhonen and Toivonen 2006; Savasere et al. 1995). Secondly, many patterns are defined based on the itemset patterns in a database. They may be called as derived patterns. For example, positive association rule and negative association rules are examples of derived patterns. A good amount of work has been reported on mining/synthesizing such derived patterns in a database (Agrawal and Srikant 1994; Han et al. 2000; Savasere et al. 1995; Wu et al. 2004). Also, solutions of many problems are based on the analysis of patterns in a database. Such applications (Wang et al. 2001; Wu et al. 2005a) process patterns in a database for the purpose of making some decisions. Thus, the mining and analysis of itemset patterns in a database is an interesting as well as important issue. Also, mining Boolean expressions induced by frequent itemsets could lead to significant nuggets of knowledge, with many potential applications in market basket data analysis, web usage mining, social network analysis and bioinformatics.

The *support* (supp) (Agrawal et al. 1993) of an itemset X in database D could be defined as the fraction of transactions in D containing all the items of X, and it is

© Springer International Publishing Switzerland 2015   31
A. Adhikari and J. Adhikari, *Advances in Knowledge Discovery in Databases*,
Intelligent Systems Reference Library 79, DOI 10.1007/978-3-319-13212-9_3

denoted by $supp(X, D)$. The importance of an itemset is judged by its support. The itemset X is *frequent* in D if $supp(X, D) \geq$ *minimum support* (α). Let $SFIS(D)$ be the set of frequent itemsets in D. Frequent itemsets determine major characteristics of a database. Wu et al. (2005b) have proposed a solution of inverse frequent itemset mining. They argued that one could efficiently generate a synthetic market basket data set from the frequent itemsets and their supports. Let X and Y be two itemsets in D. The characteristics of D are revealed more by the pair $(X, supp(X, D))$ than that of $(Y, supp(Y, D))$, if $supp(X, D) > supp(Y, D)$. Thus, it is important to study frequent itemsets more than infrequent itemsets. Hence, we propose a framework to synthesize arbitrary Boolean expressions induced by frequent itemsets in D. The proposed framework of synthesizing Boolean expressions is based on conditional patterns in D. First, we explain the concept of conditional pattern, and then we present a framework for synthesizing Boolean expressions induced by frequent itemsets in D.

In particular, let $X = \{a, b, c\}$ be a frequent itemset in D. The study of items in X might be incomplete if we have the following information about X: (i) the supports of X and its subsets, (ii) the association rules generated from X. The answers to some queries on the items of X are not immediately available from (i) and (ii). A few examples of such queries are given below:

- Find the probability that a transaction contains item a, but not items b and c with respect to frequent itemset $\{a, b, c\}$.
- Find the probability that a transaction contains items a and b but not the item c with respect to frequent itemset $\{a, b, c\}$.

The above queries correspond to a specific type of pattern in a database. These patterns are termed as conditional patterns in a database. More details about these patterns are presented in Chap. 2. It is necessary to study such patterns in a database for an effective analysis of items in the frequent itemsets.

Let $X = \{a_1, a_2, ..., a_m\}$ be a set of m binary variables, or items. Let $\vee$, $\wedge$ and $\neg$ denote the usual AND, OR and NOT operators in Boolean algebra respectively. An arbitrary *Boolean expression induced by X* could be constructed using the following steps:

(i) a_i is a Boolean expression, $i = 1, 2, ..., m$.
(ii) If a_i is a Boolean expression then $\neg a_i$ is a Boolean expression, $i = 1, 2, ..., m$.
(iii) If a_i and a_j are Boolean expressions then $(a_i \vee a_j)$ and $(a_i \wedge a_j)$ are Boolean expressions, $i, j = 1, 2, ..., m$.
(iv) Any expression obtained by applying steps (i), (ii) and (iii) finitely is a Boolean expression.

Our objective is not to give a formal definition of a well formed Boolean expression induced by X, but to understand how a Boolean expression induced by X could be constructed using a step-by-step approach. There are some other non-fundamental operators in Boolean algebra. Some examples of non-fundamental operators are NAND, NOR, and XOR. Any Boolean expression could be expressed by the set of operators $\{\neg, \wedge, \vee\}$. Thus, it is a functionally complete set of operators.

Using De Morgan's laws, one could show that $\{\neg, \wedge\}$ and $\{\neg, \vee\}$ are the minimal sets of operators by which any Boolean function could be expressed. Thus, $\{\neg, \wedge\}$ and $\{\neg, \vee\}$ are also functionally complete sets of operators. An elaborate discussion on Boolean algebra could be found in Gregg (1998).

The pattern itemset of a conditional pattern with reference to itemset $X = \{a_1, a_2, ..., a_m\}$ is of the form $b_1 \wedge b_2 \wedge \cdots \wedge b_m$, where $b_i = a_i$, or $\neg a_i$, $i = 1, 2, ..., m$. Let $\psi(X)$ be the set of all such pattern itemsets with reference to X. $\psi(X)$ is called the *generator* of Boolean expressions induced by X. $\Psi(X)$ contains $2^m - 1$ pattern itemsets. A pattern itemset of the corresponding conditional pattern is also called a *minterm*, or *standard product*. Every Boolean expression of items of X could be constructed using pattern itemsets in $\psi(X)$ (Lemma 3.3). In particular, let $X = \{a, b, c\}$. Then, $\psi(X) = \{a \wedge b \wedge c, a \wedge b \wedge \neg c, a \wedge \neg b \wedge c, a \wedge \neg b \wedge \neg c, \neg a \wedge b \wedge c, \neg a \wedge b \wedge \neg c, \neg a \wedge \neg b \wedge c\}$. The Boolean expression $\neg b \wedge c$ could be re-written as $(a \wedge \neg b \wedge c) \vee (\neg a \wedge \neg b \wedge c)$. Every Boolean expression can be expressed as a sum of some pattern itemsets in corresponding generator. A Boolean expression expressed as a sum of pattern itemsets is said to be in *canonical form*. Each pattern itemset corresponds to a set of transactions in D. In the following, we show how each pattern itemset with reference to $\{a, b, c\}$ corresponds to a set of transactions in D.

The shaded region in Fig. 3.1(a) contains the set of transactions containing the items a, b and c with respect to $\{a, b, c\}$. Thus, it corresponds to the pattern itemset of $\langle a \wedge b \wedge c, a \wedge b \wedge c \rangle$. The shaded region in Fig. 3.1(b) contains the set of transactions containing the items a, and b, but not the item c with respect to $\{a, b, c\}$. Thus, it corresponds to the pattern itemset of $\langle a \wedge b, a \wedge b \wedge c \rangle$. The shaded region in Fig. 3.1(c) contains the set of transactions containing the items a and c, but not the item b with respect to $\{a, b, c\}$. Thus, it corresponds to the pattern itemset of $\langle a \wedge c, a \wedge b \wedge c \rangle$. The shaded region in Fig. 3.1(d) contains the set of transactions containing the item a, but not the items b and c with respect to $\{a, b,$

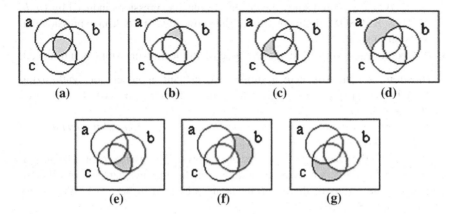

Fig. 3.1 Generator of $\{a, b, c\}$. **a** $a \wedge b \wedge c$, **b** $a \wedge b \wedge \neg c$, **c** $a \wedge \neg b \wedge c$, **d** $a \wedge \neg b \wedge \neg c$, **e** $\neg a \wedge b \wedge c$, **f** $\neg a \wedge b \wedge \neg c$, **g** $\neg a \wedge \neg b \wedge c$

c}. Thus, it corresponds to the pattern itemset of $\langle a, a \wedge b \wedge c \rangle$. The shaded region in Fig. 3.1(e) contains the set of transactions containing the items b and c, but not the item a with respect to $\{a, b, c\}$. Thus, it corresponds to the pattern itemset of $\langle b \wedge c, a \wedge b \wedge c \rangle$. The shaded region in Fig. 3.1(f) contains the set of transactions containing the item b, but not the items a and c with respect to $\{a, b, c\}$. Thus, it corresponds to the pattern itemset of $\langle b, a \wedge b \wedge c \rangle$. Finally, the shaded region in Fig. 3.1(g) contains the set of transactions containing the item c, but not the items a and b with respect to $\{a, b, c\}$. Thus, it corresponds to the pattern itemset of $\langle c, a \wedge b \wedge c \rangle$.

We number the conditional patterns with reference to itemset $\{a_1, a_2, \ldots, a_m\}$ as $1, 2, \ldots, 2^m - 1$. The numbering scheme with respect to frequent itemset $\{a, b, c\}$ is given as follows. $\langle c, a \wedge b \wedge c \rangle$ is numbered as 1, $\langle b, a \wedge b \wedge c \rangle$ is numbered as 2, $\langle b \wedge c, a \wedge b \wedge c \rangle$ is numbered as 3, and so on. Thus, the members of ψ $(\{a, b, c\})$ get numbered as follows. $\neg a \wedge \neg b \wedge c$ gets number as 1, $\neg a \wedge b \wedge \neg c$ gets number as 2, $\neg a \wedge b \wedge c$ gets number as 3, and so on.

The data in a transaction is binary in nature. We could construct 2^{2^m} Boolean expressions (functions) for m Boolean items (variables). Thus, the number of Boolean expressions induced by all frequent itemsets in a database could be large. But, there are only $2^m - 1$ conditional patterns corresponding to a frequent itemset of size m. Thus, it could be better to mine the generator of an itemset and synthesize the desired Boolean expressions afterwards. Zhao et al. (2006) have proposed BLOSOM framework for mining arbitrary Boolean expressions. The framework suffers from the following limitations:

- It does not handle NOT operator effectively.
- Let $\{a, b, c\}$ be a frequent itemset of our interest. We wish to mine some functions induced by $\{a, b, c\}$. It proposes a framework to mine minimal generators of (i) closed OR-clauses, (ii) closed AND-clauses, (iii) closed maximal min-DNF, and (iv) closed maximal min-CNF. It requires establishing a mapping from the space of minimal generators to the space of arbitrary Boolean expressions, so that we could study the desired Boolean expressions induced by $\{a, b, c\}$. Thus, BLOSOM might not provide the knowledge of Boolean expression that we wish to study.
- A specific framework for a specific type of Boolean expressions is introduced.

Therefore, we propose here a simple and elegant approach for synthesizing arbitrary Boolean expressions induced by frequent itemsets. We state the proposed problem as follows.

We are given a database D of customer transactions. Mine all the members of $\psi(X)$, for all $X \in SFIS(D)$ such that $|X| \geq 2$.

The rest of the chapter is organized as follows. We discuss related results in Sect. 3.2. In Sect. 3.3, we propose an algorithm for mining members of different generators. The results of the experiments are presented in Sect. 3.4. We discuss related work in Sect. 3.5.

3.2 Related Results

In this section, we discuss a few results related to discussion held in Sect. 3.1.

Lemma 3.1 *Let E be the event that a transaction contains at least one item of the itemset X in database D. Then, the probability of event E in D,* $P(E, D) = \sum_{Y \subseteq X, Y \neq \varphi} P\langle Y, X, D \rangle.$

Proof We state the theorem of total probability as follows (Feller 1968). For any m events $X_1, X_2, \ldots, X_m$ on database D,

$$P\left(\bigcup_{i=1}^{m} X_i, D\right) = \sum_{i=1}^{m} P(X_i, D) - \sum_{i<j;\, i,\, j=1}^{m} P(X_i \cap X_j, D) + \cdots + (-1)^{m+1}$$
$$\times P\left(\bigcup_{i=1}^{m} X_i, D\right) \tag{3.1}$$

From the definition of conditional pattern, we conclude that the events $\langle Y, X \rangle$ and $\langle Z, X \rangle$ are mutually exclusive, for $Y \neq Z$, and $Y, Z \subseteq X$. Thus, $P(\langle Y, X \rangle \cap \langle Z, X \rangle, D) = 0$, for $Y \neq Z$, and $Y, Z \subseteq X$. We apply the theorem of total probability to the conditional patterns with reference to itemset X. All the terms except the first term on the right hand side of (3.1) become zero, and the result follows. $\square$

Lemma 3.2 *Let X and Y be two itemsets in database D such that* $Y \subseteq X$. *Let* $Z = X - Y = \{a_1, a_2, \ldots, a_m\}$. *Then,* $P(Y, D) = \sum_{W \in \rho(Z)} P(\langle Y \wedge W, X \rangle, D)$, $\rho(Z) = \psi(Z) \cup \{\neg a_1 \wedge \neg a_2 \wedge \cdots \wedge \neg a_{m-1} \wedge \neg a_m\}$ *(Adhikari and Rao 2007).*

Proof The proof is based on induction on m. Now, $P(Y, D) = P(Y \wedge a_1, D) + P(Y \wedge \neg a_1, D)$. The result is true for $m = 1$. Let the result is true for $m \leq k$. We shall show that the result is true for $m = k + 1$. For $m = k$, we have $P(Y, D) = P(Y \wedge a_1 \wedge a_2 \wedge \cdots \wedge a_k, D) + P(Y \wedge \neg a_1 \wedge a_2 \wedge \cdots \wedge a_k, D) + P(Y \wedge a_1 \wedge \neg a_2 \wedge \cdots \wedge a_k, D) + P(Y \wedge \neg a_1 \wedge \neg a_2 \wedge \cdots \wedge a_k, D) + \cdots + P(Y \wedge \neg a_1 \wedge \neg a_2 \wedge \cdots \wedge \neg a_{k-1} \wedge a_k, D) + P(Y \wedge \neg a_1 \wedge \neg a_2 \wedge \cdots \wedge \neg a_{k-1} \wedge \neg a_k, D)$ [by induction hypothesis]. After incorporating a_{k+1}, we get $P(Y, D) = P(Y \wedge a_1 \wedge a_2 \wedge \cdots \wedge a_k \wedge a_{k+1}, D) + P(Y \wedge a_1 \wedge a_2 \wedge \cdots \wedge a_k \wedge \neg a_{k+1}, D) + P(Y \wedge \neg a_1 \wedge a_2 \wedge \cdots \wedge a_k \wedge a_{k+1}, D) + P(Y \wedge \neg a_1 \wedge a_2 \wedge \cdots \wedge a_k \wedge \neg a_{k+1}, D) + \cdots + P(Y \wedge \neg a_1 \wedge \neg a_2 \wedge \cdots \wedge \neg a_{k-1} \wedge \neg a_k \wedge a_{k+1}, D) + P(Y \wedge \neg a_1 \wedge \neg a_2 \wedge \cdots \wedge \neg a_k {}_{-1} \wedge \neg a_k \wedge \neg a_{k+1}, D)$. The result is true for $m = k + 1$. $\square$

Let us take an example. Let $Y = a$, and $X = a \wedge b \wedge c$. Then, $P(Y, D) = P(Y \wedge b, D) + P(Y \wedge \neg b, D) = P(Y \wedge b \wedge c, D) + P(Y \wedge b \wedge \neg c, D) + P(Y \wedge \neg b \wedge c, D) + P(Y \wedge \neg b \wedge \neg c, D)$.

Lemma 3.3 *The framework enables in synthesizing every Boolean expression induced by a frequent itemset (Adhikari and Rao 2007).*

Proof Let $X = \{a_1, a_2, \ldots, a_m\}$ be a frequent itemset. Then, $\Psi(X) = \{a_1 \wedge a_2 \wedge \cdots \wedge a_m, \neg a_1 \wedge a_2 \wedge \cdots \wedge a_m, \ldots, a_1 \wedge \neg a_2 \wedge \cdots \wedge a_m, \neg a_1 \wedge \neg a_2 \wedge \cdots \wedge a_m, \ldots, \neg a_1 \wedge \cdots \wedge \neg a_{m-1} \wedge a_m\}$, where $|\psi(X)| = 2^m - 1$. Then, the members (i.e., pattern itemsets) in $\psi(X)$ form the basic building blocks for constructing an arbitrary Boolean expression induced by X. $\psi(X)$ induces a partition (Liu 1985) of the set of transactions in D. The partition contains 2^m subsets of transactions in D. Again, each subset of transactions corresponds to a member of $\psi(X)$, except the last i.e., 2^mth subset of transactions. The last subset of transactions corresponds to the set of transactions where the Boolean expression $\neg a_1 \wedge \cdots \wedge \neg a_{m-1} \wedge \neg a_m$ is true. The support of this Boolean expression could be computed with the help the supports of the members in $\psi(X)$. Let $E(X)$ be an arbitrary Boolean expression induced by X. Thus, the support of either $E(X)$ or $\neg E(X)$ could be obtained by adding supports of some members in $\psi(X)$. Thus, it is possible to synthesize the support of every Boolean expression induced by X.   □

In particular, let $X = \{a, b, c\}$. Consider the Boolean expressions $c \vee (a \wedge \neg b)$ and $\neg a \wedge \neg c$.

The supports of the Boolean expressions in Fig. 3.2 could be computed as follows. $supp(c \vee (a \wedge \neg b), D)$ could be obtained by adding the supports of regions I, II, III, IV, and V. These regions are mutually exclusive. Each of these regions corresponds to a member of $\psi(\{a, b, c\})$. Thus, $supp(E_1, D) = supp(a \wedge b \wedge c, D) + supp(a \wedge \neg b \wedge c, D) + supp(\neg a \wedge b \wedge c, D) + supp(a \wedge \neg b \wedge \neg c, D) + supp(\neg a \wedge \neg b \wedge c, D)$. Also, $supp(E_2, D)$ could be obtained by adding the supports of regions VI and VII. But, the region VII does not correspond to any member of $\psi(\{a, b, c\})$. Now, $supp(\neg E_2, D) = supp(a \wedge b \wedge c, D) + supp(a \wedge \neg b \wedge c, D) + supp(\neg a \wedge b \wedge c, D) + supp(a \wedge \neg b \wedge \neg c, D) + supp(\neg a \wedge \neg b \wedge c, D) + supp(a \wedge b \wedge \neg c, D)$. Therefore, $supp(E_2, D) = 1 - supp(\neg E_2, D)$.

Lemma 3.4 *Let SFIS(D) be the set of frequent itemsets in database D. Then the number of distinct conditional patterns satisfying minimum support criterion is* $\sum_{X \in SFIS(D)} \left(2^{|X|} - 1\right)$.

Proof The number of non-null subsets of an itemset X is $2^{|X|} - 1$, for $X \in SFIS(D)$. Each non-null subset of X corresponds to a conditional pattern. Thus, the result follows.   □

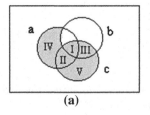 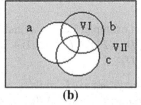

(a)   (b)

Fig. 3.2 Boolean expressions $E_1(\{a, b, c\}) = c \vee (a \wedge \neg b)$ and $E_2(\{a, b, c\}) = \neg a \wedge \neg c$ represent the *shaded areas* in this figure (**a**) and (**b**) respectively

3.3 Synthesizing Generators

Conditional patterns are derived from the frequent itemsets in a database. Let X be a frequent itemset in D. We shall express $csupp\langle Y, X, D \rangle$ in terms of the supports of the frequent itemsets in D, $Y \subseteq X$. Without any loss of generality, let $X = Y \cup Z$, where $Z = \{a_1, a_2, ..., a_m\}$. The conditional support of ith conditional pattern with reference to X is same as the support of ith member of $\psi(X)$, $i = 1, 2, ..., 2^{|X|} - 1$. Thus, the following theorem enables us to compute the supports of members of $\psi(X)$ in D, for all $X \in SFIS(D)$, and $|X| \geq 2$ (Adhikari and Rao 2007).

Lemma 3.5 *Let X, Y and Z are itemsets in database D such that $X = Y \cup Z$, where $Z = \{a_1, a_2, ..., a_m\}$. Then,*

$$P\langle Y, X, D \rangle = P(Y, D) - \sum_{i=1}^{m} P(Y \cap \{a_i\}, D) + \sum_{i<j; \, i, j=1}^{m} P(Y \cap \{a_i, a_j\}, D)$$

$$- \sum_{i<j<k; \, i, j, k=1}^{m} P(Y \cap \{a_i, a_j, a_k\}, D)$$

$$+ \cdots + (-1)^m \times P(Y \cap \{a_1, a_2, ..., a_m\}, D)$$

$$(3.2)$$

Proof We shall prove the result using the method of induction on m. For $m = 1$, $X = Y \cap \{a_1\}$. Then, $P\langle Y, X, D \rangle = P(Y, D) - P(Y \cap \{a_1\}, D)$. Thus, the result is true for $m = 1$. Let the result is true for $m = p$ (induction hypothesis). We shall prove that the result is true for $m = p + 1$. Let $Z = \{a_1, a_2, ..., a_p, a_{p+1}\}$. Due to the addition of a_{p+1}, the following observations are made: $P(Y \cap \{a_{p+1}\}, D)$ is required to be subtracted, $P(Y \cap \{a_i, a_{p+1}\}, D)$ is required to be added, for $1 \leq i \leq p$, and lastly, the term $(-1)^{p+1} \times P(Y \cap \{a_1, a_2, ..., a_{p+1}\}, D)$ is required to be added. Thus,

$$P\langle Y, X, D \rangle = P(Y, D) - \sum_{i=1}^{p+1} P(Y \cap \{a_i\}, D) + \sum_{i<j; \, i, j=1}^{p+1} P(Y \cap \{a_i, a_j\}, D)$$

$$- \sum_{i<j<k; \, i, j, k=1}^{p+1} P(Y \cap \{a_i, a_j, a_k\}, D)$$

$$+ \cdots + (-1)^{p+1} \times P(Y \cap \{a_1, a_2, ..., a_{p+1}\}, D).$$

$\square$

3.3.1 Algorithm Design

For synthesizing arbitrary Boolean expressions induced by frequent itemsets in a database, we make use of an existing frequent itemset mining algorithm (Agrawal and Srikant 1994; Han et al. 2000; Savasere et al. 1995). We synthesize only the generator of Boolean expressions induced by a frequent itemset. The generator of Boolean expressions induced by the frequent itemset X contains $2^{|X|} - 1$ pattern itemsets. The proposed algorithm synthesizes all the members of all the generators. There are two approaches of synthesizing generators of Boolean expressions induced by frequent itemsets in a database: In the first approach, we synthesize the generator from the current frequent itemset. As soon as a frequent itemset is extracted, we could call an algorithm for synthesizing members of the corresponding generator. When a frequent itemset is found, then all the non-null subsets of this frequent itemest have already been extracted. Thus, we could synthesize all the members of the generator from the frequent itemsets extracted so far. In the second approach, we synthesize members of the different generators after mining all the frequent itemsets. In this approach, all the frequent itemsets are processed after the mining task. These two approaches seem to be the same so far as the computational complexity is concerned. In this chapter, we have followed the second approach of synthesizing members of different generators. During the process of mining frequent itemsets, the frequent itemsets of smaller size get extracted before the frequent itemsets of larger size. The frequent itemsets are kept in array *SFIS*. During the processing of current frequent itemset, all the non-null subsets are available before the current itemset in SFIS.

There are $2^{|X|} - 1$ non-null subsets of an itemset X. Each non-null subset of X corresponds to a conditional pattern, and hence, it corresponds to a member of the generator of Boolean expressions induced by X. The subset X of X corresponds to a trivial conditional pattern, and gets mined during the mining of frequent itemsets in D. Thus, we need to process $2^{|X|} - 2$ subsets of X.

One could view a conditional pattern as an object with the following attributes: *pattern*, *reference*, *csupp*, and *rsupp*. We use an array *CP* to store the conditional patterns in a database. The reference attribute of the ith conditional pattern is accessed by the notation *CP(i)*.reference. Similar notations are used to access other attributes of a conditional pattern. Also, a frequent itemset could be viewed as an object with the following attributes: itemset and supp. Let N be the number of frequent itemsets in the given database D. The algorithm *synthesizingGenerators* synthesizes all the members of $\psi(X)$, $X \in SFIS(D)$ (Adhikari and Rao 2007).

procedure *synthesizingGenerators* (*N, SFIS*)
Input:
N: number of frequent itemsets in the given database
SFIS: set of frequent itemsets in the given database
Output:
Generators corresponding to the frequent itemsets
1:   $i \leftarrow 1$;
2:   $j \leftarrow 0$;
3:   **while** (i ≤ N) **do**
4:       $CP(j).rsupp \leftarrow SFIS(i).supp$; $CP(j).reference \leftarrow SFIS(i).itemset$;
5:       $sum \leftarrow 0$;
6:       **for** $k \leftarrow 1$ to $(2^{|SFIS(i).itemset|} - 1)$ **do**
7:           $tempItemset \leftarrow k$-th subset of $SFIS(i).itemset$;
8:           **if** $(SFIS(i).itemset = tempItemset)$ **then**
9:               $sum \leftarrow SFIS(i).supp$; **goto** 19;
10:         **end if**
11:         $kk \leftarrow 1$;
12:         **while** (kk ≤ i) **do**
13:             **if** $(SFIS(kk).itemset = tempItemset)$ **then**
14:                 $sum \leftarrow sum + (-1)^{|SFIS(kk).itemset| - |tempItemset|} \times SFIS(kk).supp$;
15:                 break;
16:             **end if**
17:             $kk \leftarrow kk + 1$;
18:         **end while**
19:         $CP(j).csupp \leftarrow sum$; $CP(j).pattern \leftarrow tempItemset$; $j \leftarrow j + 1$;
20:         $i \leftarrow i + 1$;
21:     **end for**
22: **end while**
23: $t \leftarrow 0$;
24: **for** $i \leftarrow 1$ to N **do**
25:     **for** $k \leftarrow 1$ to $(2^{|SFIS(i).itemset|} - 1)$
26:         display $CP(t + k)$;
27:     **end for**
28:     $t \leftarrow t + 2^{|SFIS(i).itemset|} - 1$;
29: **end for**
30: **end procedure**

Variable *i* keeps track of the current frequent itemset being processed. Variable *j* keeps track of number of conditional patterns generated. Using lines 3–22, each frequent itemset is processed. There are $2^{|X|} - 1$ non-null subsets for an itemset *X*. Each non-null subset corresponds to a conditional pattern. The generator of an itemset *X* is synthesized using lines 6–21. Let *Y* be a subset of *X*. If *Y* = *X* then the algorithm bypasses the processing of *Y* (line 8). When the algorithm synthesizes generator corresponding to a frequent itemset *X*, then it has already finished the processing of its non-null subsets. All the non-null subsets appear on or before *X* in the array *SFIS*. Thus, if a frequent itemset *X* located at position i, then we search for a subset of *X* from index 1 to *i* in array *SFIS*, since the array is sorted non-decreasing order on length of an itemset. Thus, it justifies the condition of the while loop at line 12. Formula (3.2) expresses $P\langle Y, X, D\rangle$ in terms of $P(Y \wedge Z, D)$, for all $Z \subseteq X - Y$. The coefficient of $P(Y \wedge Z, D)$ is $(-1)^{|Z|}$ in the expression of $P\langle Y, X, D\rangle$. Thus, $P\langle Y, X, D\rangle = \sum_{Z \subseteq X-Y} (-1)^{|Z|} \times P(Y \wedge Z, D)$. This formula has been applied to line 14 to calculate $P\langle Y, X, D\rangle$. We need to synthesize both interesting and non-interesting conditional patterns with reference to frequent itemsets for the solution to the proposed problem. Lines 25–27 display the generator corresponding to ith frequent itemset, for $i = 1, 2, \ldots, N$. The generator corresponding to ith frequent itemset contains $2^{|SFIS(i).itemset|} - 1$ members (i.e., pattern itemsets), $i = 1, 2, \ldots, N$.

Lemma 3.6 *Algorithm synthesizingGenerators executes in $O(N^2 \times 2^p)$ time, where N and p are the number of frequent itemsets and average size of frequent itemsets in the database respectively.*

Proof The *while*-loop at line 3 repeats N times. The *for*-loop at line 6 repeats $2^p - 1$ times. Also, the *while*-loop at line 12 repeats maximum of N times. Thus, the time complexity of lines 3–22 is $O(N^2 \times 2^p)$. The time complexity of lines 24–29 is $O(N \times 2^p)$. Therefore, the time complexity of algorithm *synthesizingGenerators* is $O(N^2 \times 2^p)$. □

3.3.2 Synthesizing First k Boolean Expressions Induced by Top p Frequent Itemsets

Using the truth table, one could determine the algebraic forms of Boolean expressions induced by a frequent itemset. A Boolean expression could be synthesized by the members of the corresponding generator. We classify the frequent itemsets in a database into different categories. The frequent itemsets of the same size are put in the same category. We sort the frequent itemsets of each category in non-increasing order by support and top frequent itemsets in each category are considered for synthesis. We do experiments for synthesizing first k Boolean expressions induced by top p frequent itemsets of each category. We have shown this phenomenon using Example 3.1.

Example 3.1 Let $\{a, b\}$ and $\{a, b, c\}$ be two frequent itemsets in D of size 2 and 3 respectively. We would like to determine first k Boolean expressions induced by $\{a, b\}$ and $\{a, b, c\}$. Let E_{ij} be the jth Boolean expression induced by the frequent itemset of size i, $j = 1, 2, \ldots, 2^i - 1$, and $i = 2, 3$. Then the truth tables for the first six Boolean expressions are given in Table 3.1.

First six Boolean expressions are based on the items of $\{a, b\}$, and the rest of the Boolean expressions are based on the items of $\{a, b, c\}$. The algebraic expressions of first six Boolean expressions are given as follows: $E_{21}(a, b) = 0$, $E_{22}(a, b) = a \wedge b$, $E_{23}(a, b) = a \wedge \neg b$, $E_{24}(a, b) = a$, $E_{25}(a, b) = \neg a \wedge b$, $E_{26}(a, b) = b$; $E_{31}(a, b, c) = 0$, $E_{32}(a, b, c) = a \wedge b \wedge c$, $E_{33}(a, b, c) = a \wedge b \wedge \neg c$, $E_{34}(a, b, c) = a \wedge b$, $E_{35}(a, b, c) = a \wedge \neg b \wedge c$, $E_{36}(a, b, c) = a \wedge c$. We express E_{ij}s in terms of members

Table 3.1 Truth tables for the first six Boolean expressions induced by {a, b} and {a, b, c}

a	b	c	E_{21}	E_{21}	E_{23}	E_{24}	E_{25}	E_{26}	E_{31}	E_{32}	E_{33}	E_{34}	E_{35}	E_{36}
0	0	0	0	0	0	0	0	0	0	0	0	0	0	0
0	0	1	0	0	0	0	0	0	0	0	0	0	0	0
0	1	0	0	0	0	0	1	1	0	0	0	0	0	0
0	1	1	0	0	0	0	1	1	0	0	0	0	0	0
1	0	0	0	0	1	1	0	0	0	0	0	0	0	0
1	0	1	0	0	1	1	0	0	0	0	0	0	1	1
1	1	0	0	1	0	1	0	1	0	0	1	1	0	0
1	1	1	0	1	0	1	0	1	0	1	0	1	0	1

of the corresponding generator. Boolean expressions E_{22}, E_{23}, and E_{25} have already been expressed in terms of the members of the concerned generator. We need not compute E_{21} and E_{31}, since $E_{21}(a, b) = E_{31}(a, b, c) = 0$. $E_{24}(a, b) = (a \wedge b) \vee (a \wedge \neg b)$, and $E_{26}(a, b) = (a \wedge b) \vee (\neg a \wedge b)$. Also, $E_{34}(a, b, c) = (a \wedge b \wedge c) \vee (a \wedge b \wedge \neg c)$, and $E_{36}(a, b, c) = (a \wedge b \wedge c) \vee (a \wedge \neg b \wedge c)$. Expressions E_{32}, E_{33} and E_{35} have already been expressed in terms of the members of concerned generator.

3.4 Experiments

Numerous experiments are carried out to study the effectiveness of the proposed approach. We present the experimental results using three real and one synthetic datasets. The dataset *retail* (Frequent itemset mining dataset repository 2004) is obtained from an anonymous Belgian retail supermarket store. The datasets *BMS-Web-Wiew-*1 and *BMS-Web-Wiew-*2 can be found from KDD CUP 2,000 (Frequent itemset mining dataset repository). The dataset *T10I4D100K* (Frequent itemset mining dataset repository) was generated using the generator from IBM Almaden Quest research group. We present some characteristics of these datasets in Table 3.2. Among these four datasets the first three datasets are real and the forth one is synthetic.

Let *NT*, *AFI*, *ALT*, and *NI* denote the number of transactions, average frequency of an item, average length of a transaction and number of items in the corresponding dataset respectively. For the purpose of synthesizing Boolean expressions , we have implemented apriori algorithm (Agrawal and Srikant 1994) since it is simple and easy to implement. In Tables 3.3, 3.4, 3.5 and 3.6, we present first six Boolean expressions induced top five frequent itemsets from *retail*, *BMS-Web-Wiew-*1, *BMS-Web-Wiew-*2 and *T10I4D100K* respectively.

The Boolean functions E_{22} and E_{32} are not shown, since they are all equal to 0. The Boolean expressions induced by frequent itemsets of size one are not studied here.

*BMS-Web-Wiew-*1 and *BMS-Web-Wiew-*2 do not report any frequent itemsets of size greater than two.

Table 3.2 Dataset characteristics

Dataset	NT	ALT	AFI	NI
retail	88,162	11.31	99.67	10,000
BMS-Web-Wiew-1	149,639	2.00	155.71	1,922
BMS-Web-Wiew-2	358,278	2.00	7,165.56	100
T10I4D100K	100,000	11.10	1,276.12	870

Table 3.3 (a) First six Boolean expressions induced by top five frequent itemsets of size 2 in *retail* at $\alpha = 0.05$. (b) First six Boolean expressions induced by top five frequent itemsets of size 3 in *retail* at $\alpha = 0.05$

Frequent itemset	$supp(E_{22}, D)$	$supp(E_{23}, D)$	$supp(E_{24}, D)$	$supp(E_{25}, D)$	$supp(E_{26}, D)$
(a)					
{39, 48}	0.3306	0.2562	0.5868	0.1582	0.4888
{39, 41}	0.1295	0.4573	0.5868	0.0422	0.1717
{38, 39}	0.1173	0.0603	0.1776	0.4694	0.5868
{41, 48}	0.1023	0.0694	0.1717	0.3865	0.4888
{32, 39}	0.0959	0.0793	0.1752	0.4909	0.5868
Frequent itemset	$supp(E_{32}, D)$	$supp(E_{33}, D)$	$supp(E_{34}, D)$	$supp(E_{35}, D)$	$supp(E_{36}, D)$
(b)					
{39, 41, 48}	0.0836	0.0459	0.1295	0.2470	0.3306
{38, 39, 48}	0.0692	0.0481	0.1173	0.0209	0.0901
{32, 39, 48}	0.0613	0.0346	0.0959	0.0299	0.0911
{1, 39, 48}	0.0449	0.0215	0.0663	0.0170	0.0618
{5, 39, 48}	0.0432	0.0197	0.0629	0.0162	0.0594

Table 3.4 First six Boolean expressions induced by top five frequent itemsets of size 2 in *BMS-Web-Wiew*-1 at $\alpha = 0.01$

Frequent itemset	$supp(E_{22}, D)$	$supp(E_{23}, D)$	$supp(E_{24}, D)$	$supp(E_{25}, D)$	$supp(E_{26}, D)$
{5, 7}	0.0139	0.2353	0.2491	0.2012	0.2151
{1, 5}	0.0137	0.1761	0.1899	0.2355	0.2491
{3, 5}	0.0131	0.1953	0.2083	0.2361	0.2491
{5, 9}	0.0129	0.2363	0.2491	0.2014	0.2142
{1, 7}	0.0124	0.1774	0.1899	0.2027	0.2151

T10I4D100K reports only one frequent itemset of size 3. We observe that the proposed framework is simple and elegant. It enables us to synthesize arbitrary Boolean expressions induced by frequent itemsets in a dataset.

Also, we have conducted experiments to study the relationship between the size of a dataset and the execution time required for mining generators. The execution time required for mining generators in a dataset increases as the number of

Table 3.5 First six Boolean expressions induced by top five frequent itemsets of size 2 in *BMS-Web-Wiew-2* at $\alpha = 0.009$

Frequent itemset	$supp(E_{22}, D)$	$supp(E_{23}, D)$	$supp(E_{24}, D)$	$supp(E_{25}, D)$	$supp(E_{26}, D)$
{1, 3}	0.0236	0.1695	0.1932	0.1710	0.1946
{1, 7}	0.0229	0.1702	0.1932	0.1741	0.1970
{3, 7}	0.0228	0.1719	0.1946	0.1743	0.1970
{3, 5}	0.0220	0.1726	0.1946	0.1572	0.1793
{1, 9}	0.0220	0.1712	0.1932	0.1512	0.1732

Table 3.6 (a) First six Boolean expressions induced by top five frequent itemsets of size 2 in *T10I4D100K* at $\alpha = 0.01$. (b) First six Boolean expressions induced by frequent itemsets of size 3 in *T10I4D100K* at $\alpha = 0.01$

Frequent itemset	$supp(E_{22}, D)$	$supp(E_{23}, D)$	$supp(E_{24}, D)$	$supp(E_{25}, D)$	$supp(E_{26}, D)$
(a)					
{217, 346}	0.0134	0.0405	0.0539	0.0214	0.0347
{789, 829}	0.0119	0.0335	0.0454	0.0690	0.0809
{368, 829}	0.0119	0.0665	0.0785	0.0690	0.0809
{368, 682}	0.0119	0.0665	0.0785	0.0319	0.0438
{39, 825}	0.0119	0.0307	0.0426	0.0237	0.0356
Frequent itemset	$supp(E_{32}, D)$	$supp(E_{33}, D)$	$supp(E_{34}, D)$	$supp(E_{35}, D)$	$supp(E_{36}, D)$
(b)					
{39, 704, 825}	0.0104	0.0004	0.0111	0.0015	0.0119

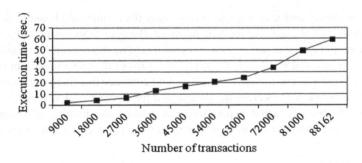

Fig. 3.3 Execution time versus the number of transactions from *retail* at $\alpha = 0.05$

transactions contained in a dataset increases. We observe this phenomenon in Figs. 3.3 and 3.4.

We have also conducted experiments to find the execution time for synthesizing generators in a dataset. The time required (only) for synthesizing generators for each of the above datasets is 0 ms at the respective value of α as shown in Tables 3.3, 3.4, 3.5, and 3.6.

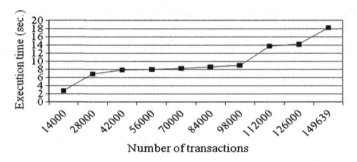

Fig. 3.4   Execution time versus the number of transactions from *BMS-Web-Wiew*-1 at α = 0.01

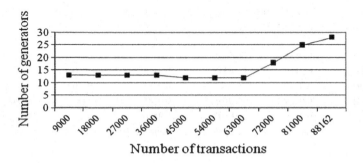

Fig. 3.5   Number of generators versus the number of transactions from *retail* at α = 0.05

Also, we have conducted experiments to study the relationship between the size of a dataset and the number of generators of Boolean expressions induced by frequent itemsets of size greater than or equal to 2. The experiments are conducted on datasets *retail* and *BMS-Web-Wiew*-1. The results of the experiments are shown in Figs. 3.5 and 3.6. From these graphs we could conclude that there is no universal relationship between the size of the dataset and the number of generators in it.

We have also conducted experiments for finding the number of generators corresponding to frequent itemsets of size greater than or equal to 2 in a dataset at a

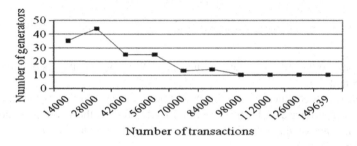

Fig. 3.6   Number of generators versus the number of transactions from *BMS-Web-Wiew*-1 at α = 0.01

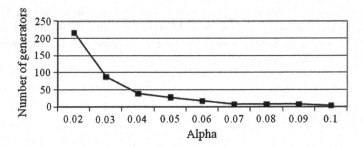

Fig. 3.7 Number of generators versus α for *retail*

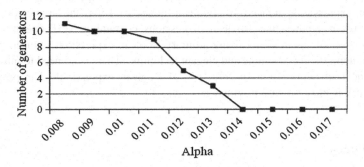

Fig. 3.8 Number of generators versus α for *BMS-Web-Wiew-*1

given α. The number of generators in a dataset decreases as α increases. We observe this phenomenon in Figs. 3.7 and 3.8.

3.5 Related Work

Mining conjunctive Boolean expressions (Agrawal and Srikant 1994; Han et al. 2000; Savasere et al. 1995) have been well studied within the context of frequent itemset mining. Several implementations of mining conjunctive Boolean expressions (Zhao et al. 2006) have been reported. The maximum-entropy approach to support estimation of a general Boolean expression is proposed by Pavlov et al. (2000). Itemset mining typically results in large amounts of redundant itemsets. Several approaches such as closed itemsets, non-derivable itemsets and generators have been suggested for reducing the amount of itemsets losslessly. Muhonen and Toivonen (2006) have proposed a pruning method based on combining techniques for closed and non-derivable itemsets that allows further reductions of itemsets. This reduction is done without loss of information, that is, the complete collection of frequent itemsets can still be derived from the collection of closed non-derivable itemsets. Shima et al. (2004) have proposed a technique of mining closed and minimal monotone disjunctive normal forms. The proposed technique for synthesizing Boolean expression is somewhat different from that of BLOSOM or, the technique proposed by Shima et al. Within the association rule context, there has been previous work on mining negative rules (Wu et al. 2004).

3.6 Conclusion

Frequent itemsets determine major characteristics of a database. Thus, we could analyze the characteristics of a database in more detail by mining arbitrary Boolean expressions. This chapter proposes an algorithm for synthesizing generators of Boolean expressions induced by frequent itemsets. Thus, the framework also enables us in synthesizing Boolean expressions containing NOT operator. The generators enable us in synthesizing arbitrary Boolean expressions induced by the frequent itemsets. It is a simple and elegant technique. There is no need to introduce a specific framework for a specific type of Boolean expressions. The proposed framework is effective and promising.

References

Adhikari A, Rao PR (2007) A framework for synthesizing arbitrary Boolean expressions induced by frequent itemsets. In: Proceedings of Indian international conference on artificial intelligence, pp 5–23

Agrawal R, Srikant R (1994) Fast algorithms for mining association rules. In: Proceedings of 20th very large databases (VLDB) conference, pp 487–499

Agrawal R, Imielinski T, Swami A (1993) Mining association rules between sets of items in large databases. In: Proceedings of ACM SIGMOD conference management of data, pp 207–216

Feller W (1968) An introduction to probability theory and its applications, vol 1, 3rd edn. Wiley, New York

FIMI 2004, http://fimi.cs.helsinki.fi/src/

Frequent itemset mining dataset repository (2004), http://fimi.cs.helsinki.fi/data

Gregg JR (1998) Ones and zeros: understanding Boolean algebra, digital circuits, and the logic of sets. Wiley-IEEE Press, New York

Han J, Pei J, Yiwen Y (2000) Mining frequent patterns without candidate generation. In: Proceedings of ACM SIGMOD conference management of data, pp 1–12

Liu CL (1985) Elements of discrete mathematics, 2nd edn. McGraw-Hill, New York

Muhonen J, Toivonen H (2006) Closed non-derivable itemsets. In: Proceedings of PKDD, pp 601–608

Pavlov D, Mannila H, Smyth P (2000) Probabilistic models for query approximation with large sparse binary data sets. In: Proceedings of 16th conference on uncertainty in artificial intelligence, pp 465–472

Savasere A, Omiecinski E, Navathe S (1995) An efficient algorithm for mining association rules in large databases. In: Proceedings of the 21st international conference on very large data bases, pp 432–443

Shima Y, Mitsuishi S, Hirata K, Harao M, Suzuki E, Arikawa S (2004) Extracting minimal and closed monotone DNF formulas. In: Proceedings of international conference on discovery science, vol 3245, Springer, Berlin, pp 298–305

Wang K, Zhou S, He Y (2001) Hierarchical classification of real life documents. In: Proceedings of the 1st (SIAM) international conference on data mining, pp 1–16

Wu X, Zhang C, Zhang S (2004) Efficient mining of both positive and negative association rules. ACM Trans Inf Syst 22(3):381–405

Wu X, Wu Y, Wang Y, Li Y (2005a) Privacy-aware market basket data set generation: a feasible approach for inverse frequent set mining. In: Proceedings of SIAM international conference on data mining, pp 103–114

Wu X, Zhang C, Zhang S (2005b) Database classification for multi-database mining. Inf Syst 30 (1):71–88

Zhao L, Zaki MJ, Ramakrishnan M (2006) BLOSOM: a framework for mining arbitrary Boolean expressions. In: Proceedings of KDD, pp 827–832

Chapter 4
Measuring Association Among Items in a Database

Measuring association among variables is an important step for finding solutions to many data mining problems. An existing metric might not be effective to serve as a measure of association among a set of items in a database. In this chapter, we propose two measures of association, A_1 and A_2. We introduce the notion of associative itemset in a database. We express the proposed measures in terms of supports of itemsets. In addition, we provide theoretical foundations of our work. We present experimental results on both real and synthetic databases to show the effectiveness of A_2.

4.1 Introduction

The analysis of relationships among variables is a fundamental task being at the heart of many data mining problems. For instance, association rules (Agrawal et al. 1993) find relationships between sets of items in a database of transactions. Such rules express buying patterns of customers e.g., finding how the presence of one item affects the presence of another and so forth.

Many measures of association have been reported in the literature of data mining, machine learning, and statistics. They could be categorized into two groups. Some measures deal with a set of objects, or could be generalized to deal with a set of objects. On the other hand, remaining measures could not be generalized. Confidence (Agrawal et al. 1993), conviction (Brin et al. 1997) are examples of the second category of measures. On the other hand, measures such as Jaccard (Tan et al. 2002) could be generalized to find association among a set of items in a database. We shall see later why measures such as support (Agrawal et al. 1993), generalized Jaccard, all-confidence (Omiecinski 2003) have not been effective in measuring association among a set of items in a database.

Various problems could be addressed using association among a set of items in market basket data. For example, a company might be interested in analyzing items that are purchased frequently. Let the items P, Q, and R be purchased frequently, and a few specific problems are stated below involving these items.

© Springer International Publishing Switzerland 2015

A. Adhikari and J. Adhikari, *Advances in Knowledge Discovery in Databases*,
Intelligent Systems Reference Library 79, DOI 10.1007/978-3-319-13212-9_4

(i) Some items (products) could be high profit making. Naturally, the company would like to promote them. There are various ways one could promote an item. An indirect way of promoting an item P is to promote items that are highly associated with it. The implication of high association between P and Q is that if Q is purchased by a customer then P is likely to be purchased by the same customer at the same time. Thus, P gets indirectly promoted.

(ii) Again, an item say R, could be low-profit making. Thus, it is important to know how it promotes the sales of other items. Otherwise, the company could stop dealing with R.

To solve the above problems, one could cluster the frequent items in a database. In the context of (i), one could promote item P indirectly, by promoting other items in the class containing P. In the context of (ii), the company could keep on dealing with R if the class size containing R is reasonably large. Thus, a suitable metric for capturing association among a set of items could enable us to cluster frequent items in a database. In general, many corporate decisions could be taken effectively by incorporating knowledge inherent in data. Later, we shall show that a measure of association based on a 2×2 contingency table might not be effective in clustering a set of items in a database. Thus, we need a new measure for capturing association among a set of items in a database.

In this chapter, we propose two measures of association among a set of items in a database. The first measure could be considered as a generalized Jaccard measure for capturing association among items. On the other hand, the second measure of association is based on a weighting model. We provide theoretical foundations of the work. For the purpose of measuring association among a set of items, we express the proposed measures in terms of supports of itemsets. The main contributions of this work are given as follows:

1. We propose two measures of association among a set of items in a database.
2. We introduce the notion of associative itemset in a database.
3. We provide theoretical foundations of the work.
4. We express second measure in terms of supports of itemsets.
5. We provide experimental results to show the effectiveness of the second measure of association.

In the following section, we study some existing measures and explain why these measures are not suitable for capturing association among a set of items in a database.

This chapter is organized as follows. We discuss related work in Sect. 4.2. In Sect. 4.3, we propose two new measures of association among a set of items in a database. We discuss various properties of the proposed measures in Sect. 4.4. Also, we express second measure in terms of supports of itemsets. In Sect. 4.5, we mention an application of the proposed measures of association. Experimental results are provided in Sect. 4.6 to show the effectiveness of the second measure of association.

4.2 Related Work

Tan et al. (2002) have described several key properties of twenty one interestingness measures proposed in statistics, machine learning and data mining literature. One needs to examine these properties in order to select right interestingness measure for a given application domain. Many measures such as Odds ratio, Yule's Q, Yule's Y, Kappa, J-Measure, Certainty factor have not been relevant in the context of measuring association among items in a database.

Hershberger and Fisher (2005) discuss measures such as Phi, Pearson's contingency coefficient, Cramer's V coefficient for making association analysis of two variables. These measures of association are used with chi-square and other contingency table analyses. When using the chi-square statistic, these coefficients can be helpful in interpreting the relationship between two variables once statistical significance has been established. In general, they might not be suitable for measuring association among a set of items.

Agrawal et al. (1993) have proposed support measure in the context of finding association rules in a database. To find support of itemset, it requires counting frequency of the itemset in the given database. An itemset in a transaction could be a source of association among items in the itemset. But, support of an itemset does not consider the frequencies of it subsets. As a result, support of an itemset might not be a good measure of association among items in an itemset.

Piatetsky-Shapiro (1991) has proposed leverage measure in the context of mining strong rules in a database. It captures both strength and volume of the effect in a single value. It is difference between the observed frequency with which the antecedent and consequent co-occur and the frequency that would be expected if the two were independent. Thus, it has limited use in the context of measuring association among a set of items. Adhikari and Rao (2007) have proposed a measure called OA, to measure overall association between two items in a database. It might not be suitable for measuring of association among a set of items in a database.

Aggarwal and Yu (1998) have proposed collective strength of an itemset as a measure. Collective strength is based on the concept of violation of an itemset. An itemset X is said to be in violation of a transaction, if some items of X are present in the transaction and others are not. Collective strength of an itemset X has been defined as follows.

$$C(X) = \frac{1 - v(X)}{1 - E(v(X))} \times \frac{E(v(X))}{v(X)},$$

where $v(X)$ is the violation rate of itemset X. It is the fraction of transactions in violation of itemset X. $E(v(X))$ is the expected violation rate of itemset X. The major concern regarding computation of $C(X)$ is that the computation of $E(v(X))$ is based on statistical independence of items of X.

Cosine (Han and Kamber 2001) and correlation (Han and Kamber 2001) are used to measure association between two objects. They might not be suitable as a measure of association among items of an itemset.

Confidence and conviction are used to measure strength of association between itemsets in some sense. They might not be useful in the current context, since we are interested in capturing association among items of an itemset. In the following section, we introduce two measures for capturing association among a set of items in a database.

4.3 New Measures of Association

Before we propose our measures of association, we mention a few definitions and notations used frequently in this chapter.

A set of items in a database is called an *itemset*. Every itemset in a database is associated with a statistical measure, called *support*. Support of an itemset X in database D is the fraction of transactions in D containing X, denoted by $S(X, D)$. In general, let $S(E, D)$ be the support of Boolean expression E defined on the trans-actions in database D. An itemset X is called *frequent* in D if $S(X, D) \geq \alpha$, where α is user defined level of *minimum support*. If X is frequent then Y is also frequent, for $\phi \neq Y \subseteq X$, since $S(Y, D) \geq S(X, D)$. Each item in a frequent itemset is called a *frequent item*. Let $|X|$ denote the number of items of itemset X. Let X be $\{x_1, x_2, \ldots, x_m\}$. The following notations are used frequently in this chapter:

- $S_X \langle Y, D \rangle$: support of Boolean expression that a transaction in D contains all the items of Y, but not items of $X - Y$, for $\neq Y \subseteq X$
- $S(\cup_{i=1}^m \{x_i\}, D)$: support of Boolean expression that a transaction in D contains at least one item of X
- $S(\cap_{i=1}^m \{x_i\}, D)$: support of Boolean expression that a transaction in D contains all the items of X

A measure of association gives numerical estimate of the magnitude of the sta-tistical dependence among n items, for an integer $n \geq 2$. Highly associated items are likely to be purchased together. In other words, items of itemset X are highly asso-ciated, if one of the items of X is purchased then the remaining items of X are also likely to be purchased in the same transaction. One could define association among a set of items in many ways. Our first measure of association A_1 is defined as follows.

Definition 4.1 Let $X = \{x_1, x_2, \ldots, x_m\}$ be an itemset in database D. Let δ be the *minimum level of association*. The measure of association A_1 is defined as follows.

$$A_1(X, D) = \begin{cases} S(X, D)/S(\cup_{i=1}^m \{x_i\}, D), & \text{for } |X| \geq 2 \\ \delta, & \text{for } |X| = 1 \end{cases} \qquad (4.1)$$

Measure A_1 is the proportion of the number of transactions containing all the items of X and the number of transactions containing at least one of the items of X. The association among items of an itemset and the number of association rules generated from the itemset are positively correlated, provided the support of the itemset is high. If the association among items of an itemset is more then it is expected to generate more association rules and vice versa. Palshikar et al. (2005) have proposed heavy itemsets for mining association rules. An itemset X is *heavy* for given support and confidence values, if all possible association rules made up of items only of X are present. Thus, items of heavy itemsets are expected to have high association among themselves.

A transaction in a database D provides the following information regarding association among items of X: (i) A transaction that contains all the items of X contributes maximum value towards overall association among items of X. We attach weight 1.0 to each such transaction. (ii) A transaction that contains k items of X contributes some value towards overall association among the items of X, for $2 \leq k \leq |X|$. We attach weight $k/|X|$ to each such transaction. (iii) A transaction that does not contain any item of X contributes no information regarding association among the items of X. (iv) A transaction that contains only one item of X contributes maximum value towards overall dispersion among the items of X. At a given X, we attach a weight to each transaction that contributes some value towards overall association among the items of X. Our second measure of association A_2 is defined as follows.

Definition 4.2 Let $X = \{x_1, x_2, ..., x_m\}$ be an itemset in database D. Let δ be the *minimum level of association*. The measure of association A_2 is defined as follows.

$$A_2(X, D) = \begin{cases} \sum\limits_{Y \subseteq X, |Y| \geq 2} \left\{ \dfrac{S_X\langle Y, D\rangle}{S(\bigcup_{i=1}^{m} \{x_i\}, D)} \times \dfrac{|Y|}{|X|} \right\}, & \text{for } |X| \geq 2 \\ \delta, & \text{for } |X| = 1 \end{cases} \quad (4.2)$$

A_2 could be considered as conditional expectation of $|Y|/|X|$ at given X, for $Y \subseteq X$, and $|Y| \geq 2$. Thus, A_2 could be expressed as follows.

$$A_2(X, D) = \sum_{Y \subseteq X, |Y| \geq 2} \left\{ CS_X(Y, D) \times \frac{|Y|}{|X|} \right\}, \quad (4.3)$$

where

$$CS_X(Y, D) = \frac{S_X\langle Y, D\rangle}{S(\bigcup_{i=1}^{m} \{x_i\}, D)}.$$

Variable $|Y|/|X|$ has conditional support $CS_X(Y, D)$, for $Y \subseteq X$, and $|Y| \geq 2$. Conditional support $CS_X(Y, D)$ refers to the fraction of transactions containing Y among the transactions containing at least one item of X.

4.4 Properties of A_1 and A_2

A_1 and A_2 could be used to measure association among a set of items in a database. Thus, each of these measures could be considered as a *generalized measure of association*. The following corollary is obtained from Definitions 4.1 and 4.2.

Corollary 4.12.1 *Let* $X = \{x_1, x_2\}$ *be an itemset in database D. Then,*

$$A_1(X,D) = A_2(X,D) = S(\{x_1\} \cap \{x_2\}, D)/S(\{x_1\} \cup \{x_2\}, D), \quad \text{for } |X| = 2. \quad (4.4)$$

The measure in (4.4) has been reported as a measure of similarity between two objects (Wu et al. 2005; Xin et al. 2005). To judge goodness of the proposed measures, we state below *monotone property* of a measure of association.

Property 4.1 Given an itemset X, if a subset Y occurs more frequently in the transactions containing at least two items of X then the items of X have stronger association, for $Y \subseteq X$, and $|Y| \geq 2$.

A_2 satisfies monotone property, for every $Y \subseteq X$, and $|Y| \geq 2$. But, A_1 satisfies monotone property, only for $Y = X$, and $|Y| \geq 2$. Therefore, A_2 is more appealing measure of association than A_1. The following example verifies that A_2 measures association among items of an itemset more accurately than A_1.

Example 4.1 Consider the database $D = \{\{a, b, c, d\}, \{a, c\}, \{a, h\}, \{b, c\}, \{b, c, d\}, \{b, d, e\}, \{b, e\}, \{c, d, e\}\}$. Here, $A_1(\{b, c, d\}, D) = 0.2857$, and $A_2(\{b, c, d\}, D) = 0.5714$. By observing the presence of items b, c, and d in different transactions in D, one could conclude that the association among items of $\{b, c, d\}$ is closer to 0.5714 than 0.2857.

Moreover, the Examples 4.2 and 4.3 presented later, also show that A_2 is more appealing than A_1 in capturing association among items in a database. In the context of monotone property of a measure of association, we discuss here the effectiveness of all-confidence measure. For an itemset X, all-confidence measure has been defined as follows: all-confidence $(X) = S(X, D)/\text{maximum}(S(\{x\}, D))$, for $x \in X$. Measure all-confidence satisfies monotone property, only for $Y = X$, where $Y \subseteq X$, and $|Y| \geq 2$. Thus, measure all-confidence might not be effective in capturing association among items in an itemset at all the situations.

In Sect. 4.6, experimental results are provided using measure A_2. The following two lemmas are useful to show some interesting properties of A_2.

Lemma 4.1 *Let* $X = \{x_1, x_2, \ldots, x_m\}$ *be an itemset in database D. Also, let* $T(i) = \sum_{Y \subseteq X, |Y|=m-i} S_X\langle Y, D\rangle$, $i = 0, 1, \ldots, m-1$. *Then T(i) can be expressed as follows.*

$$\left\{ \sum_{Y\subseteq X, |Y|=m-i} S(Y, D) -\ {}^{m-i+1}C_{m-i} \times \sum_{Y\subseteq X, |Y|=m-i+1} S(Y, D) +\ {}^{m-i+2}C_{m-i} \right.$$

$$\times \sum_{Y\subseteq X, |Y|=m-i+2} S(Y, D) + \cdots + (-1)^{i-1} \times^{m-1}C_{m-i}$$

$$\left. \times \sum_{Y\subseteq X, |Y|=m-1} S(Y, D) + (-1)^i \times {}^mC_{m-i} \times S(X, D) \right\} \qquad (4.5)$$

Proof For $i = k$, $|Y| = m - k$. $S_X\langle Y, D \rangle$ could be expressed by itemsets of size greater than or equal to $m - k$. There are ${}^mC_{m-k}$ distinct $\langle Y \rangle$s, and thus, we have ${}^mC_{m-k}$ distinct expressions for $S_X\langle Y, D \rangle$, one for each $\langle Y \rangle$. Each expression of $S_X\langle Y, D \rangle$ contains a $S(X, D)$. Thus, the last term of $T(k)$ is ${}^mC_{m-k} \times S(X, D)$. The second last term contains itemsets of size $m - 1$. Not all expressions of $S_X\langle Y, D \rangle$ contain a particular itemset Y_1 of size $m - 1$. Y_1 is present in the expression of $S_X\langle Y, D \rangle$, if $Y_1 \subseteq Y$. The number of expressions of $S_X\langle Y, D \rangle$ that contain Y_1 is ${}^{m-1}C_{m-k}$. Other terms could be obtained in a similar way.   □

We verify Lemma 4.1 with the help of following example. Let $X = \{x_1, x_2, x_3\}$. Then,

$$T(1) = S_X\langle \{x_1, x_2\}, D \rangle + S_X\langle \{x_1, x_3\}, D \rangle + S_X\langle \{x_2, x_3\}, D \rangle$$
$$= S(\{x_1, x_2\}, D) - S(\{x_1, x_2, x_3\}, D) + S(\{x_1, x_3\}, D)$$
$$\quad - S(\{x_1, x_2, x_3\}, D) + S(\{x_2, x_3\}, D) - S(\{x_1, x_2, x_3\}, D)$$
$$= \sum_{|Y|=3-1, Y\subseteq X} S(Y, D) + (-1)^1 \times {}^3C_{3-1} \times S(X, D).$$

Again,

$$T(2) = S_X\langle \{x_1\}, D \rangle + S_X\langle \{x_2\}, D \rangle + S_X\langle \{x_3\}, D \rangle$$
$$= S(\{x_1\}, D) - S(\{x_1, x_2\}, D) - S(\{x_1, x_3\}, D) + S(\{x_1, x_2, x_3\}, D)$$
$$\quad + S(\{x_2\}, D) - S(\{x_1, x_2\}, D) - S(\{x_2, x_3\}, D) + S(\{x_1, x_2, x_3\}, D)$$
$$\quad + S(\{x_3\}, D) - S(\{x_1, x_3\}, D) - S(\{x_2, x_3\}, D) + S(\{x_1, x_2, x_3\}, D)$$
$$= \sum_{|Y|=3-2, Y\subseteq X} S(Y, D) + (-1)^1 \times {}^{3-1}C_{3-2}$$
$$\times \sum_{|Y|=3-1, Y\subseteq X} S(Y, D) + (-1)^2 \times {}^3C_{3-2} \times S(X, D).$$

Thus, the expressions of $T(1)$ and $T(2)$ verify Lemma 4.1, and it has been used to prove Lemma 4.2.

Lemma 4.2 *Let $X = \{x_1, x_2, ..., x_m\}$ be an itemset in database D. Then*

$$\sum_{Y \subseteq X, 1 \leq |Y| \leq |X|} \left\{ S_X \langle Y, D \rangle \times \frac{|Y|}{|X|} \right\} = \frac{1}{m} \times \sum_{i=1}^{m} S(\{x_i\}, D) \qquad (4.6)$$

Proof

$$\sum_{\substack{Y \subseteq X, |Y|=|X| \text{down to } 1}} \left\{ S_X \langle Y, D \rangle \times \frac{|Y|}{|X|} \right\} = \sum_{i=0}^{m-1} \left(\frac{m-i}{m} \right) \times T(i), \qquad (4.7)$$

$$\text{where } i = m - |Y|$$

$$= S(X,D) + \left(\frac{m-1}{m} \right) \times \left(\sum_{Y \subseteq X, |Y|=m-1} S(Y,D) - {}^{m}C_{m-1} \times S(X,D) \right) + \cdots + \left(\frac{m-2}{m} \right)$$

$$\times \left(\sum_{Y \subseteq X, |Y|=m-2} S(Y,D) - {}^{m-1}C_{m-2} \times \left(\sum_{Y \subseteq X, |Y|=m-1} S(Y,D) \right) + {}^{m}C_{m-2} \times S(X,D) \right)$$

$$- \left(\frac{m-3}{m} \right) \times \left(\sum_{Y \subseteq X, |Y|=m-3} S(Y, D) - {}^{m-2}C_{m-3} \times \left(\sum_{Y \subseteq X, |Y|=m-2} S(Y,D) \right) + {}^{m-1}C_{m-3} \right.$$

$$\times \left(\sum_{Y \subseteq X, |Y|=m-1} S(Y,D) \right) \left. - {}^{m}C_{m-3} \times S(X,D) \right) + \left(\frac{m-(m-1)}{m} \right) \times \left(\sum_{Y \subseteq X, |Y|=1} S(Y,D) \right)$$

$$- {}^{2}C_1 \times \left(\sum_{Y \subseteq X, |Y|=2} S(Y,D) \right) + \cdots \pm {}^{m}C_1 \times S(X,D) \right), \quad \text{(Lemma 4.1)} \qquad (4.8)$$

$$= S(X,D) \times \left\{ 1 - {}^{m-1}C_1 + {}^{m-1}C_2 - \cdots \pm {}^{m-1}C_{m-1} \right\}$$

$$+ \left(\frac{m-1}{m} \right) \times \left(\sum_{Y \subseteq X, |Y|=m-1} S(Y,D) \right) \times \left\{ 1 - {}^{m-2}C_1 + {}^{m-2}C_2 + \cdots \pm {}^{m-2}C_{m-2} \right\}$$

$$+ \left(\frac{m-2}{m} \right) \times \left(\sum_{Y \subseteq X, |Y|=m-2} S(Y,D) \right) \times \left\{ 1 - {}^{m-3}C_1 + {}^{m-3}C_2 + \cdots \pm {}^{m-3}C_{m-3} \right\}$$

$$+ \cdots + \left(\frac{2}{m} \right) \times \left(\sum_{Y \subseteq X, |Y|=2} S(Y,D) \right) \times \left\{ 1 - {}^{1}C_1 \right\} + \left(\frac{1}{m} \right) \times \left(\sum_{Y \subseteq X, |Y|=1} S(Y,D) \right) \qquad (4.9)$$

$$= \left(\frac{1}{m} \right) \times \left(\sum_{Y \subseteq X, |Y|=1} S(Y,D) \right), \text{since the coefficient of } \left(\frac{p}{m} \right) \times \left(\sum_{Y \subseteq X, |Y|=p} S(Y, D) \right) \text{ is zero, for } 2 \leq p \leq m. \qquad \square$$

We verify Lemma 4.2 with the help of following example. Let $X = \{x_1, x_2, x_3\}$. Then,

$$\sum_{Y \subseteq X, 1 \le |Y| \le |X|} \left\{ S_X\langle Y, D\rangle \times \frac{|Y|}{|X|} \right\} = \frac{1}{3}\{S_X\langle\{x_1\}, D\rangle + S_X\langle\{x_2\}, D\rangle + S_X\langle\{x_3\}, D\rangle\}$$

$$+ \frac{2}{3}\{S_X\langle\{x_1, x_2\}, D\rangle + S_X\langle\{x_1, x_3\}, D\rangle + S_X\langle\{x_2, x_3\}, D\rangle\}$$

$$+ S_X\langle\{x_1, x_2, x_3\}, D\rangle$$

$$= \frac{1}{3}\{S_X(\{x_1\}, D) - S_X(\{x_1, x_2\}, D)$$

$$- S_X(\{x_1, x_3\}, D) + S_X(\{x_1, x_2, x_3\}, D) + \cdots\}$$

$$+ \frac{2}{3}\{S_X(\{x_1, x_2\}, D) - S_X(\{x_1, x_2, x_3\}, D) + \cdots\}$$

$$+ S_X(\{x_1, x_2, x_3\}, D)$$

$$= \frac{1}{3}\{S_X(\{x_1\}, D) + S_X(\{x_2\}, D) + S_X(\{x_3\}, D)\}.$$

We use Lemma 4.2 for proving Lemma 4.3.

Lemma 4.3 *Let $X = \{x_1, x_2, \ldots, x_m\}$ be an itemset in database D, for an integer $m \ge 2$. Then,*

$$A_2(X, D) = \frac{1}{m \times S(\bigcup_{i=1}^{m} \{x_i\}, D)} \times \left[\sum_{i=1}^{m} \{S(\{x_i\}, D) - S_X\langle\{x_i\}\rangle, D\} \right],$$

(4.10)

where $x_i \in X$.

Proof

$$A_2(X, D) = \frac{1}{S(\bigcup_{i=1}^{m} \{x_i\}, D)}$$

$$\times \left[\sum_{Y \subseteq X, |Y| \ge 1} S_X\langle Y, D\rangle \times \frac{|Y|}{|X|} - \frac{\sum_{i=1}^{m} S_X\langle\{x_i\}, D\rangle}{m} \right] \quad (4.11)$$

$$= \frac{1}{S(\bigcup_{i=1}^{m} \{x_i\}, D)} \times \left[\frac{\sum_{i=1}^{m} S(\{x_i\}, D)}{m} - \frac{\sum_{i=1}^{m} S_X\langle\{x_i\}, D\rangle}{m} \right], \quad \text{(Lemma 4.2).} \quad (4.12)$$

$\square$

Lemma 4.3 gives a simple expression for A_2. A few corollaries of Lemma 4.3 are given below.

Corollary 4.3.1 *Let $X = \{x_1, x_2, \ldots, x_m\}$ be an itemset in database D. If all the items in X have equal support then*

$$A_2(X, D) = q/S\left(\cup_{i=1}^{m}\{x_i\}, D\right),\qquad\qquad(4.13)$$

where

$$q = S(\{x_i\}, D) - S_X\langle\{x_i\}, D\rangle,\quad\text{for } i = 1, 2, \ldots, m.$$

Corollary 4.3.2 *Let* $X = \{x_1, x_2, \ldots, x_m\}$ *be an itemset in database D. Then,*

$$A_2(X, D) = \sum_{i=1}^{m}[q_i/m],\qquad\qquad(4.14)$$

where

$$q_i = [S(\{x_i\}, D) - S_X\langle\{x_i\}, D\rangle]/S\left(\cup_{i=1}^{m}\{x_i\}, D\right).$$

q_i/m *is the contribution of item* x_i *towards overall association among items in X, for* $i = 1, 2, \ldots, m.$

Based on measure A_2, we define an *associative itemset* as follows.

Definition 4.3 Let $X = \{x_1, x_2, \ldots, x_m\}$ be an itemset in database D. Also, let δ be the *minimum level of association*. X is associative at level δ if $A_2(X, D) \geq \delta$.

From Definition of A_2, an itemset of size 1 is associative at level δ. If $\delta > \alpha$ then a frequent itemset X might not be associative at level δ. This is because of the fact that the association among items of X might lie in $[\alpha, \delta)$. In many applications, one might be interested in the itemsets that are frequent as well as associative. In the next section, we mention one such application. The following two examples illustrate the difference between associative itemsets and frequent itemsets.

Example 4.2 We consider the following three transactional databases. Let $D_1 = \{\{a, b, c, d\}, \{a, c\}, \{a, h\}, \{b, c\}, \{b, c, d\}, \{b, d, e\}, \{b, e\}, \{c, d, e\}\}$, $D_2 = \{\{a, b, c, e\}, \{a, b, f\}, \{a, d\}, \{a, g, h\}, \{b, d, g\}, \{b, f, g\}, \{b, g, i\}, \{b, j\}, \{c, d\}\}$ and $D_3 = \{\{a, b, c, e\}, \{a, b, g, i\}, \{a, b, j\}, \{a, c, d, e\}, \{b, d, g\}, \{c, d\}, \{f, g\}, \{g, h\}\}$. Let $\alpha = 0.2$, and $\delta = 0.4$. In database D_1, $S(\{a, b\}, D_1) = 1/8$, and $A_1(\{a, b\}, D_1) = A_2(\{a, b\}, D_1) = 1/7$. Thus, the itemset $\{a, b\}$ is not frequent and also not associative. In database D_2, $S(\{a, b\}, D_2) = 2/9$, and $A_1(\{a, b\}, D_2) = A_2(\{a, b\}, D_2) = 1/4$. Thus, the itemset $\{a, b\}$ is frequent but not associative. In database D_3, $S(\{a, b\}, D_3) = 3/8$, and $A_1(\{a, b\}, D_3) = A_2(\{a, b\}, D_3) = 3/5$. Thus, the itemset $\{a, b\}$ is frequent as well as associative.

An associative itemset at level α might not necessarily be frequent. This is because of the fact that the subsets of the itemset might be available frequently in different transactions.

Example 4.3 Consider the database D_1 of Example 4.1. Let $\alpha = 0.2$, and $\delta = 0.4$. $S(\{c, d, e\}, D_1) = 0.125$, and $A_2(\{c, d, e\}, D_1) = 3/7$. Thus, the itemset $\{c, d, e\}$ is associative, but not frequent.

Thus, the support of an itemset could not effectively measure association among items of an itemset. In Lemma 4.4, we shall prove that association among items of a frequent itemset is always greater than or equal to α under A_2. Thus, a frequent itemset is associative at level α.

Lemma 4.4 *Let $X = \{x_1, x_2, \ldots, x_m\}$ be a frequent itemset in database D, for an integer $m \geq 2$. Then, $A_2(X, D) \geq \alpha$.*

Proof

$$A_2(X, D) = \sum_{Y \subseteq X, |Y| \geq 2} \left\{ \frac{S_X\langle Y, D \rangle}{S(\cup_{i=1}^{m}\{x_i\}, D)} \times \frac{|Y|}{|X|} \right\}$$

$$= \frac{S(X, D)}{S(\cup_{i=1}^{m}\{x_i\}, D)} + \sum_{Y \subset X, |Y| \geq 2} \frac{S_X\langle Y, D \rangle}{S(\cup_{i=1}^{m}\{x_i\}, D)} \times \frac{|Y|}{|X|} \tag{4.15}$$

$S(X, D) \geq \alpha$ implies $S(X, D)/S(\cup_{i=1}^{m}\{x_i\}, D) \geq \alpha$, since $0 < S(\cup_{i=1}^{m}\{x_i\}, D) \leq 1$.

Thus,

$$A_2(X, D) \geq \alpha + \sum_{Y \subset X, |Y| \geq 2} \left\{ \frac{S_X\langle Y, D \rangle}{S(\cup_{i=1}^{m}\{x_i\}, D)} \times \frac{|Y|}{|X|} \right\}. \tag{4.16}$$

$\square$

In Lemma 4.5, we discuss an important property of A_2. Association among items of an itemset under A_2 lies in $(0, 1]$.

Lemma 4.5 *Let $X = \{x_1, x_2, \ldots, x_m\}$ be a frequent itemset in database D, for an integer $m \geq 1$. Then $0 < A_2(X, D) \leq 1$.*

Proof From Definition 4.2, we get $0 < A_2(X, D) \leq 1$, for $m = 1$. Thus, we need to prove the result for $m \geq 2$. Also from Definition 4.2, we get $A_2(X, D) > 0$, for $m \geq 2$. We shall use the method of induction on $|X|$ to show that $A_2(X, D) \leq 1$, for $m \geq 2$. For $m = 2$, $X = \{x_1, x_2\}$.
Then,

$$A_2(X, D) = \frac{S(\{x_1\}, D) - S_X\langle\{x_1\}, D\rangle + S(\{x_2\}, D) - S_X\langle\{x_2\}, D\rangle}{2 \times \{S(\{x_1\}, D) + S(\{x_2\}, D) - S(\{x_1\} \cap \{x_2\}, D)\}}, \tag{4.17}$$
(Lemma 4.3)

Also,

$$S_X\langle\{x_i\},D\rangle = S(\{x_i\},D) - S(\{x_1\}\cap\{x_2\},D), \quad i = 1,2.$$

Thus,

$$A_2(X,D) = \frac{S(\{x_1,x_2\},D)}{S(\{x_1\},D) + S(\{x_2\},D) - S(\{x_1\}\cap\{x_2\},D)} \tag{4.18}$$

We have, $S(\{x_1\}\cap\{x_2\},D) \leq S(\{x_1\},D)$ and $S(\{x_2\},D) - S(\{x_1\}\cap\{x_2\},D) \geq 0$. So, $A_2(X,D) \leq 1$. Thus, the result is true for $m = 2$. We assume that the result is true for $m \leq k - 1$. Now, we shall prove that it is true for $m = k$.

$$A_2(X,D) = \frac{1}{k \times S\left(\bigcup_{i=1}^{k}\{x_i\},D\right)} \times \left[\sum_{i=1}^{k}\{S(\{x_i\},D) - S_X\langle\{x_i\},D\rangle\}\right] \tag{4.19}$$

$$= \frac{1}{k \times S\left(\bigcup_{i=1}^{k}\{x_i\},D\right)} \times \left[\sum_{i=1}^{k-1}\{S(\{x_i\},D) - S_X\langle\{x_i\},D\rangle\}\right]$$
$$+ \frac{1}{k \times S\left(\bigcup_{i=1}^{k}\{x_i\},D\right)} \times \{S(\{x_k\},D) - S_X\langle\{x_k\},D\rangle\} \tag{4.20}$$

$$= \frac{1}{k \times S\left(\bigcup_{i=1}^{k}\{x_i\},D\right)} \times \left[c_1 \times (k-1) \times S\left(\bigcup_{i=1}^{k-1}\{x_i\},D\right) + c_2 \times S(\{x_k\},D)\right],$$
$$0 \leq c_1, c_2 \leq 1, \quad \text{(induction hypothesis)} \tag{4.21}$$

$$\leq \frac{c_1 \times (k-1)}{k} + \frac{c_2}{k} \leq \frac{(k-1)}{k} + \frac{1}{k} = 1. \tag{4.22}$$

$\square$

Let $X = \{x_1, x_2, \ldots, x_m\}$ be a frequent itemset in D. Let $Y \subset X$ such that $|Y| = |X| - 1$, and $|X| \geq 3$. Let $Y = \{x_1, x_2, \ldots, x_{m-1}\}$. We try to establish the relationship between $A_2(X, D)$ and $A_2(Y, D)$. Using formula (4.20) we get,

$$A_2(X,D) \times m \times S\left(\bigcup_{i=1}^{m}\{x_i\},D\right) = A_2(Y,D) \times (m-1) \times S\left(\bigcup_{i=1}^{m}\{x_i\},D\right)$$
$$+ S(\{x_m\},D) - S_X\langle\{x_m\},D\rangle \tag{4.23}$$

or, $A_2(X, D) = A_2(Y, D) \times K_1 + K_2$, where

$$K_1 = \frac{(m - 1) \times S\left(\bigcup_{i=1}^{m-1} \{x_i\}, D\right)}{m \times S\left(\bigcup_{i=1}^{m} \{x_i\}, D\right)}, \quad \text{and} \quad K_2 = \frac{S(\{x_m\}, D) - S_X\langle\{x_m\}, D\rangle}{m \times S\left(\bigcup_{i=1}^{m} \{x_i\}, D\right)}.$$

(4.24)

We note that, $0 < K_1, K_2 \leq 1$. There does not exist any fixed relationship between $A_2(X, D)$ and $A_2(Y, D)$, for all $Y \subset X$ such that $|Y| = |X| - 1$, and $|X| \geq 3$. We consider the following example.

Example 4.4 Consider the database $D_4 = \{\{a, b, c, d\}, \{a, b, d\}, \{a, b, e\}, \{a, b, f\}, \{a, f\}, \{a, g\}, \{d, i\}, \{i, j\}\}$. $A_2(\{a, b\}, D_4) = 0.66667$, $A_2(\{a, c\}, D_4) = 0.16667$, $A_2(\{a, b, c\}, D_4) = 0.5$. We observe that, $A_2(\{a, b, c\}, D_4) < A_2(\{a, b\}, D_4)$, and $A_2(\{a, b, c\}, D_4) > A_2(\{a, c\}, D_4)$.

We wish to express A_2 in terms of supports of different itemsets. Given an itemset X, the Lemma 4.6 expresses $S_X\langle\{x_i\}, D\rangle$ in terms of the support of itemset $Y \subseteq X$, for $x_i \in X$.

Lemma 4.6 *Let* $X = \{x_1, x_2, \ldots, x_m\}$ *be an itemset in database D, for an integer* $m \geq 1$. *Then*

$$S_X\langle\{x_i\}, D\rangle = S(\{x_i\}, D) - \sum_{j=1; j \neq i}^{m} S(\{x_i\} \cap \{x_j\}, D)$$

$$+ \sum_{j,k=1; j<k; j,k \neq i}^{m} S(\{x_i\} \cap \{x_j\} \cap \{x_k\}, D) - \cdots + (-1)^{m-1} \times S\left(\bigcap_{i=1}^{m} \{x_i\}, D\right) 1, 2, \ldots, m.$$

(4.25)

Proof We shall prove the result using the method of induction on m. The result trivially follows for $m = 1$. For $m = 2$, $X = \{x_1, x_2\}$. Then, $S_X\langle\{x_i\}, D\rangle = S(\{x_i\}, D) - S(\{x_1\} \cap \{x_2\}, D)$, $i = 1, 2$. Hence, the result is true for $m = 2$. Assume that the result is true for $m = p$. We shall prove that the result is true for $m = p + 1$. Let $X = \{x_1, x_2, \ldots, x_{p+1}\}$. Due to the addition of x_{p+1}, the following observations are made. $S(\{x_i\} \cap \{x_{p+1}\}, D)$ is required to be subtracted, for $1 \leq i \leq p$. $S(\{x_i\} \cap \{x_j\} \cap \{x_{p+1}\}, D)$ is required to be added, for $1 \leq i < j \leq p$. Finally, the term $(-1)^p \times S(\{x_1\} \cap \{x_2\} \cap \cdots \cap \{x_{p+1}\}, D)$ is required to be added. Thus,

$$S_X\langle\{x_i\}, D\rangle = S(\{x_i\}, D) - \sum_{j=1, j\neq i}^{p+1} S(\{x_i\} \cap \{x_j\}, D)$$

$$+ \sum_{j,k=1; j<k; j, k\neq i}^{p+1} S(\{x_i\} \cap \{x_j\} \cap \{x_k\}, D) \qquad (4.26)$$

$$+ \cdots + (-1)^p \times S(\{x_1\} \cap \{x_2\} \cap \cdots \cap \{x_{p+1}\}, D)$$

Thus, the induction step follows. Hence, the result follows. $\qquad\qquad\square$

Lemma 4.7 expresses association among items of X in terms of supports of different subsets of X.

Lemma 4.7 *Let $X = \{x_1, x_2, \ldots, x_m\}$ be an itemset in database D, for $m \geq 2$. Then*

$$A_2(X, D)$$

$$= \frac{\sum_{i=1}^{m}\left[\sum_{j=1, j\neq i}^{m} S(\{x_i\} \cap \{x_j\}, D) - \sum_{j,k=1; j, k\neq i}^{m} S(\{x_i\} \cap \{x_j\} \cap \{x_k\}, D) + \cdots \pm S(\{x_1\} \cap \cdots \cap \{x_m\}, D)\right]}{m \times \left[\sum_{i=1}^{m} S(\{x_i\}, D) - \sum_{i,j=1; i<j}^{m} S(\{x_i\} \cap \{x_j\}, D) + \cdots \pm S(\{x_1\} \cap \{x_2\} \cap \cdots \cap \{x_m\}, D)\right]}$$

$$(4.27)$$

Proof We state the theorem of total probability as follows (Papoulis 1984). For any m events $x_1, x_2, \ldots, x_m$, we have

$$S\left\{\bigcup_{i=1}^{m} \{x_i\}, D\right\} = \sum_{i=1}^{m} S(\{x_i\}, D) - \sum_{1\leq i<j\leq m} S(\{x_i\} \cap \{x_j\}, D)$$

$$+ \cdots \pm S(\{x_1\} \cap \{x_2\} \cap \cdots \cap \{x_m\}, D) \qquad (4.28)$$

Result follows using Lemmas 4.3 and 4.6. $\qquad\qquad\square$

Result (4.28) expresses measure A_2 in terms of supports of different itemsets. A few corollaries of Lemma 4.7 are given below.

Corollary 4.7.1

$$A_2(X, D) = \frac{S(\{x_1\} \cap \{x_2\}, D)}{S(\{x_1\}, D) + S(\{x_2\}, D) - S(\{x_1\} \cap \{x_2\}, D)}, \quad \text{for} \quad m = 2.$$

$$(4.29)$$

Corollary 4.7.2 *For m = 3, $A_2(X, D) = E_1/E_2$, where*

$$E_1 = 2 \times \{ S(\{x_1\} \cap \{x_2\}, D) + S(\{x_1\} \cap \{x_3\}, D) + S(\{x_2\} \cap \{x_3\}, D)\}$$
$$- 3 \times S(\{x_1\} \cap \{x_2\} \cap \{x_3\}, D) \quad \text{and}$$
$$E_2 = 3 \times \{ S(\{x_1\}, D) + S(\{x_2\}, D) + S(\{x_3\}, D) - S(\{x_1\} \cap \{x_2\}, D) - S(\{x_1\} \cap \{x_3\}, D)\}$$
$$3 \times \{-S(\{x_2\} \cap \{x_3\}, D) + S(\{x_1\} \cap \{x_2\} \cap \{x_3\}, D)\}$$

$$(4.30)$$

Thus, $A_2(X, D)$ could be computed when supports of subsets of X are available. In Lemma 4.8, we study some properties of measure A_1.

Lemma 4.8 *Let X be an itemset in database D. Then the measure of association A_1 satisfies the following properties.* (i) $0 < A_1(X, D) \leq 1$, (ii) $A_1(X, D) \leq A_1(Y, D)$, *for* $Y \subseteq X$, *and* $|Y| \geq 2$, *and* (iii) $A_1(X, D) \geq \alpha$, *if X is a frequent itemset.*

Proof

(i) $A_1(X, D) > 0$, since $S(X, D) > 0$. Also, $A_1(X, D) \leq 1$, since $S(\cup_{y \in X}\{y\}, D) \geq S(X, D)$.

(ii) Let $X = \{x_1, x_2, ..., x_m\}$. We consider an itemset $Y \subseteq X$ such that $|Y| \geq 2$. Then, $A_1(X, D) \leq A_1(Y, D)$, since $S(X, D) \leq S(Y, D)$ and $S(\cup_{z \in X}\{z\}, D) \geq S(\cup_{z \in Y}\{z\}, D)$.

(iii) $A_1(X, D) = S(X, D)/S(\cup_{x_i \in X}\{x_i\}, D) \geq S(X, D) \geq \alpha$, since $0 < S(\cup_{y \in X}\{y\}, D) \leq 1$. $\qquad\square$

The proposed measures of association could be used in many applications. In Sect. 4.5, we mention one such application.

4.4.1 Capturing Association

Let $X = \{x_1, x_2, ..., x_k\}$ be an itemset in the given database D. In finding association among items in X, the following procedure (Adhikari and Rao 2008; Wu et al. 2005) could be followed. An algorithm finds association between every pair of items. The items in X form a class, if $^kC_2$ association values corresponding to $^kC_2$ pairs of items are close. The level of association among the items in this class could be assumed as the minimum of $^kC_2$ association values. If the number of items in a class is more than two, then such technique might fail to estimate correctly the association among the items in an itemset. Then the accuracy of estimating association among items in X becomes low. Instead of that, we could have estimated association among items in X using A_2 and the difference in similarity using measure A_2 could be estimated as $DS(X, D) = A_2(\{x_1, x_2, ..., x_k\}, D) - minimum$ $\{EMS(x_i, x_j, D) : 1 \leq i < j \leq k\}$, where *EMS* is an existing measure of similarity for finding the similarity between two items in D. Due to the *monotone property* of A_2, the amount of difference could be significant (see Tables 4.5, 4.6 and 4.7). In particular, $DS(X, D)$ becomes 0, if $k = 2$ (Corollary 4.12.1).

4.5 An Application: Clustering Frequent Items in a Database

We have observed that the measure A_2 is effective in finding association among items of an itemset in a database. For the purpose of clustering frequent items in a database, one could find associations among items of each frequent itemset of size greater than 1. Items of a highly associative itemset could be put in the same class. Thus, one could cluster the frequent items in a given database using A_2.

Adhikari and Rao (2008) have proposed a technique for clustering multiple databases. If a cluster contains a class of size greater than 2, one might cluster frequent items in a database with higher degree of accuracy, since A_2 possesses monotone property. In the context of clustering data, an overview of different clustering techniques is given by Jain et al. (1999).

During the clustering, there are two approaches of measuring association among items of each itemset in a database. In the first approach, one could synthesize association among items of the current frequent itemset. As soon as a frequent itemset is found during the mining process, one could call an algorithm of finding association among items of the current frequent itemset. When a frequent itemset is extracted, then all the non-null subsets of the frequent itemest might have been available (Agrawal et al. 1993). Thus, one could synthesize association among items of the current frequent itemset. Also, one could synthesize association among items of each frequent itemsets after the mining task. In the second approach, all the frequent itemsets could be processed at the end of the mining task. These two approaches seem to be the same so far as the computational complexity is concerned.

4.6 Experimental Results

Experimental results are based on different sources of data. We present the experimental results using six databases. The databases *retail* (Frequent itemset mining dataset repository 2004), *mushroom* (Frequent itemset mining dataset repository 2004) are real. The dataset *ecoli* is a subset of *ecoli database* (UCI ML repository content summary)[1] and has been processed for the purpose of conducting experiment. Also, we have omitted non-numeric fields of *ecoli database* for the

---

[1] UCI ML repository content summary, http://www.ics.uci.edu/mlearn/MLSummary.html.

Table 4.1 Dataset characteristics

D	NT(D)	ALT(D)	AFI(D)	NI(D)
retail	88,162	11.31	80.46	10,000
mushroom	8,124	23.00	1570.18	119
ecoli	336	7.00	25.57	92
check	5	2.80	2.33	6
random500	10,000	6.47	109.40	500
random1000	10,000	12.49	111.86	1,000

purpose of conducting experiments. The fifth and sixth databases are generated synthetically for the purpose of conducting experiments. The fourth database *check* is artificial. The database *check* contains the following transactions: {34, 47, 62}, {34, 55, 62, 102}, {47, 62}, {47, 62, 90}, {55, 102}. We have introduced database *check* for the purpose of verifying results. We present some characteristics of these databases in Table 4.1.

Let $NT(D)$, $ALT(D)$, $AFI(D)$, and $NI(D)$ denote the number of transactions, average length of a transaction, average frequency of an item, and number of items in database D, respectively. There are many interesting algorithms (Agrawal and Srikant 1994; Han et al. 2000; Savasere et al. 1995) to mine frequent itemsets in a database. We implement apriori algorithm (Agrawal et al. 1993) for mining frequent itemsets in a database, since it is simple and easy to implement.

4.6.1 Top Associative Itemsets Among Frequent Itemsets

We have conducted experiments on different datasets to mine itemsets that are frequent as well as associative. Databases *retail, mushroom, ecoli, random500, random1000* and *check* are mined at αs 0.03, 0.1, 0.5, 0.001, 0.002 and 0.1 respectively. Top 10 associative itemsets among frequent itemsets in these databases are given in Tables 4.2, 4.3 and 4.4.

Some itemsets in *mushroom* database are highly frequent. Thus, associations among items in these itemsets are significantly high. On the other hand, the items in *random500* and *random1000* databases are sparsely distributed.

4.6.2 Finding the Difference in Similarity

We have conducted experiments on different databases to mine itemsets that are frequent as well as associative. For the purpose of comparison, we take similarity measure $sim_1(x_i, x_j, D) = S(\{x_1\} \cap \{x_2\}, D)/S(\{x_1\} \cup \{x_2\}, D)$ (Wu et al. 2005).

Table 4.2 Top 10 associative itemsets in databases *retail* and *mushroom*

retail		mushroom	
Itemset	A_2(Itemset)	Itemset	A_2(Itemset)
{39, 41, 48}	0.39183	{85, 86}	0.97538
{38, 39, 48}	0.38044	{34, 85, 86}	0.97530
{32, 39, 48}	0.36929	{34, 85}	0.97415
{38, 39, 41}	0.23977	{85, 86, 90}	0.96570
{39, 41}	0.21057	{34, 85, 90}	0.96455
{38, 170}	0.19350	{34, 86, 90}	0.94847
{41, 48}	0.18763	{36, 85, 86}	0.93763
{38, 39}	0.18498	{34, 36, 85}	0.93755
{36, 38}	0.17723	{85, 90}	0.92171
{38, 110}	0.17396	{34, 36, 86}	0.92114

Table 4.3 Top 10 associative itemsets in databases *ecoli* and *check*

ecoli		check	
Itemset	A_2(Itemset)	Itemset	A_2(Itemset)
{48, 50}	0.94152	{47, 62}	0.75000
{44, 48, 50}	0.68469	{47, 62, 90}	0.58333
{37, 48, 50}	0.68264	{34, 55, 102}	0.55556
{40, 48, 50}	0.68051	{34, 47, 62}	0.50000
{48, 50, 54}	0.67758	{34, 62}	0.50000
{42, 48, 50}	0.67650	{55, 62, 102}	0.33333
{48, 50, 51}	0.66764	{34, 62, 102}	0.33333
{48, 49, 50}	0.65511	{34, 55, 62}	0.33333
{47, 48, 50}	0.65024	{47, 90}	0.33333
{44, 48}	0.14873	{34, 102}	0.33333

Table 4.4 Top 10 associative itemsets in databases *random500* and *random1000*

random500		random1000	
Itemset	A_2(itemset)	Itemset	A_2(itemset)
{15, 49, 145}	0.00511	{213, 446, 610}	0.00781
{49, 167, 440}	0.00503	{337, 385, 780}	0.00714
{15, 167, 223}	0.00423	{229, 850, 919}	0.00689
{15, 78, 222}	0.00402	{167, 480, 674}	0.00645
{15, 49}	0.00398	{519, 689, 881}	0.00617
{15, 167}	0.00345	{337, 385}	0.00581
{49, 167}	0.00323	{337, 780}	0.00489
{167, 223}	0.00298	{167, 674}	0.00433
{44, 334}	0.00286	{446, 610}	0.00410
{300, 445}	0.00276	{689, 881}	0.00380

Table 4.5 Top 10 associative itemsets along with their difference in similarity values for *retail*, and *mushroom*

retail		mushroom	
Itemset	DS(Itemset, *retail*)	Itemset	DS(Itemset, *mushroom*)
{39, 41, 48}	0.20419	{85, 86}	0
{38, 39, 48}	0.22089	{34, 85, 86}	0.00115
{32, 39, 48}	0.22196	{34, 85}	0
{38, 39, 41}	0.09351	{85, 86, 90}	0.08861
{39, 41}	0	{34, 85, 90}	0.06448
{38, 170}	0	{34, 86, 90}	0.07138
{41, 48}	0	{36, 85, 86}	0.12196
{38, 39}	0	{34, 36, 85}	0.12490
{36, 38}	0	{85, 90}	0
{38, 110}	0	{34, 36, 86}	0.10849

We present top 10 associative itemsets along with their difference in similarities are given in Tables 4.5, 4.6 and 4.7.

Some itemsets in the *mushroom* database are highly frequent. Thus, associations among items in these itemsets are significantly high. The itemsets in *random500* and *random1000* are sparsely distributed. As a result, the associations among items in these itemsets are not significant.

Table 4.6 Top 10 associative itemsets along with their difference in similarity values for *ecoli*, and *check*

ecoli		check	
Itemset	DS(Itemset, *ecoli*)	Itemset	DS(Itemset, *check*)
{48, 50}	0	{34, 47, 62}	0.50000
{44, 48, 50}	0.79279	{47, 62}	0
{37, 48, 50}	0.54808	{47, 62, 90}	0.25000
{40, 48, 50}	0.54119	{34, 55, 102}	0.22222
{48, 50, 54}	0.55296	{34, 62}	0
{42, 48, 50}	0.53718	{55, 62, 102}	0.13333
{48, 50, 51}	0.52654	{34, 62, 102}	0.13333
{48, 49, 50}	0.54366	{34, 55, 62}	0.13333
{47, 48, 50}	0.54048	{47, 90}	0
{44, 48}	0	{34, 102}	0

Table 4.7 Top 10 associative itemsets along with their difference in similarity values for *random500* and *random1000*

random500		random1000	
Itemset	DS(Itemset, *random500*)	Itemset	DS(Itemset, *random1000*)
{15, 49, 145}	0.00027	{213, 446, 610}	0.00044
{49, 167, 440}	0.00041	{337, 385, 780}	0.00067
{15, 167, 223}	0.00039	{229, 850, 919}	0.00038
{15, 78, 222}	0.00024	{167, 480, 674}	0.00030
{15, 49}	0	{519, 689, 881}	0.00011
{15, 167}	0	{337, 385}	0
{49, 167}	0	{337, 780}	0
{167, 223}	0	{167, 674}	0
{44, 334}	0	{446, 610}	0
{300, 445}	0	{689, 881}	0

4.6.3 Execution Time for Measuring Association

We have studied execution time for measuring association among items in each frequent itemset of size greater than one. In the first case, execution time is studied with respect to number of transactions in a database. As the number of transactions in a database increases, the number of frequent itemsets is likely to increase. Therefore, the execution time increases as the size of a database increases. We observe this phenomenon in Figs. 4.1, 4.2, 4.3, 4.4 and 4.5.

In the second case, the execution time is studied with respect to α. As we increase α, the number of frequent itemsets is likely to decrease. Therefore, the execution time decreases as α increases. We observe this phenomenon in Figs. 4.6, 4.7, 4.8, and 4.9.

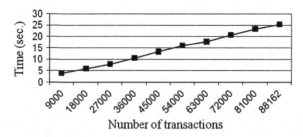

Fig. 4.1 Execution time versus database size at $\alpha = 0.03$ (*retail*)

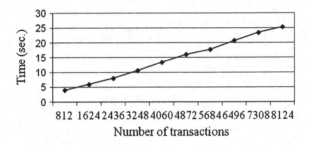

Fig. 4.2   Execution time versus database size at α = 0.1 (*mushroom*)

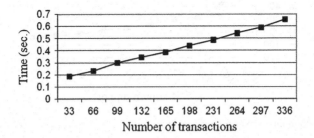

Fig. 4.3   Execution time versus database size at α = 0.1 (*ecoli*)

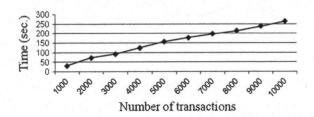

Fig. 4.4   Execution time versus database size at α = 0.001 (*random500*)

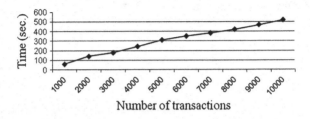

Fig. 4.5   Execution time versus database size at α = 0.002 (*random1000*)

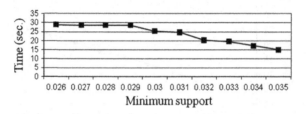

Fig. 4.6 Execution time versus α (*retail*)

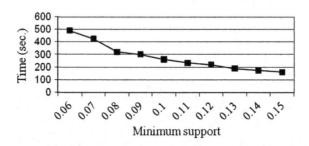

Fig. 4.7 Execution time versus α (*mushroom*)

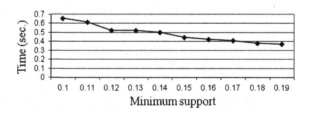

Fig. 4.8 Execution time versus α (*ecoli*)

Fig. 4.9 Execution time versus α (*random500*)

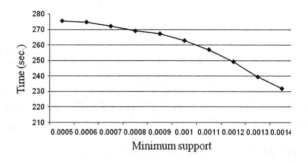

4.7 Conclusion

In this chapter, we propose two measures of association among items in an itemset in a database. An existing measure might not be effective in capturing association among items in an itemset of size greater than 2. Many research problems could boil down to capturing association among items in an itemset. We have given an example of one such application in Sect. 4.5.

We have introduced the notion of associative itemset in a database. We have provided many useful lemmas and examples to make the foundations of the proposed measures strong and clear. Using monotone property of a measure of association, we show that A_2 measures association among items in an itemset more accurately than A_1.

For the purpose of computing A_2, we express it in terms of supports of itemsets. Experimental results show that the difference in similarity among a set of items obtained using measure A_2 could be significantly different than the similarity obtained using an existing similarity measure. Thus, a solution of a problem obtained using measure A_2 could be more effective. The measure of association A_2 is effective in capturing statistical association among items in a database.

References

Adhikari A, Rao PR (2007) Study of select items in multiple databases by grouping. In: Proceedings of 3rd Indian international conference on artificial intelligence, pp 1699–1718

Adhikari A, Rao PR (2008) Efficient clustering of databases induced by local patterns. Decis Support Syst 44(4):925–943

Aggarwal C, Yu P (1998) A new framework for itemset generation. In: Proceedings of PODS, pp 18–24

Agrawal R, Imielinski T, Swami A (1993) Mining association rules between sets of items in large databases. In: Proceedings of ACM SIGMOD conference management of data, pp 207–216

Agrawal R, Srikant R (1994) Fast algorithms for mining association rules. In: Proceedings of the international conference on very large databases, pp 487–499

Brin S, Motwani R, Ullman JD, Tsur S (1997) Dynamic itemset counting and implication rules for market basket data. In: Proceedings of SIGMOD conference, pp 255–264

Frequent itemset mining dataset repository (2004) http://fimi.cs.helsinki.fi/data

Han J, Pei J, Yiwen Y (2000) Mining frequent patterns without candidate generation. In: Proceedings of ACM SIGMOD conference management of data, pp 1–12

Han J, Kamber M (2001) Data mining: concepts and techniques. Morgan Kauffmann Publishers, Burlington

Hershberger SL, Fisher DG (2005) Measures of association, Encyclopedia of statistics in behavioral science. Wiley, London

Jain AK, Murty MN, Flynn PJ (1999) Data clustering: a review. ACM Comput Surv 31 (3):264–323

Omiecinski ER (2003) Alternative interest measures for mining associations in databases. IEEE Trans Knowl Data Eng 15(1):57–69

Palshikar GK, Kale MS, Apte MM (2005) Association rule mining using heavy itemsets. In: Proceedings of international conference on management of data, pp 148–155

Papoulis A (1984) Probability, random variables and stochastic processes, 2 edn. McGraw-Hill, New York

Piatetsky-Shapiro G (1991) Discovery, analysis, and presentation of strong rules. In: Proceedings of knowledge discovery in databases, pp 229–248

Savasere A, Omiecinski E, Navathe S (1995) An efficient algorithm for mining association rules in large databases. In: Proceedings of the 21st international conference on very large data bases, pp 432–443

Tan P-N, Kumar V, Srivastava J (2002) Selecting the right interestingness measure for association patterns. In: Proceedings of SIGKDD conference, pp 32–41

Wu X, Zhang C, Zhang S (2005) Database classification for multi-database mining. Inf Syst 30 (1):71–88

Xin D, Han J, Yan X, Cheng H (2005) Mining compressed frequent-pattern sets. In: Proceedings of the 31st VLDB conference, pp 709–720

Chapter 5
Mining Association Rules Induced by Item and Quantity Purchased

Most of the real market basket data are non-binary in the sense that an item could be purchased multiple times in the same transaction. In this case, there are two types of occurrences of an itemset in a database: the number of transactions in the database containing the itemset, and the number of occurrences of the itemset in the database. Traditional support-confidence framework might not be adequate for extracting association rules in such a database. In this chapter, we introduce three categories of association rules. We introduce a framework based on traditional support-confidence framework for mining each category of association rules. We present experimental results based on two databases.

5.1 Introduction

Pattern recognition (Agrawal et al. 1993; Adhikari et al. 2014) and interestingness measures (Tan et al. 2002) are two important as well as interesting topics at the heart of many data mining problems. Association analysis using association rules (Agrawal et al. 1993; Antonie and Zaïane 2004) has been studied well on binary data. An association rules in a database are expressed in the form of a forward implication, $X \rightarrow Y$, between two itemsets X and Y in a database such that $X \cap Y = \phi$. Association rule mining in a binary database is based on support-confidence framework established by Agrawal et al. (1993). Most of the real life databases are non-binary, in sense that an items could be purchased multiple times in a transaction. It is necessary to study the applicability of traditional support-confidence framework for mining association rules in these databases.

A positive association rule in a binary database *BD* expresses positive association between itemsets X and Y, called the *antecedent* and *consequent*, of the association rule respectively. Each itemset in *BD* is associated with a statistical measure, called *support* (Agrawal et al. 1993). Support of itemset X in *BD* is the fraction of transactions in *BD* containing X, denoted by $supp(X, BD)$. The interestingness of an association rule is expressed by its support and confidence measures. The support and confidence of an association rule $r: X \rightarrow Y$ in a binary database *BD* are defined as

© Springer International Publishing Switzerland 2015
A. Adhikari and J. Adhikari, *Advances in Knowledge Discovery in Databases*,
Intelligent Systems Reference Library 79, DOI 10.1007/978-3-319-13212-9_5

follows: $supp(r, BD) = supp(X \cap Y, BD)$, and $conf(r, BD) = supp(X \cap Y, BD)/supp(X, BD)$. An association rule r in BD is *interesting* if $supp(r, BD) \geq$ *minimum support*, and $conf(r, BD) \geq$ *minimum confidence*. The parameters, minimum support and minimum confidence, are user input given to an association rule mining algorithm.

Association rules in a binary database have limited usage, since in a real transaction an item might be present multiple times. Let *TIMT* be the type of a database such that a Transaction in the database might contain an Item Multiple Times. In this chapter, a database refers to a *TIMT* type database, if the type of the database is unspecified. Then, the question comes to our mind whether the traditional support-confidence framework still works for mining association rules in a *TIMT* type database. Before answering to this question, first we take the following Example 5.1 of a *TIMT* type database *DB*.

Example 5.1 Let $DB = \{\{A(300), B(500), C(1)\}, \{A(2), B(3), E(2)\}, \{A(3), B(2), E(1)\}, \{A(2), E(1)\}, \{B(3), C(2)\}\}$, where $x(\eta)$ denotes item x purchased η numbers at a time in the corresponding transaction. The number of occurrences of itemset $\{A, B\}$ in the first transaction is equal to minimum $\{300, 500\}$, i.e., 300. Thus, the total number of occurrences of $\{A, B\}$ in DB is 304. Also, $\{A, B\}$ has occurred in 3 out of 5 transactions in DB. Thus, the following attributes of itemset X are important consideration for making association analysis of items in X: number of occurrences of X in DB, and number of transactions in DB containing X.

The chapter is organized as follows. In Sect. 5.2, we study association rules in a *TIMT* type database and introduce three categories of association rules. In Sect. 5.3, we introduce a framework based on traditional support-confidence framework for mining each category of association rules. We study the properties of proposed interestingness measures in Sect. 5.4. In Sect. 5.5, we discuss a method for mining association rules in a *TIMT* type database. Experimental results are provided in Sect. 5.6. We discuss related work in Sect. 5.7.

5.2 Association Rules in a TIMT Type Database

We are given a *TIMT* type database *DB*. A transaction in *DB* containing p items could be stored as follows: $\{i_1(n_1), i_2(n_2), \ldots, i_p(n_p)\}$, where item i_k is purchased n_k numbers at a time in the transaction, for $i = 1, 2, \ldots, p$. Each itemset X in a transaction is associated with the following two attributes: transaction-itemset frequency (*TIF*), and transaction-itemset status (*TIS*). These two attributes are defined as follows: $TIF(X, \tau, DB) = m$, if X occurs m times in transaction τ in DB. $TIS(X, \tau, DB) = 1$, for $X \in \tau, \tau \in DB$, and $TIS(X, \tau, DB) = 0$, otherwise. Also, each itemset X in DB is associated with the following two attributes: transaction frequency (*TF*), and database frequency (*DF*). These two attributes are defined as follows: $TF(X, DB) = \sum_{\tau \in DB} TIS(X, \tau, DB)$, and $DF(X, DB) = \sum_{\tau \in DB} TIF(X, \tau, DB)$.

Table 5.1 Distributions of itemset $\{A, B\}$ in transactions of different databases

Database	Trans #1	Trans #2	Trans #3	Trans #4	Trans #5
DB_1	1,000	0	0	0	0
DB_2	1	1	1	1	0
DB_3	500	400	400	0	200

When an item is purchased multiple times in a transaction then the existing generalized frameworks (Steinbach and Kumar 2004; Steinbach et al. 2004) might not be adequate for mining association rules, since they are based on a binary database. The following Example 5.2 shows why the traditional support-confidence framework is not adequate for mining association rules in a *TIMT* type database.

Example 5.2 Let there are three *TIMT* type databases DB_1, DB_2, and DB_3 containing five transactions each. $DB_1 = \{\{A(1000), B(2000), C(1)\}, \{A(5), C(2)\}, \{B (4), E(2)\}, \{E(2), F(1)\}, \{F(2)\}\}$; $DB_2 = \{\{A(1), B(1), C(2)\}, \{A(1), B(1), E(2)\}, \{A (1), B(1), F(1)\}, \{A(1), B(1), G(2)\}, \{H(3)\}\}$; $DB_3 = \{\{A(500), B(600)\}, \{A(700), B (400), E(1)\}, \{A(400), B(600), E(3)\}, \{G(3)\}, \{A(200), B(500), H(1)\}\}$. The numbers of occurrences of itemset $\{A, B\}$ in transactions of different databases are given.

In Table 5.1, we observe the following points regarding itemset $\{A, B\}$: (i) It has high database frequency, but low transaction frequency in DB_1, (ii) In DB_2, it has high transaction frequency, but relatively low database frequency, and (iii) It has high transaction frequency and high database frequency in DB_3.

Based on the above observations, it might be required to consider database frequencies and transaction frequencies of $\{A\}$, $\{B\}$ and $\{A, B\}$ to study association between items A and B. Thus, we could have the following categories of association rules in a *TIMT* type database: (I) Association rules induced by transaction frequency of an itemset, (II) Association rules induced by database frequency of an itemset, and (III) Association rules induced by both transaction frequency and database frequency of an itemset. The goal of this chapter is to provide frameworks for mining association rules under different categories in a *TIMT* type database.

5.3 Frameworks for Mining Association Rules

Each framework is based on traditional support-confidence framework for mining association rules in a binary database.

5.3.1 Framework for Mining Association Rules Under Category I

Based on the number of transactions containing an itemset, we define *transaction-support* (*tsupp*) of the itemset in a database as follows.

Definition 5.1 Let X be an itemset in *TIMT* type database *DB*. Transaction-support of X in *DB* is given as follows: $tsupp(X, DB) = TF(X, DB)/|DB|$.

Let X and Y be two itemsets in *DB*. An itemset X is *transaction-frequent* in *DB* if $tsupp(X, DB) \geq \alpha$, where α is user-defined *minimum transaction-support* level. We define transaction-support of association rule $r: X \rightarrow Y$ in *DB* as follows: $tsupp(r, DB) = tsupp(X \cap Y, DB)$. We define *transaction-confidence* (*tconf*) of association rule r in *DB* as follows: $tconf(r, DB) = tsupp(X \cap Y, DB)/tsupp(X, DB)$. An association rule r in *DB* is *interesting* with respect to transaction frequency of an itemset if $tsupp(r, DB) \geq$ *minimum transaction-support*, and $tconf(r, DB) \geq \beta$, where β is user-defined *minimum transaction-confidence* level.

5.3.2 Framework for Mining Association Rules Under Category II

Let $X = \{x_1, x_2, \ldots, x_k\}$ be an itemset in database *DB*. Also, let τ be a transaction in *DB*. Let item x_i be purchased η_i numbers at a time in τ, $i = 1, 2, \ldots, k$. Then, $TIF(X, \tau, DB) =$ minimum $\{\eta_1, \eta_2, \ldots, \eta_k\}$. Based on the frequency of an itemset in a database, we define *database-support* (*dsupp*) of an itemset as follows.

Definition 5.2 Let X be an itemset in *TIMT* type database *DB*. Database-support of X in *DB* is given as follows: $dsupp(X, DB) = DF(X, DB)/|DB|$.

An item in a transaction could occur more than once. Thus, $dsupp(X, DB)$ could be termed as the *multiplicity* of itemset X in *DB*. An important characteristic of a database is the average multiplicity of an item (*AMI*) in the database. Let m be the number of distinct items in *DB*. We define *AMI* in a *TIMT* type database *DB* as follows: $AMI(DB) = \sum_{i=1}^{m} dsupp(x_i, DB)/m$, where x_i is the ith item in *DB*, $i = 1$, $2, \ldots, m$. An itemset X is *database-frequent* in *DB* if $dsupp(X, DB) \geq \gamma$, where γ is user-defined *minimum database-support* level. Let Y be another itemset in *DB*. We define *database-support* of association rule $r: X \rightarrow Y$ in *DB* as follows: $dsupp(r, DB) = dsupp(X \cap Y, DB)$. Also, we define *database-confidence* (*dconf*) of association rule $r: X \rightarrow Y$ in *DB* as follows: $dconf(r, DB) = dsupp(X \cap Y, DB)/dsupp(X, DB)$. An association rule $r: X \rightarrow Y$ is *interesting* with respect to database frequency of an itemset if $dsupp(r, DB) \geq$ *minimum database-support* and $dconf(r, DB) \geq \delta$, where δ is user defined *minimum database-confidence* level.

5.3.3 Framework for Mining Association Rules Under Category III

An association rule r in *TIMT* type database *DB* is *interesting* with respect to both transaction frequency and database frequency of an itemset if $tsupp(r, DB) \geq \alpha$, $tconf(r, DB) \geq \beta$, $dsupp(r, DB) \geq \gamma$, and $dconf(r, DB) \geq \delta$. The parameters α, β, γ, and δ are defined in Sects. 5.3.1 and 5.3.2. The parameters are user defined inputs given to a category III association rule mining algorithm. Based on the framework, we extract association rules in a database in Example 5.3.

Example 5.3 Let *DB* = {{$A(1), B(1)$}, {$A(2), B(3), C(2)$}, {$A(1), B(4), E(1)$}, {$A(3)$, $E(1)$}, {$C(2), F(2)$}}. Let $\alpha = 0.4$, $\beta = 0.6$, $\gamma = 0.5$, and $\delta = 0.5$. Transaction-frequent and database-frequent itemsets are given in Tables 5.2 and 5.3, respectively. Interesting association rules under category III are given in Table 5.4.

5.3.4 Dealing with Items Measured in Continuous Scale

We discuss here the issue of handling items that are measured in continuous scale. Consider the item milk in a departmental store. Let there are four types of milk packets: 0.5, 1, 1.5, and 2 kl. The minimum packaging unit could be considered as 1 unit. Thus, 3.5 kl of milk could be considered as 7 units of milk.

Table 5.2 Transaction-frequent itemsets in *DB*

itemset	A	B	C	E	AB	AE
tsupp	0.8	0.6	0.4	0.4	0.6	0.4

Table 5.3 Database-frequent itemsets in *DB*

itemset	A	B	C	E	F	AB	AC	AE	BC	CF	ABC
dsupp	1.4	1.6	0.8	0.4	0.4	0.8	0.4	0.4	0.4	0.4	0.4

Table 5.4 Interesting association rules in *DB* under category III

$r: X \to Y$	$tsupp(r, DB)$	$tconf(r, DB)$	$dsupp(r, DB)$	$dconf(r, DB)$
$A \to B$	0.6	0.75	0.8	0.57143
$B \to A$	0.6	1.0	0.8	0.5

5.4 Properties of Different Interestingness Measures

Transaction-support and transaction-confidence measures are the same as the traditional support and confidence measures of an itemset in a database, respectively. Thus, they satisfy all the properties that are satisfied by traditional measures.

Property 5.1 $0 \leq tsupp(Y, DB) \leq tsupp(X, DB) \leq 1$, *for itemsets X and Y ($\supseteq$ X) in DB*.

Transaction-support measure satisfies anti-monotone property [11] of traditional support measure.

Property 5.2 $tconf(r, DB)$ *lies in* $[tsupp(r, DB), 1]$, *for an association rule r in DB*.

If an itemset X is present in transaction τ in DB then $TIF(X, \tau, DB) \geq 1$. Thus, we have the following property.

Property 5.3 $tsupp(X, DB) \leq dsupp(X, DB) < \infty$, *for itemset X in DB*.

In this case, database-confidence of an association rule r might not lie in [*dsupp (r, DB)*, 1.0], since database-support of an itemset in DB might be greater than 1 (see Table 5.3). But, the database confidence of an association rule r in DB satisfies the following property.

Property 5.4 $dconf(r, DB)$ *lies in* $[0, 1]$, *for an association rule r in DB*.

5.5 Mining Association Rules

Many interesting algorithms have been proposed for mining positive association rules in a binary database (Agrawal and Srikant 1994; Han et al. 2000). Thus, there are several implementations (FIMI 2004) of mining positive association rules in a binary database. In the context of mining association rules in a *TIMT* type database, we shall implement apriori algorithm (Agrawal and Srikant 1994), since it is simple and easy to implement. For mining association rules in a *TIMT* type database, we could apply apriori algorithm. For mining association rules under category III, the pruning step of interesting itemset generation requires testing on two conditions: minimum transaction-support and minimum database-support. The interesting association rules under category III satisfy the following two additional conditions: minimum transaction-confidence and minimum database-confidence.

Table 5.5 Association rules in different databases under category I

Database	(α, β)	(antecedent, consequent, *tsupp*, *tconf*)			
retail	(0.05, 0.2)	(48, 39, 0.33, 0.69)	(39, 48, 0.33, 0.58)	(41, 39, 0.13, 0.76)	(39, 41, 0.13, 0.23)
ecoli	(0.1, 0.3)	(48, 50, 0.96, 0.98)	(50, 48, 0.96, 0.96)	(44, 48, 0.14, 0.40)	(40, 50, 0.14, 0.41)

Table 5.6 Association rules in different databases under category II

Database	(γ, δ)	(antecedent, consequent, *dsupp*, *dconf*)			
retail	(0.07, 0.4)	(48, 39, 0.7255, 0.5073)	(39, 48, 0.7255, 0.4208)	(41, 39, 0.2844, 0.5584)	(38, 39, 0.2574, 0.4848)
ecoli	(0.1, 0.2)	(48, 50, 2.7583, 0.9231)	(50, 48, 2.7583, 0.9176)	(44, 48, 0.3005, 1.0000)	(40, 50, 0.2828, 1.0000)

5.6 Experiments

We present the experimental results using real databases *retail* (Frequent itemset mining dataset repository 2004) and *ecoli* (UCI ML repository content summary).[1] The database *ecoli* is a subset of *ecoli database*. Due to unavailability of *TIMT* type database, we have applied a preprocessing technique. If an item is present in a transaction then the number of occurrences of the item is generated randomly between 1 and 5. Thus, a binary transactional database gets converted into a *TIMT* type database. Top 4 association rules under category I (sorted on transaction-support) and under category II (sorted on database-support) are given in Tables 5.5 and 5.6, respectively.

Also, we have obtained execution times for extracting association rules at different database sizes. As the size of a database increases, the execution time also increases. We have observed such phenomenon in Figs. 5.1 and 5.2.

Also, we have obtained execution times for extracting association rules at different minimum database-supports. As the value of minimum database-support increases, the number of interesting itemsets decreases. So, the time required to extract category II association rules also decreases. We have observed such phenomenon in Figs. 5.3 and 5.4.

---

[1] http://www.ics.uci.edu/~mlearn/MLSummary.html.

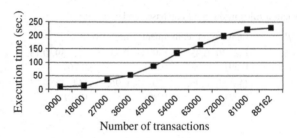

Fig. 5.1 Execution time versus size of database at $\alpha = 0.05$, $\gamma = 0.07$, $\beta = 0.2$, and $\delta = 0.4$ (*retail*)

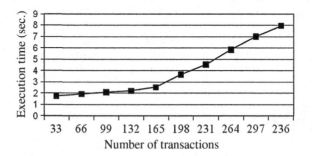

Fig. 5.2 Execution time versus size of database at $\alpha = 0.1$, $\gamma = 0.1$, $\beta = 0.3$, and $\delta = 0.3$ (*ecoli*)

Fig. 5.3 Execution time
versus γ at $\alpha = 0.04$, $\beta = 0.2$,
and $\delta = 0.4$ (*retail*)

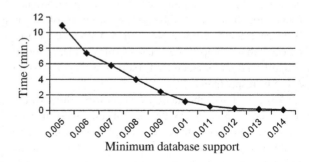

Fig. 5.4 Execution time
versus γ at $\alpha = 0.08$, $\beta = 0.3$,
and $\delta = 0.2$ (*ecoli*)

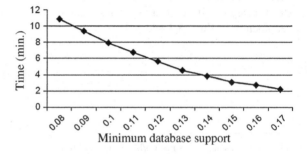

The graph of execution time versus α at a given tuple (γ, β, δ) is similar to the graph of execution time versus γ at a given tuple (α, β, δ). As the value of minimum transaction-support increases, the number of interesting itemsets decreases. So, the time required to extract category I association rules also decreases.

5.7 Related Work

Agrawal and Srikant (1994) have proposed apriori algorithm that uses breadth-first search strategy to count the supports of itemsets. The algorithm uses an improved candidate generation function, which exploits the downward closure property of support and makes it more efficient than earlier algorithm (Agrawal et al. 1993). Han et al. (2000) have proposed data mining method frequent pattern growth (*FP-growth*) which uses an extended prefix-tree (*FP*-tree) structure to store the database in a compressed form. In the context of interestingness measures, Tan et al. (2002) have presented an overview of twenty one interestingness measures proposed in statistics, machine learning and data mining literature. An algorithm for mining calendar-based patterns in non-binary dataset is designed by Adhikari (2014). Also, an algorithm for finding periodicity of patterns is presented. Some association analysis such as overall association between items in non-binary data is discussed in Adhikari et al. (2011).

5.8 Conclusion

The traditional support-confidence framework has limited usage in association analysis of items, since a real life transaction might contain an item multiple times. The traditional support-confidence framework is based on the frequency of an itemset in a binary database. In the case of a *TIMT* type database, there are two types of frequency of an itemset: transaction frequency, and database frequency. Due to these reasons, we get three categories of association rules in a *TIMT* type database. We have introduced a framework for mining each category of association rules. The proposed frameworks are effective for studying association among items in real life market basket data.

References

Adhikari A, Adhikari J, Pedrycz W (2014) Data analysis and pattern recognition in multiple databases. Springer, Switzerland
Adhikari A, Ramachandrarao P, Pedrycz W (2011) Study of select items in different data sources by grouping. Knowl Inf Syst 27(1):23–43

Adhikari J (2014) Mining calendar-based periodic patterns from nonbinary transactions. J Intell
    Syst, pp 1–5

Agrawal R, Imielinski T, Swami A (1993) Mining association rules between sets of items in large
    databases. In: Proceedings of ACM SIGMOD conference management of data, pp 207–216

Agrawal R, Srikant R (1994) Fast algorithms for mining association rules. In: Proceedings of 20th
    very large databases (VLDB) conference, pp 487–499

Antonie M-L, Zaïane OR (2004) Mining positive and negative association rules: an approach for
    confined rules. In: Proceedings of PKDD, pp 27–38

FIMI (2004) http://fimi.cs.helsinki.fi/src/

Frequent Itemset Mining Dataset Repository (2004). http://fimi.cs.helsinki.fi/data

Han J, Pei J, Yiwen Y (2000) Mining frequent patterns without candidate generation. In:
    Proceedings of ACM SIGMOD conference management of data, pp 1–12

Steinbach M, Kumar V (2004) Generalizing the notion of confidence. In: Proceedings of ICDM
    conference, pp 402–409

Steinbach M, Tan P-N, Xiong H, Kumar V (2004) Generalizing the notion of support. In:
    Proceedings of KDD, pp 689–694

Tan P-N, Kumar V, Srivastava J (2002) Selecting the right interestingness measure for association
    patterns. In: Proceedings of SIGKDD conference, pp 32–41

Chapter 6
Mining Patterns in Different Related Databases

Effective data analysis using multiple databases requires highly accurate patterns. As the local pattern analysis might extract patterns of low quality from multiple databases, it becomes necessary to improve mining multiple databases. In this chapter, we present an idea of multi-database mining by making use of local pattern analysis. We elaborate on the existing specialized and generalized techniques which are used for mining multiple large databases. In the sequel, we discuss a certain generalized technique, referred to as a pipelined feedback model, which is of particular relevance for mining multiple large databases. It significantly improves the quality of the synthesized global patterns. We define two types of error occurring in multi-database mining techniques. Experimental results are provided and they are reported for both real-world and synthetic databases. They help us assess the effectiveness of the pipelined feedback model.

6.1 Introduction

Many large companies operate from a number of branches usually located at different geographical regions. Each branch collects data continuously and local data become stored locally. The collection of all branch databases might be large. Many corporate decisions of a multi-branch company are based on data stored over the branches. The challenges are to make meaningful decisions which are based on large volume of distributed data. This creates not only risk but also offers opportunities. One of the risks is a significant amount investment on hardware and software to deal with multiple large databases. The use of inefficient data mining techniques has to be taken into account and in many scenarios this shortcoming could be very detrimental to the quality of results.

Based on the number of data sources, patterns in multiple databases could be classified into three categories. These are local patterns, global patterns and patterns that are neither local nor global. A pattern based on a single database is called a

© Springer International Publishing Switzerland 2015

A. Adhikari and J. Adhikari, *Advances in Knowledge Discovery in Databases*,
Intelligent Systems Reference Library 79, DOI 10.1007/978-3-319-13212-9_6

local pattern. Local patterns are useful for local data analysis, and locally restricted decision making activities (Adhikari and Rao 2008b; Wu et al. 2005; Zhang et al. 2004b). On the other hand, global patterns are based on all the databases taken into consideration. They are useful for data analyses of global nature (Adhikari and Rao 2008a; Wu and Zhang 2003) and global decision making problems. The intent of this chapter is to introduce and analyze a certain global model of data mining, referred to as a pipelined feedback model (PFM) (Adhikari et al. 2007b) which is used for mining/synthesizing global patterns in multiple large databases.

In Sect. 6.2, we formalize the idea of multi-database mining using local pattern analysis. Next, we discuss existing generalized multi-database mining techniques (Sect. 6.3). We analyze the existing specialized multi-database mining techniques in Sect. 6.4. The pipelined feedback model for mining multiple large databases is covered in Sect. 6.5. We also define a way in which an error associated with the model is quantified (Sect. 6.6). In Sect. 6.7, we provide experimental results using both synthetic and real-world databases.

6.2 Multi-database Mining Using Local Pattern Analysis

Consider a large company that deals with multiple large databases. For mining multiple databases, we are faced with three scenarios viz., (i) Each of the local databases is small, so that a single database mining technique (SDMT) could mine the union of all databases. (ii) At least one of the local databases is large, so that a SDMT could mine every local database, but fail to mine the union of all local databases. (iii) At least one of the local databases is very large, so that a SDMT fails to mine at least one local database. We are faced with challenges when handling the cases (ii) and (iii) and these challenges are inherently present because of the large size of some databases.

The first question which comes to our mind is whether a traditional data mining technique (Agrawal and Srikant 1994; Han et al. 2000; Coenen et al. 2004) could provide a sound solution when dealing with multiple large databases. To apply a "traditional" data mining technique we need to amass all the branch databases together. In such cases, a traditional data mining technique might not offer a good solution due to the following reasons:

- It might not be suitable as it requires heavy investment on hardware and software to deal with a large volume of data.
- A single computer might take unreasonable amount of time to mine a huge amount of data.
- It is difficult to identify local patterns if a traditional data mining technique is applied to the collection of all local databases.

In light of these problems and associated constraints, as encountered so far there have been attempts to deal with multi-database mining in a different way. Zhang et al. (2003) designed a multi-database mining technique (MDMT) using local pattern analysis. Multi-database mining using local pattern analysis could be classified into two categories viz., the techniques that analyze local patterns, and the techniques that analyze approximate local patterns. A multi-database mining technique using local pattern analysis could be viewed as a two-step process, denoted symbolically as M + S. Its essence can be explained as follows:

- Mine each local database using a SDMT by applying the model M (Step 1)
- Synthesize patterns using the algorithm S (Step 2)

We use the notation of M + S to stress a character of a multi-database mining technique in which we first use the model of mining (M) being followed by the synthesizing algorithm S.

One could apply sampling techniques (Babcock et al. 2003) for taming large volume of data. If an itemset is frequent in a large dataset then it is likely to be frequent in a sample dataset. Thus, one can mine patterns approximately in a large dataset by analyzing patterns in a representative sample dataset.

In addition to generalized multi-database mining techniques, there exist also specialized multi-database mining techniques. In what follows, we discuss some of the existing multi-database mining techniques.

6.3   Generalized Multi-database Mining Techniques

There is a significant variety of techniques that can be used in the multi-database mining applications.

6.3.1   Local Pattern Analysis

Under this model of mining multiple databases, each branch requires to mine the database using a traditional data mining technique. Afterwards, each branch is required to forward the pattern base to the central office. Then the central office processes the locally processed pattern bases collected from different branches to synthesize the global patterns and subsequently to support decision-making activities. Zhang et al. (2003) designed a multi-database mining technique (MDMT) using local pattern analysis. In Chap. 10, we presented an extended version of this model (Adhikari and Rao 2008a). It improves the quality of synthesized global patterns in multiple databases. In addition, it supports a systematic approach to synthesize the global patterns.

6.3.2 Partition Algorithm

For the purpose of mining multiple databases, one could apply *partition algorithm* (PA) proposed by Savasere et al. (1995). The algorithm is designed for mining a very large database by partitioning. The algorithm works as follows. It scans the database twice. The database is divided into disjoint partitions, where each partition is small enough to fit in memory. In a first scan, the algorithm reads each partition and computes locally frequent itemsets in each partition using apriori algorithm (Agrawal and Srikant 1994). In the second scan, the algorithm counts the supports of all locally frequent itemsets toward the complete database. In this case, each local database could be considered as a partition. Though partition algorithm mines frequent itemsets in a database exactly, it might be an expensive solution to mining multiple large databases, since each database is required to be scanned twice. During the time of the second scanning, all the local patterns obtained at the first scan are analyzed. Thus, partition algorithm used for mining multiple databases could be considered as another type of local pattern analysis.

6.3.3 IdentifyExPattern Algorithm

Zhang et al. (2004a) have proposed algorithm, *IdentifyExPattern* (IEP) for identifying global exceptional patterns in multi-databases. Every local database is mined separately at *random order* (RO) using a SDMT to synthesize global exceptional patterns. For identifying global exceptional patterns in multiple databases, the following pattern synthesizing approach has been proposed. A pattern in a local database is assumed as absent, if it does not get reported. Let $supp_a(p, DB)$ and $supp_s(p, DB)$ be the actual (i.e., apriori) support and synthesized support of pattern p in database DB. Let D be the union of all local databases. Then support of pattern p has been synthesized in D based on the following expression:

$$supp_s(p, D) = \frac{1}{num(p)} \sum_{i=1}^{num(p)} (supp_a(p, D_i) - \alpha)/(1 - \alpha) \qquad (6.1)$$

where $num(p)$ is the number of databases that report p at a given minimum support level (α).

The size (i.e., the number of transactions) of a local database and support of an itemset in a local database seem to be important parameters that are used to determine the presence of an itemset in a database, since the number of transactions containing the itemset X in a database D_1 is equal to $supp(X, D_1) \times size(D_1)$. The major concern in this investigation is that the algorithm does not consider the size of a local database to synthesize the global support of a pattern. Using the IEP algorithm, the global support of a pattern has been synthesized using only supports of the pattern present in local databases.

6.3.4 RuleSynthesizing Algorithm

Wu and Zhang (2003) have proposed *RuleSynthesizing* (RS) algorithm for synthesizing high frequency association rules in multiple databases. Using this technique, every local database is mined separately at *random order* (RO) using a SDMT for synthesizing high frequency association rules. A pattern in a local database is assumed as absent, if it does not get reported. Based on the association rules present in different databases, the authors have estimated weights of different databases. Let w_i be the weight of the ith database, $i = 1, 2, ..., n$. Without any loss of generality, let the association rule r be extracted from first m databases, for $1 \leq m \leq n$. Here, $supp_a(r, D_i)$ has been assumed as 0, for $i = m + 1, m + 2, ..., n$. Then the support of r in D has been determined in the following way:

$$supp_s(r, D) = w_1 \times supp_a(r, D_1) + \cdots + w_m \times supp_a(r, D_m) \qquad (6.2)$$

Algorithm *RuleSynthesizing* offers an indirect approach for synthesizing association rules in multiple databases. Thus the time complexity of the algorithm is reasonably high. The algorithm executes in $O(n^4 \times maxNosRules \times totalRules^2)$ time, where n, *maxNosRules*, and *totalRules* are the number of data sources, the maximum among the numbers of association rules extracted from different databases, and the total number of association rules in different databases, respectively.

6.4 Specialized Multi-database Mining Techniques

For finding solution to a specific application, it might be possible to devise a better multi-database mining technique. In this section, we elaborate in detail on three specific multi-database mining techniques.

6.4.1 Mining Multiple Real Databases

We have proposed algorithm *Association-Rule-Synthesis* (ARS) for synthesizing association rules in multiple real data sources (Adhikari and Rao 2008a). The algorithm uses the model shown in Fig. 6.1. While synthesizing an association rule, it uses a specific method which is explained as follows: For real databases, the trend of the customers' behaviour exhibited in a single database is usually present in other databases. In particular, a frequent itemset in one database is usually present in some transactions of other databases even if it does not get extracted. The proposed estimation procedure captures such trend and estimates the support of a missing association rule. Without any loss of generality, let the itemset X be extracted from first m databases, for $1 \leq m \leq n$. Then trend of X in first m databases could be expressed as follows:

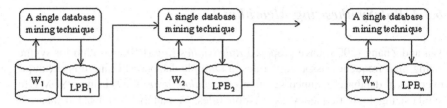

Fig. 6.1 Pipelined feedback model of mining multiple databases

$$trend^{1,\,m}(X|\alpha) = \frac{1}{\sum_{i=1}^{m}|D_i|} \times \sum_{i=1}^{m}(supp_a(X,D_i) \times |D_i|) \qquad (6.3)$$

The number of transactions in a database could be considered as its weight. In (6.3), the trend of X in first m databases is estimated as a weighted sum of supports in the first m databases. We can use the detected trend of X encountered in the first m databases for synthesizing support of X in D. We estimate the support of X in database D_j by computing the expression $\alpha \times trend^{1,\,m}(X|\alpha)$, $j = k + 1$, $k + 2$, ..., n. Then the synthesized support of X is determined as follows:

$$supp_s(X, D) = \frac{trend^{1,\,m}(X|\alpha)}{\sum_{i=1}^{n}|D_i|} \times \left[(1-\alpha) \times \sum_{i=1}^{m}|D_i| + \alpha \times \sum_{i=1}^{n}|D_i|\right] \qquad (6.4)$$

Association-Rule-Synthesis algorithm might return approximate global patterns, since an itemset might not get extracted from all the databases.

6.4.2 Mining Multiple Databases for the Purpose of Studying a Set of Items

Many important decisions are based on a set of specific items called the *select items*. We have proposed a technique for mining patterns of select items in multiple databases (Adhikari and Rao 2007a). A detailed discussion of this technique is provided in Chap. 12.

6.4.3 Study of Temporal Patterns in Multiple Databases

Adhikari et al. (2009) have proposed a technique for clustering items in multiple databases based on their level of stability where a certain stability measure is used to quantify this feature. The technique is presented in Chap. 9.

6.5 Mining Multiple Databases Using Pipelined Feedback Model (PFM)

Before applying the pipelined feedback model, one needs to prepare data warehouses at different branches of a multi-branch organization. In Fig. 10.1, we have shown how to preprocess data warehouse at each branch. Let W_i be the data warehouse corresponding to the ith branch, $i = 1, 2, ..., n$. Then the local patterns for the ith branch are extracted from W_i, $i = 1, 2, ..., n$. We mine each data warehouse using any SDMT technique. In Fig. 6.1, we present a model of mining multiple databases (Adhikari et al. 2007b).

In PFM, W_1 is mined using a SDMT and as result a local pattern base LPB_1 becomes extracted. While mining W_2, all the patterns in LPB_1 are extracted irrespective of their values of interestingness measures like, minimum support and minimum confidence. Apart from these patterns, some new patterns that satisfy user-defined threshold values of interestingness measures are also extracted. In general, while mining W_i, all the patterns in W_{i-1} are mined irrespective of their values of interestingness measures, and some new patterns that satisfy user-defined threshold values of interestingness measures, $i = 2, 3, ..., n$. Due to this nature of mining each data warehouse, the technique is called a feedback model. Thus, $|LPB_{i-1}| \leq |LPB_i|$, for $i = 2, 3, ..., n$. There are $n!$ arrangements of pipelining for n databases. All the arrangements of data warehouses might not produce the same result of mining. If the number of local patterns increases, one gets more accurate global patterns which leads to a better analysis of local patterns. An arrangement of data warehouses would produce near optimal result if the cardinality $|LPB_n|$ is maximal. Let $size(W_i)$ be the size of W_i (in bytes), $i = 1, 2, ..., n$. We will adhere to the following rule of thumb regarding the arrangements of data warehouses for the purpose of mining. The number of patterns in W_i is greater than or equal to the number of patterns in W_{i-1} when $size(W_i) \geq size(W_{i-1})$, $i = 2, 3, ..., n$. For the purpose of increasing the number of local patterns, W_{i-1} precedes W_i in the pipelined arrangement of mining data warehouses if $size(W_{i-1}) \geq size(W_i)$, $i = 2, 3, ..., n$. Finally, we analyze the patterns in LPB_1, LPB_2, ..., and LPB_n to synthesize global patterns, or analyze local patterns.

Let W be the collection of all branch data warehouses. For synthesizing global patterns in W we discuss here a simple pattern synthesizing (SPS) algorithm. Without any loss of generality, let the itemset X be extracted from first m databases, for $1 \leq m \leq n$. Then the synthesized support of X in W comes in the form.

$$supp_s(X, W) = \frac{1}{\sum_{i=1}^{n} |W_i|} \times \sum_{i=1}^{m} [supp_a(X, W_i) \times |W_i|]. \qquad (6.5)$$

6.5.1 Algorithm Design

In this section, we present an algorithm for mining multiple large databases. The method is based on the pipelined feedback model discussed above.

Algorithm 6.1 Mine multiple data warehouses using pipelined feedback model.
procedure *PipelinedFeedbackModel* ($W_1, W_2, ..., W_n$)
Input: $W_1, W_2, ..., W_n$
Output: local pattern bases
01: **for** i = 1 to n **do**
02:    **if** W_i does not fit in memory **then**
03:       partition W_i into $W_{i1}, W_{i2}, ...,$ and W_{ip_i} for an integer p_i;
04:    **else** $W_{i1} = W_i$;
05:    **end if**
06: **end for**
07: sort data warehouses on size in non-increasing order and the data warehouses are renamed as
    $DW_1, DW_2, ..., DW_N$, where $N = \sum_{i=1}^{n} p_i$;
08: **let** $LPB_0 = \phi$;
09: **for** i = 1 to N **do**
10:    mine DW_i using a SDMT with input LPB_{i-1};
11: **end for**
12: return $LPB_1, LPB_2, ..., LPB_N$;
end procedure

In the algorithm, the usage of LPB_{i-1} during mining DW_i has been explained above. Once a pattern has been extracted from a data warehouse, then it also gets extracted from the remaining data warehouses. Thus, the algorithm *Pipelined-FeedbackModel* improves the quality of synthesized patterns as well as contributes significantly to an analysis of local patterns.

6.6 Error Evaluation

To evaluate the quality of MDMT: PFM + SPS, one needs to quantify the error produced by the method. First, in an experiment we mine frequent itemsets in multiple databases using PFM, and afterwards synthesize global patterns using the SPS algorithm. One needs to find how the global synthesized support differs from the exact (apriori) support of an itemset.

PFM improves mining multiple databases significantly over local pattern analysis. In the PFM, we have $LPB_{i-1} \subseteq LPB_i$, for i = 2, 3, ..., n. Then, patterns in $LPB_i - LPB_{i-1}$ are generated from databases $D_i, D_{i+1}, ..., D_n$. We assume $supp_a(X, D_j) = 0$, for each $X \in LPB_i - LPB_{i-1}$, and j = 1, 2, ..., $i - 1$. Thus, the error of mining X could be defined as follows:

$$E(X|\text{PFM, SPS}) = \left| supp_a(X, D) - \frac{1}{\sum_{j=1}^{n}|D_j|} \times \sum_{j=i}^{n} \left[supp_a(X, D_j) \times |D_j| \right] \right|, \tag{6.6}$$

for $X \in LPB_i - LPB_{i-1}$ and $i = 2, 3, \ldots, n$.

Also, E(X|PFM, SPS) = 0, for $X \in LPB_1$.

When a frequent itemset is reported from D_1 then it gets reported from every databases using PFM algorithm. Thus, E(X|PFM, SPS) = 0, for $X \in LPB_1$.

Otherwise, an itemset X is not reported from all the databases. It is synthesized using SPS algorithm. Then the synthesized support is subtracted from its apriori support for finding the error of mining X.

There are several ways one could define the error of an experiment. In particular, one could concentrate on the following definitions.

1. *Average error* (AE)

$$AE(D, \alpha) = \frac{1}{|LPB_1 + \sum_{i=2}^{n}(LPB_i - LPB_{i-1})|} \sum_{X \in \left[LPB_1 + \cup \left\{ \bigcup_{i=2}^{n}(LPB_i - LPB_{i-1}) \right\} \right]} E(X|\text{PFM, SPS})] \tag{6.7}$$

2. *Maximum error* (ME)

$$ME(D, \alpha) = \text{maximum}\{E(X|\text{PFM, SPS}), \text{ for } X \in LPB_1 \\ \cup \left\{ \bigcup_{i=2}^{n}(LPB_i - LBP_{i-1}) \right\} \} \tag{6.8}$$

where $supp_a(X_i, D)$ is obtained by mining D using a traditional data mining technique, $i = 1, 2, \ldots, m$. $supp_s(X_i, D)$ is obtained by SPS, for $i = 1, 2, \ldots, m$.

6.7 Experiments

We have carried out a series of experiments to study and quantify the effectiveness of the PFM. We present experimental results using three synthetic databases and two real-world databases. The synthetic databases are *T10I4D100K* (*T*) (Frequent itemset mining dataset repository 2004), *random500* (*R1*) and *random1000* (*R2*). The databases *random500* and *random1000* are generated synthetically for the purpose of conducting experiments. The real databases are *retail* (*R*) (Frequent itemset mining dataset repository 2004) and *BMS-Web-Wiew-1* (*B*) (Frequent itemset mining dataset repository 2004). The main characteristics of these datasets are displayed in Table 6.1.

Table 6.1 Database characteristics

D	NT	ALT	AFI	NI
T	100,000	11.10	1276.12	870
R	88,162	11.31	99.67	10,000
B	149,639	2.00	155.71	1,922
R1	10,000	6.47	109.40	500
R2	10,000	12.49	111.86	1,000

Table 6.2 Input database characteristics

D	NT	ALT	AFI	NI	DB	NT	ALT	AFI	NI
T_0	10,000	11.06	127.66	866	T_5	10,000	11.14	128.63	866
T_1	10,000	11.133	128.41	867	T_6	10,000	11.11	128.56	864
T_2	10,000	11.07	127.64	867	T_7	10,000	11.10	128.45	864
T_3	10,000	11.12	128.44	866	T_8	10,000	11.08	128.56	862
T_4	10,000	11.14	128.75	865	T_9	10,000	11.08	128.11	865
R_0	9,000	11.24	12.07	8,384	R_5	9,000	10.86	16.71	5,847
R_1	9,000	11.21	12.27	8,225	R_6	9,000	11.20	17.42	5,788
R_2	9,000	11.34	14.60	6,990	R_7	9,000	11.16	17.35	5,788
R_3	9,000	11.49	16.66	6,206	R_8	9,000	12.00	18.69	5,777
R_4	9,000	10.96	16.04	6,148	R_9	7,162	11.69	15.35	5,456
B_0	14,000	2.00	14.94	1,874	B_5	14,000	2.00	280.00	100
B_1	14,000	2.00	280.00	100	B_6	14,000	2.00	280.00	100
B_2	14,000	2.00	280.00	100	B_7	14,000	2.00	280.00	100
B_3	14,000	2.00	280.00	100	B_8	14,000	2.00	280.00	100
B_4	14,000	2.00	280.00	100	B_9	23,639	2.00	472.78	100
$R1_0$	1,000	6.37	10.73	500	$R1_5$	1,000	6.34	10.68	500
$R1_1$	1,000	6.50	11.00	500	$R1_6$	1,000	6.62	11.25	500
$R1_2$	1,000	6.40	10.80	500	$R1_7$	1,000	6.42	10.83	500
$R1_3$	1,000	6.52	11.05	500	$R1_8$	1,000	6.58	11.16	500
$R1_4$	1,000	6.30	10.60	500	$R1_9$	1,000	6.65	11.30	500
$R2_0$	1,000	6.42	5.43	996	$R2_5$	1,000	6.44	5.46	997
$R2_1$	1,000	6.41	5.44	995	$R2_6$	1,000	6.48	5.50	996
$R2_2$	1,000	6.56	5.58	995	$R2_7$	1,000	6.48	5.49	997
$R2_3$	1,000	6.53	5.54	998	$R2_8$	1,000	6.54	5.56	996
$R2_4$	1,000	6.50	5.54	991	$R2_9$	1,000	6.50	5.56	988

Let NT, AFI, ALT, and NI denote the number of transactions, average frequency of an item, average length of a transaction, and number of items in a database, respectively. Each of the above databases is split into 10 databases for the purpose of carrying out experiments. The databases obtained from T, R, B, $R1$ and $R2$ are named as T_i, R_i, B_i, $R1_i$ and $R2_i$, respectively, for $i = 0, 1, \ldots, 9$. The databases T_i, R_i,

Table 6.3 Error obtained for the first three databases for selected value of α

Database	T10I4D100K		retail		BMS-Web-Wiew-1	
α	0.05		0.11		0.19	
Error type	AE	ME	AE	ME	AE	ME
MDMT: RO + IEP	0.01	0.04	0.01	0.06	0.05	0.15
MDMT: RO + RS	0.01	0.04	0.01	0.06	0.02	0.13
MDMT: RO + ARS	0.01	0.04	0.01	0.06	0.02	0.11
MDMT: PFM + SPS	0	0.05	0.01	0.06	0	0
MDMT: RO + PA	0	0	0	0	0	0

Table 6.4 Error reported for the last two databases for selected value of α

Database	random500		random1000	
α	0.005		0.004	
Error type	AE	ME	AE	ME
MDMT: RO + IEP	0.01	0.01	0.01	0.01
MDMT: RO + RS	0.01	0.01	0	0.01
MDMT: RO + ARS	0.01	0.01	0.01	0.01
MDMT: PFM + SPS	0.01	0.01	0	0
MDMT: RO + PA	0	0	0	0

B_i, $R1_i$, $R2_i$ are called input databases (*DBs*), for $i = 0, 1, \ldots, 9$. Some characteristics of these input databases are presented in the Table 6.2. In Tables 6.3 and 6.4, we include some outcomes to quantify how the proposed technique improves the results of mining. We have completed experiments using other MDMTs on these databases for the purpose of comparing them with MDMT: PFM + SPS.

Figures 6.2, 6.3, 6.4, 6.5 and 6.6 show average error versus different values of α. From these graphs, we conclude that AE normally increases as α increases. The number of databases reporting a pattern decreases when the values of α increase. Thus, the AE of synthesizing patterns normally increases as α increases. In case of Fig. 6.4, the graphs for MDMT: PFM + SPS and MDMT: RO + PA are similar to those with the X-axis.

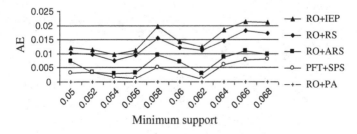

Fig. 6.2 AE versus α for experiments conducted for database T

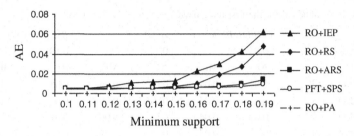

Fig. 6.3 AE versus α for experiments using database R

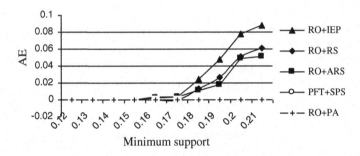

Fig. 6.4 AE versus α for experiments using database B

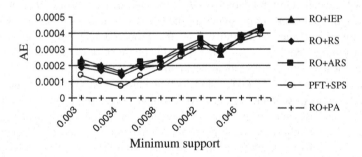

Fig. 6.5 AE versus α for experiments using database *R1*

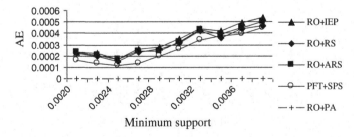

Fig. 6.6 AE versus α for experiments using database *R2*

6.8 Conclusions

In this chapter, we have discussed several generalized as well as specialized multi-database mining techniques. For a particular problem at hand, one technique could be more suitable than the others. However, we cannot claim that there is a single method of universal nature which outperforms all other techniques. Instead, a choice of the method has to be problem-driven. We have formalized the idea of multi-database mining using local pattern analysis by considering an underlying two-step process. We have also presented the pipelined feedback model which is particularly suitable for mining multiple large databases. It improves significantly the accuracy of mining multiple databases as compared to an existing technique that scans each database only once. The pipelined feedback model could also be used for mining a large database by dividing it into a series of sub-databases. Experimental results obtained with the use of the MDMT: PFM + SPS are promising and underline the usefulness of the method studied here.

References

Adhikari A, Rao PR (2007a) Study of select items in multiple databases by grouping. In: Proceedings of the 3rd Indian international conference on artificial intelligence, pp 1699–1718

Adhikari A, Rao PR (2008a) Synthesizing heavy association rules from different real data sources. Pattern Recogn Lett 29(1):59–71

Adhikari A, Rao PR (2008b) Efficient clustering of databases induced by local patterns. Decis Support Syst 44(4):925–943

Adhikari A, Rao PR, Adhikari J (2007b) Mining multiple large databases. In: Proceedings of the 10th international conference on information technology, pp 80–84

Adhikari J, Rao PR, Adhikari A (2009) Clustering items in different data sources induced by stability. Int Arab J Inf Technol 6(4):66–74

Agrawal R, Srikant R (1994) Fast algorithms for mining association rules. In: Proceedings of international conference on very large data bases, pp 487–499

Babcock B, Chaudhury S, Das G (2003) Dynamic sample selection for approximate query processing. In: Proceedings of ACM SIGMOD conference management of data, pp 539–550

Coenen F, Leng P, Ahmed S (2004) Data structure for association rule mining: T-trees and P-trees. IEEE Trans Knowl Data Eng 16(6):774–778

Frequent itemset mining dataset repository (2004) http://fimi.cs.helsinki.fi/data

Han J, Pei J, Yiwen Y (2000) Mining frequent patterns without candidate generation. In: Proceedings of ACM SIGMOD conference on management of data, pp 1–12

Savasere A, Omiecinski E, Navathe S (1995) An efficient algorithm for mining association rules in large databases. In: Proceedings of the 21st international conference on very large data bases, pp 432–443

Wu X, Zhang S (2003) Synthesizing high-frequency rules from different data sources. IEEE Trans Knowl Data Eng 14(2):353–367

Wu X, Zhang C, Zhang S (2005) Database classification for multi-database mining. Inf Syst 30 (1):71–88

Zhang S, Wu X, Zhang C (2003) Multi-database mining. IEEE Comput Intell Bull 2(1):5–13

Zhang C, Liu M, Nie W, Zhang S (2004a) Identifying global exceptional patterns in multi-database mining. IEEE Comput Intell Bull 3(1):19–24

Zhang S, Zhang C, Yu JX (2004b) An efficient strategy for mining exceptions in multi-databases. Inf Sci 165(1–2):1–20

Chapter 7
Mining Icebergs in Different Time-Stamped Data Sources

Many organizations possess large databases collected over a long period of time. Analysis of such databases might be strategically important for further growth of the organizations. For instance, it might be of interest to learn about interesting changes in sales over time. In this chapter, we introduce a new pattern, called notch, of an item in time-stamped databases. Based on this notion, we propose a special kind of notch, called a generalized notch and subsequently, a specific type of generalized notch, called an iceberg, in time-stamped databases. We design an algorithm for mining interesting icebergs in time-stamped databases. We also present experimental results obtained for both synthetic and real-world databases.

7.1 Introduction

Many organizations collect transactional data continuously over a long period of time. A database grown over a long period of time might contain useful as well as interesting temporal patterns. By taking into account time aspect, many interesting applications/patterns such as surprising patterns (Keogh et al. 2002), discords (Keogh et al. 2006), calendar based patterns (Mahanta et al. 2008) have been recently reported. Surprising patterns, anomaly detection and discords could be considered as exceptional patterns occurring in a time series. These exceptional patterns form important as well as interesting contributions to temporal data mining. In this chapter, we study another kind of exceptional patterns in transactional time-stamped data. We define exceptional patterns and discuss how to mine them from time-stamped databases.

Though an analysis of time series data (Box et al. 2003; Brockwell and Richard 2002; Keogh 1997; Tsay 1989) has been intensively studied, the analysis of time-stamped data still calls for more research. Specifically, in the context of multiple time-stamped databases, little work has been reported so far. Therefore, there arises an urgent need to study multiple time-stamped databases. In Example 7.1, we observe an interesting type of temporal pattern in multiple time-stamped databases that needs to be fully analyzed.

The support of an itemset (Agrawal et al. 1993) is defined as the fraction of transactions containing the itemset. It has been used extensively in identifying

© Springer International Publishing Switzerland 2015   97
A. Adhikari and J. Adhikari, *Advances in Knowledge Discovery in Databases*,
Intelligent Systems Reference Library 79, DOI 10.1007/978-3-319-13212-9_7

different types of patterns in a database. Some examples are association rule (Agrawal et al. 1993), negative association rule (Wu et al. 2004), and conditional pattern (Adhikari and Rao 2008). Nonetheless the support measure has a limited use in discovering some other types of patterns in a database. We illustrate this issue using the following example.

Example 7.1 Consider a company that maintains customers' transactions on a yearly basis. Many important problems can be studied given such yearly databases. Let item A be of our interest. In view of analyzing item A over the years, let us consider the sales series of A from the year 1970–1979:

(0.9, 1,000, 1970), (0.31, 2,700, 1971), (0.36, 2,500, 1972), (0.29, 3,450, 1973), (0.37, 1,689, 1974), (0.075, 7,098, 1975), (0.073, 8,900, 1976), (0.111, 6,429, 1977), (0.09, 9,083, 1978), (0.07, 10,050, 1979).

The individual components of each triple refer to the support of A, the number of transactions in the yearly database and the corresponding year, respectively.

The sales series of item A is depicted in Fig. 7.1. There is a significant downfall of sales from 1972 and rise in sales from the year 1975. Year 1975 is an important point (Pratt and Fink 2002) for the company. It is a significant down-to-up change in the sales series. It is not surprising to observe a significant up-to-down change in a sales series of an item. Such patterns in time-stamped series are interesting as well as important to investigate. They could reveal the sources of customers' purchase behavior that might provide an important knowledge to the organization.

At this point, one might be interested in knowing how a time series data differs from a time-stamped data. Transactional data are time-stamped data collected over time at no particular frequency (Leonard and Wolfe 2005). Whereas, time series data are time-stamped data collected over time at a particular frequency. For example, point of sales data could be considered as time-stamped data, but sales per month/year could be considered as time series data. One could convert transactional data into time series data for the purpose of specific data analyses. The frequency associated with time series data varies from problem to problem. For future planning of business activities, one might need to analyze the past data of customer transactions collected over a long period of time. While analyzing the past data it is useful as well as important to figure out the abrupt changes in sales of an item along with time. Existing algorithms, as mentioned above, fail to detect these changes.

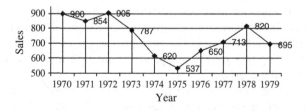

Fig. 7.1 Sales of item A reported in consecutive years

Therefore, in this chapter our objective is to define such exceptional patterns and design an algorithm to extract such patterns from time-stamped databases.

For the purpose of studying patterns in time-stamped databases one may need to handle multiple databases over time. One could call these time variant databases as *time databases*. In this context, the choice of time granularity is an important issue as the characteristics of temporal patterns is heavily dependent on this parameter. It is quite reasonable to consider the time granularity as one year, since a season re-appears on a yearly basis and the customers' purchase behaviour might vary from season to season.

Consider an established company having data collected over fifty consecutive years. The company might be interested in knowing the performance of different items over the years. Such analysis might help the company in devising the future strategies. Our objective is to identify abrupt changes in sales of each item (as defined in Sects. 7.4 and 7.5) over the years as depicted in Fig. 7.1.

The chapter is organized as follows. We discuss related work in Sect. 7.2. In Sect. 7.3, we introduce a new temporal pattern, called notch, of an item. Based on this pattern, we propose the concepts of generalized notch (Sect. 7.4) and iceberg notch (Sect. 7.5). We present another view of sales series in Sect. 7.6. In Sect. 7.7 we design an algorithm for mining icebergs in time-stamped databases. Experimental results are presented in Sect. 7.8.

7.2 Related Work

Temporal sequences appear in a vast range of domains ranging from engineering to medicine and finance, and the ability to model and extract information from them becomes crucial from a conceptual as well as applied perspective. Identifying exceptional patterns in time-stamped databases deserves much attention. In Sects. 7.4 and 7.5, we will propose two exceptional patterns in time-stamped databases.

There are mainly two general directions of temporal data mining (Roddick and Spillopoulou 1999). One concerns the discovery of casual relationships among temporally oriented events. Another one deals with the discovery of similar patterns within the same time sequence or among different time sequences. Sequences of events describe the behavior and actions of users or systems that can be collected in several domains. The proposed problem falls under the first category of the problems, since we are interested in identifying exceptional patterns by comparing sales in different years.

Agrawal et al. (1995) introduced the *shape definition language* (SDL), which used limited vocabulary such as {*Up, up, stable, zero, down, Down*} to describe different gradients in the series. The similarity of two time series is proportional to the length of the longest common sequence of words in their *SDL* representation. Such coarse information might not be always helpful. We define two exceptional patterns viz. a generalized notch and an iceberg notch.

Perng et al. (2000) proposed the landmark model where perceptually important points of a time series are used in its representation. The perceptual importance depends on the specific type of the time series. In general, sound choices for landmarks are local maxima and minima as well as inflection points. The advantage of using the landmark-based method is that this time representation is invariant to amplitude scaling and time warping. Some of the local maxima and minima might lead to higher level of exceptionality. Here we are concerned with defining similar exceptionalities in time-stamped databases.

There has been a significant amount of work on discovering temporal patterns of interest in sequence databases and time series databases. Temporal data mining is concerned with the analyses of data with an intention of finding patterns and regularities from a set of temporal data. In this context sequential association rule (Agrawal and Srikant 1995), periodical association rule (Li and Deogun 2005), calendar association rule (Li et al. 2003) calendar-based periodic pattern (Mahanta et al. 2008) and up-to-date pattern (Hong et al. 2009) are some of the interesting temporal patterns reported in the recent time.

As noted in Sect. 7.1, support history of an item provides important information of an item over time. We have proposed an algorithm for clustering items in multiple databases based on their support history (Adhikari et al. 2009). We have introduced the notion of stability of an item based on its support history.

Lomet et al. (2008) integrated a temporal indexing technique, the TSB-tree, into Immortal DB to serve as the core access method. The TSB-tree provides high performance access and update for both current and historical data.

Keogh et al. (2006) proposed an algorithm for finding unusual time series where the notion of time discords is introduced. A time discord is a subsequence of a longer time series that is maximally different from all other subsequences of the series. Discords can be used to detect anomalies in an efficient way.

Many algorithms are designed incrementally to support time-dependent analysis of data. We have proposed algorithms incrementally to study an overall influence of a set of items on another set of items in time databases (Adhikari and Rao 2010).

Castellana et al. (2007) proposed a new approach to perform change detection analyses based on a combination of supervised and unsupervised techniques is presented. Experimental results were reported for experiments carried out for image data. Wang et al. (2010) examined an unsupervised search method to discover motifs from multivariate time series data. The algorithm first scans the entire series to construct a list of candidate motifs in linear time. The list is then used to populate a sparse self-similarity matrix for further processing to generate the final selections. In contrast, the algorithm to be proposed is based on time-stamped data.

7.3 Notches in Sales Series

The change in sales of an item could be defined by the change of its support over time. The support of an item results in a *support history* (Böttcher et al. 2008) of the item in time databases. An analysis of a support history could be important in understanding customers' behavior (Adhikari et al. 2009). While dealing with the support history, the size of a database is an important issue. Support 0.129 from a database containing 1,000 transactions might be less important than the support 0.091 from a database containing 100,000 transactions. Thus, a mere analysis of the support history over time might not be effective in an application. One needs to analyze the supports along with the sizes in time databases. In Example 7.1, we observe that the support of A has been decreased from year 1970 to 1971. But the sales of A have been increased from the year 1970 to 1971. Hence, a negative change in support of an item might imply a positive change in frequency of the item. Thus, one needs to be careful in dealing with support history of an item in different databases.

Let us consider a company that has been operating for the last k years. For the purpose of studying temporal patterns, yearly databases could be constructed based on a time-stamp. Each of these databases corresponds to a specific period of time. Let D be the collection of customer transactions over k years. For the purpose of defining a temporal pattern we divide D into k yearly databases. Let DT_i be the database corresponding to the ith year, $i = 1, 2, ..., k$. Each of these time databases is mined using a traditional data mining technique (Agrawal and Srikant 1994; Han et al. 2000). Mining time-stamped databases could help business leaders make better decisions by listening to their suppliers and/or customers after analyzing their transactions collected over time (Leonard and Wolfe 2005).

Over the years, an item may exhibit many consecutive data points having similar sales. As opposed to similar data patterns considering each data point, a limited yet meaningful number of points, may play a dominant role in many decision making problems. These meaningful data points could be defined in various ways, like average, peak, or slope of lines (Pratt and Fink 2002). In the context of the proposed problem, such compression of data points seems to be irrelevant. Given the sales series of an item, one might be interested in identifying abrupt changes in the sales series. The goal of our considerations is to define a new type of pattern based on abrupt variation of sales of an item over the years and to design an algorithm to mine such patterns in time databases.

Over the years, there may exist many ups and downs in sales of an item. One might be interested in identifying abrupt changes of sales in different years. It might be helpful to figure out the causes behind it and to take actions accordingly. Let $s_i(A)$ be the sales of item A for the ith year, $i = 1, 2, ..., k$. We define the change in sales series of item A at year i as follows (Singh and Stuart 1998).

The change in sales series at year i is *increasing* if

$$s_{i-1}(A) < s_i(A) < s_{i+1}(A), \quad i = 2, 3, ..., k - 1. \tag{7.1}$$

The change of sales series at year i is *decreasing* if

$$s_{i-1}(A) > s_i(A) > s_{i+1}(A), \quad i = 2, 3, ..., k - 1. \tag{7.2}$$

The change of sales series at year i is *altering* if

$$s_{i-1}(A) < s_i(A) \text{ and } s_i(A) > s_{i+1}(A), \qquad\qquad (7.3)$$

or,

$$s_{i-1}(A) > s_i(A) \text{ and } s_i(A) < s_{i+1}(A), \quad i = 2, 3, \ldots, k - 1. \qquad (7.4)$$

The notion of *strict extrema* (Fink and Gandhi 2007) at a year corresponding to an item is defined as follows.

Let $s_i(A)$ be the amount of sales of an item A at year i, $i = 2, 3, \ldots, k - 1$. There exists a *strict extrema* at year i for the item A if the change of support history of A at year i is altering. Based on the concept of strict extrema, we define a notch as follows.

Definition 7.1 There exists a *notch* at year i for the item A if there is a strict extrema at year i in the sales series of item A.

Let $\Delta s_i(A)$ be the difference in sales of item A between the years i and $i - 1$. Now we propose a few definitions as follows.

Definition 7.2 Let there exists a notch at year i for item A. The notch at year i for item A is *downward* if $\Delta s_i(A) < 0$ and $\Delta s_{i+1}(A) > 0$.

Definition 7.3 Let there exists a notch at year i for item A. The notch at year i for item A is *upward* if $\Delta s_i(A) > 0$ and $\Delta s_{i+1}(A) < 0$.

Itemset (Agrawal et al. 1993) could be considered as a basic type of pattern present in a transactional database. Many important as well as interesting patterns are based on itemset patterns. Similarly an upward/downward notch could be considered as a basic type of pattern in time databases. In Sect. 7.4, we illustrate how the notion of notch could help analyzing a special type of trend in time databases. Thus, it becomes important to mine notches in time databases.

7.3.1 Identifying Notches

Notches in sales series of an item could be considered as basic building blocks to construct temporal patterns. Later we introduce the notion of two interesting temporal patterns viz., generalized notches and iceberg notches in time databases. One could simply scan the sales series of an item to identify its interesting notches. Let n and k be the number of items in time databases and the number of time-stamped (yearly) databases, respectively. Then the time complexity of identifying notches is $O(n \times k)$.

7.4 Generalized Notch

Based on the concept of notch, we present here the notion of a generalized notch in time databases. Let us refer to Fig. 7.1. There are two downward notches in the years 1971 and 1975 having sales 854 and 537, respectively. The concept of notch can be generalized based on strict extrema as clarified in Sect. 7.3. One could notice in Fig. 7.1 that the downward notch in the year 1975 is wider than that of 1971. The width of a downward generalized notch is based on the two consecutive local maxima within which the downward generalized notch is enclosed. The width of the downward generalized notch in 1975 is 1978 − 1972 = 6. Similarly, the width of an upward generalized notch is based on the two consecutive local minimums within which the upward generalized notch is enclosed. The width of the upward generalized notch in 1972 is 1975 − 1971 = 4. Based on the above discussion, we define a width of a generalized notch as follows (Adhikari et al. 2011).

Definition 7.4 Let there be a generalized downward (upward) notch in the year i. Also, let the generalized downward (upward) notch be enclosed with the local maximums (minimums) in the years i_1 and i_2 ($>i_1$). The width of the generalized notch in the year i is equal to $i_2 - i_1$.

The width of a generalized notch could be divided into left width and right width. The left width and right width are equal to $(i - i_1)$ and $(i_2 - i)$, respectively. In case of downward generalized notch the sales value gradually decreases, and then attains the minimum value, and then it gradually increases. Thus, the change of sales value between two consecutive years seems to be an important characteristic of a generalized notch. In this regard, one might be interested in the change of sales for a year as compared to its previous year. Also, the sales at year i as compare to sales of year i_1 and i_2 are important characteristics of a generalized notch. Accordingly, one could define left-height and right-height of the generalized notch in the year i for an item A as follow: $left-\text{height}(A, i) = |\text{sales}(A, i_1) - \text{sales}(A, i)|$, and $right-\text{height}(A, i) = |sales(A, i) - sales(A, i_2)|$. We define the height of a generalized notch as follows (Adhikari et al. 2011).

Definition 7.5 Let there exists a generalized downward (upward) notch in the year i. Also, let the generalized downward (upward) notch be enclosed with the local maximums (minimums) i_1 and i_2 ($> i_1$). The height of the generalized notch in the year i is equal to *maximum* {*left-height* (A, i), *right-height* (A, i)}.

Based on the concept of generalized notch, we focus on the notion of iceberg.

7.5 Iceberg Notch

An analysis of sales series of items is an important issue. In view of performing this task, one could analyze the sales series for each item. In analyzing a sales series in-depth, it is evident that an existence of a notch might be an indication of a bigger notch. This represents an exceptionality of sales of an item. Based on such an exception, we define iceberg in time databases as follows (Adhikari et al. 2011).

Definition 7.6 An *iceberg* notch is a generalized notch that satisfies the following conditions: (i) The height of the generalized notch is greater than or equal to α, and (ii) The width of the generalized notch is greater than or equal to β. Both α and β are user-defined thresholds.

An iceberg notch is a generalized notch having a larger width and a lager height. The concept of iceberg in data management is not new. For example, iceberg queries (Han and Kamber 2001) are commonly used in data mining, particularly in market basket analysis.

Let us illustrate the concept of an iceberg using an example. Let the value of α and β be set to 300 and 5, respectively. Also, let the values of i, i_1 and i_2 be 1975, 1972, and 1978, respectively (with respect to Fig. 7.1). We observe that $leftHeight(A, 1975) = |sales(A, 1972) - sales(A, 1975)| = |905 - 537| = 368$ and $rightHeight(A, \quad 1975) = |sales(A, 1975) - sales(A, 1978)| = |537 - 820| = 283$, respectively. Therefore, the height of the iceberg is *maximum* {368, 283} i.e. $368 \geq \alpha$. Also, $i_2 - i_1 = (1978 - 1972) = 6 \geq \beta$. So, there exists an interesting downward iceberg notch in the year 1975. We also observe an upward notch in the year 1972 with height $368 \geq \alpha$ and width $(1975 - 1971) = 4 < \beta$. So, the upward notch in the year 1972 is not an iceberg. We state the problem as follows.

Let there are k time databases. Let I be the set of all items in time databases. There exists a sales series of each item in I, consisting of k sales values, one for each time database. Mine interesting iceberg notches in sales series for each item in I.

7.6 Sales Series

A sales series of an item might provide interesting information about the item. It is basically the same as the support history of the item. As noted above, in many problems, it is preferable to analyze sales series rather than looking at the support history of an item. Many temporal patterns might originate by analyzing such types of temporal series.

7.6.1 Another View at Sales Series

Each data in a sales series can be mapped onto a member in the set {−1, 0, 1} by comparing the data with the previous data in the same series. Thus, a time-stamped series could be mapped into a ternary series. This provides a simplified view of the original time-stamped series data. Such simplified view might provide some useful information. The procedure for mapping a time-stamped series into a ternary series is illustrated in Example 7.2.

Example 7.2 Consider the sales data given in Example 7.1. The sales of item A in 1971 decreased from the sales in 1970. We note this behavior by putting -1 in the ternary series of item A corresponding to year 1971. The sales of item A in 1972 increased over the sales in 1971. We note this behaviour by putting $+1$ in the ternary series of item A corresponding to year 1972. If the sales of item A in any year remains same as that of previous year then we note this behaviour by putting 0 in the ternary series. Thus, we obtain the ternary series (TS) of item A in the following form:

Year	1970	1971	1972	1973	1974	1975	1976	1977	1978	1979
$TS(A)$		-1	$+1$	-1	-1	-1	$+1$	$+1$	$+1$	-1

In the above series, one can observe the existence of a generalized notch. The width of a downward generalized notch can be obtained from a run of -1's and the subsequent run of $+1$'s. The width is obtained by adding the number of -1's in the first run and the number of $+1$'s in the second run. A similar procedure can be followed for finding width of an upward generalized notch. In $TS(A)$ we observe a downward generalized notch in 1975. The width of this notch is equal to $3 + 3 = 6$. Also, there exists an upward generalized notch in 1978 having width of 4.

A slightly different procedure could also be followed for obtaining a ternary series corresponding to a sales series of an item. Let x and y be the sales for the year 1970 and 1971, respectively. Let δ be the level of significance of difference in sales. We put $+1$ in the ternary series of the item corresponding to year 1971, if $y - x > \delta$. We put -1 in the ternary series of the item corresponding to year 1971, if $x - y > \delta$. We put 0 in the ternary series of the item corresponding to year 1971, if $|x - y| \leq \delta$. The method of obtaining a ternary series using this procedure might be useful in many situations, since a small change in sales value might be insignificant in many situations. This procedure seems to be more realistic than the previous one.

7.6.2 Other Applications of Icebergs

We have mined icebergs from binary transactional datasets. In binary transactions the database can be viewed as records with Boolean attributes, where each attribute corresponds to a single item. Further, in the record of a transaction, the attribute corresponding to an item is true if and only if the transaction contains the item, otherwise false. But this type of datasets has got limited usage, since in a real life the items are often purchased multiple times. Therefore, in reality most of the market basket data are non-binary. Nowadays researchers from the data mining community are more concerned with qualitative aspect of attributes (e.g.

significance, utility) as compared to considering only quantitative ones (e.g. number of appearances in a database etc.), because qualitative properties are required in order to fully exploit the attributes present in the dataset. By using this measure, icebergs can also be discovered. This utility measure calculates the actual profit value of an item in a non-binary transaction database. Thus, the same approach could be followed to mine the icebergs from the total utility value of the itemsets.

Icebergs could also be mined from other type of datasets such as rainfall data and crops production. To mine icebergs from rainfall data, yearly total rainfall is accumulated for a particular region. Similarly, total amount of particular crops are summed up for a specific crop in a region.

7.7 Mining Icebergs in Time-Stamped Databases

Let there are n items in time databases. For each item in time databases there exists a time-stamped series containing k data. In this section we are interested in identifying icebergs in each time-stamped series. For mining icebergs in time databases, we make use of an existing frequent itemset mining algorithm (Agrawal and Srikant 1994; Han et al. 2000). For the requirement of proposed problem one needs to mine the frequencies of each item in the time databases.

7.7.1 Non-incremental Approach

Based on the discussion held in the previous sections, we design an algorithm for mining icebergs in time databases. Let n and k be the number of items in time databases and the number of time databases, respectively. In this algorithm, we use a two-dimensional array F for storing frequencies of all items in time databases. F consists of n rows and $k + 1$ columns. The first column contains the items in time databases. For example, $F(i)(1)$ contains the ith item in time databases, $i = 1, 2, ...,$ n. The ith row of F contains ith item and its frequencies in k time databases. For example, $F(i)(j)$ contains the frequency of ith item in $(j - 1)$th database, $j = 2, 3, ...,$ $k + 1$. Therefore, we need to check the existence of a generalized notch using the values in the columns from 2 to $k + 1$.

For the purpose of computing interestingness of a generalized notch, we determine the change of sales of a local minimum (maximum) with respect to its previous and next local maximum (minimum). During the process of mining icebergs, the generalized notches are kept in array GN. A generalized notch can be described by the following attributes: left year (*leftYear*), right year (*rightYear*), item (*item*), year of occurrence (*year*), type of generalized notch (*type*), change of sales at the year of occurrence with respect to the previous local extremum (*leftHeight*), change of sales at the year of occurrence with respect to the next local extremum (*rightHeight*), width of generalized notch (*width*) and the sales at the year of occurrence (*sales*). The goal of the proposed algorithm is to find all the interesting icebergs for each item in time databases. The algorithm is given as follows (Adhikari et al. 2011).

Algorithm 7.1. Mine icebergs in time-stamped databases.
procedure *MineIcebergs* (*k*, *F*, *α*, *β*)
Inputs: *k*, *F*, *α*, *β*
k: number of yearly databases
F: array of frequencies of items in yearly databases
α: user-defined threshold of height of a generalized notch
β: user-defined threshold of width of a generalized notch
Outputs:
Interesting icebergs in time databases
01: **let** *index* = 1;
02: **for** *i* = 1 to *n* **do**
03:   **let** *j* = 2; **let** *left* = 2; **let** *flat* = false; **let** *prevDown* = false; **let** *prevUp* = false;
04:   **while not** end of the sales series corresponding to *i*-th item **do**
05:     **if** there is a downward trend **and** *prevUp* is false **then**
06:       find *mid*, *leftWidth*; **let** *prevDown* = true;
07:       compute *leftHeight*; **go to** 04;
08:     **end if** {05}
09:     **if** there is an upward trend **and** *prevDown* is true **then**
10:       find *left*, *mid*, *right*, *leftWidth*, *rightWidth*; **let** *prevDown* = false;
11:       **let** *leftHeight* = *rightHeight*;
12:       compute *rightHeight*;
13:       *GN*(*index*).*type* = *down*; **go to** 30;
14:     **end if** {09}
15:     **if** there is an upward trend **and** *prevDown* is false **then**
16:       **let** *prevUp* = true; find *mid*, *leftWidth*;
17:       compute *leftHeight*; **go to** 04;
18:     **end if** {15}
19:     **if** there is a downward trend **and** *prevUp* is true **then**
20:       find *left*, *mid*, *right*, *leftWidth*, *rightWidth*;
21:       **let** *prevUp* = false; **let** *prevDown* = true;
22:       **let** *leftHeight* = *rightHeight*;
23:       compute *rightHeight*;
24:       *GN*(*index*).*type* = *up*; **go to** 30;
25:     **end if** {19}
26:     **if** the sales of the *j*-th and (*j*+1)-th year remain same **then**
27:       find *left*; **let** *flat* = true; **let** *prevDown* = false; **let** *prevUp* = false;
28:       **go to** 04;
29:     **end if**
30:     **if** *flat* is false **then**
31:       compute *height* as defined in Definition 5;
32:       **if** the current generalized notch satisfies the criteria *α* and *β* **then**
33:         store it in *GN*(*index*); increase *index* by 1;
34:       **end if**
35:       **if** the current generalized notch is downward **then**
36:         **let** *prevDown* = false; **let** *prevUp* = true;
37:       **else if** the current generalized notch is upward **then**
38:         **let** *prevDown* = true; **let** *prevUp* = false;
39:       **end if**
40:     **end if** {35}
41:     **end if** {30}
42:   **end while** {04}
43: **end for** {02}
44: display icebergs from *GN*;
end procedure

The lines 2–43 are repeated for each item in time databases. In each repetition, the interesting icebergs corresponding to an item are identified. The variable *index* is used to index array *GN*. The variable *j* is used to keep track of current sales data of the item under consideration. The starting value of *j* is 2, since the sales data for the first year of an item is kept staring from column number 2 of array *F*. We use three Boolean variables viz., *flat*, *prevUp* and *prevDown*. While identifying downward (or, upward) generalized notches, we first go through its left leg of a generalized notch. After reaching its minimum/maximum value, if the next point also attains the same value then *flat* becomes true. The width of a generalized notch is determined by the following years: left year (*left*), middle year (*mid*), and right year (*right*). Accordingly, the width of a generalized notch has two components: left width (*leftWidth*) and right width (*rightWidth*). After identifying the left leg of a downward (or, upward) generalized notch, *prevDown* (or, *prevUp*) becomes true. After storing the details of the current generalized notch *index* gets increased by 1 (line 33). We identify generalized notches for each item in time databases. For this purpose, we introduce a *for*-loop at line 02 which ends at line 43. Some lines, e.g. lines 40, 41, 42, 43, are ended with a number enclosed in curly brackets, to mark the ends of composite statement starting with the line number kept in curly bracket.

Lines 2 and 4 repeat for n and k times respectively. In other words, the sales series corresponding to each item is processed for identifying icebergs. Thus, the time complexity of lines 1–43 is $O(n \times k)$. Again, the time complexity of line 44 can not be more than $O(n \times k)$, since the number of interesting icebergs is always less than $n \times k$.

Theorem 7.1 Correctness of the *MineIcebergs* algorithm.

Proof Consider that there are n items in k time-stamped databases. Each sales series is processed using lines 2–43. For the purpose of mining interesting icebergs, each sales series is checked completely by applying a *while*-loop shown in lines 4–42. A sales series can start with one of the following three ways: (a) showing a downward trend, (b) showing an upward trend (c) remained at a constant level. The algorithm handles each of these cases separately.

Case (a) has been checked at the line numbered as 05. Once a downward trend changes we again go back to *while*-loop a line 04 for finding one of the following two possibilities: an upward trend and a constant sales.

Case (b) has been checked at the line number 15. Once the upward trend changes we again go back to *while*-loop a line 04 for finding one of the following two possibilities: a downward trend and a constant sales.

Case (c) has been checked at the line number 26. Once the flatness changes we again go back to *while*-loop a line 04 for finding one of the following two possibilities: a downward trend and an upward trend.

Once the left leg of a downward generalized notch is detected in lines 5–8, its right leg is detected in lines 9–14. When the left leg of an upward generalized notch is detected in lines 15–18, its right leg is detected in lines 19–25. After detecting a generalized notch at lines 13 and 24, we go to line 30 for detecting its interestingness and re-initializing required Boolean variables.

Thus, the above algorithm considers all the possibilities that would arise in each sales series.   □

7.7.2 Incremental Approach

An incremental approach seems to be a natural way of designing an algorithm for mining icebergs in time-stamped databases. Every year we need to mine a yearly database, and the frequencies of all the items are computed. As a result, one more term in the frequency series of each item gets added. Thus, one needs to mine only the current database for the purpose of making an analysis based on the entire database. As a result one can mine icebergs using a cost-effective and faster algorithm.

Let ΔD be the database for the current year. In this algorithm, we use a two-dimensional array ΔF for storing frequencies of all items in the current yearly database. It consists of n rows and 2 columns. The first column contains the items in the current database and the second column contains the frequencies of the items. Similar to Algorithm 7.1, array F contains the frequencies of items of previous databases. Array ΔF is merged with the array F, and the frequency are stored in $(k + 2)$th column. It can be considered as a preprocessing task of the proposed incremental mining algorithm. Currently, F consists of n rows and $k + 2$ columns.

For the purpose of computing interesting icebergs incrementally, we use array GN, where all the interesting icebergs for all items are stored for k years. The attributes of a generalized notch have already been described in the previous section. The goal of the proposed algorithm is to find all the interesting icebergs for each item in time databases incrementally. We have made use of procedure *MineIcebergs* while designing an incremental algorithm for mining icebergs. The incremental algorithm is given below.

Algorithm 7.2. Mine icebergs incrementally in time-stamped databases.
procedure *MineIcebergsIncrementally* (F, ΔF, GN, α, β)
Inputs: F, ΔF, GN, α, β
F: array of frequencies of items in the previous yearly databases
ΔF: array of frequencies of items in the current yearly database
GN: array of in interesting icebergs in previous yearly database
α: user-defined threshold of height of a generalized notch
β: user-defined threshold of width of a generalized notch
Outputs:
Interesting icebergs in time databases
01: copy frequencies of items in ΔF into $(k+2)$-th column of F
02: **for** $j = 1$ to n **do**
03:     **if** there exists an iceberg of j-th item in GN **then**
04:         go to the last iceberg I of j-th item in GN;
05:         **if** *rightYear* of I is equal to $(k+1)$ **then**
06:             **if** (I is upward) and (sales of $(k+2)$-th year is less than *rightYear*) **then**
07:                 update current iceberg;
08:             **else** call *MineIcebergs* ($k+3$-*year*, $F(year, k+2)$, α, β);
09:             **end if**
10:             **if** (I is downward) and (sales of $(k+2)$-th year is greater than *rightYear*) **then**
11:                 update current iceberg;
12:             **else** call *MineIcebergs* ($k+3$-*year*, $F(year, k+2)$, α, β);
13:             **end if**
14:         **else**
15:             go to the last iceberg I of j-th item in GN;
16:             call *MineIcebergs* ($k+3$-*year*, $F(year, k+2)$, α, β);
17:         **end if** {05}
18:     **else** call *MineIcebergs* ($k+1$, $F(2, k+2)$, α, β);
19:     **end if** {03}
20: **end for** {02}
21: display icebergs from GN;
end procedure

Line 1 merges ΔF with F so that F now contains n rows and $k + 2$ columns. Line numbers 2–20 repeats for each item present in the database. Lines 3–4 are used to check whether an iceberg of the current item is present in GN, and go to the last iceberg if it has been found. Lines 5–13 are used to update the last iceberg of current item with the help of current data kept in $(k + 2)$th column. If the last iceberg of the current item occurs in the middle, then we go to the last iceberg and then call *MineIcebergs* (lines 15–16). In case if there is no iceberg for the current item then we call *MineIcebergs* using all the data kept in columns 2 to $(k + 2)$. Lines 2–20 execute for n times. Among these statements, line 18 has the maximum time complexity i.e., $O(n \times k)$. Therefore, the complexity of *MineIcebergsIncrementally* is $O(n^2 k)$.

The incremental algorithm appears to be more time consuming than the initial algorithm *MineIcebergs*. Both the algorithms are executed in the main memory and do not require any secondary memory access. These algorithms are required to follow a data mining algorithm that actually reads data from the secondary storage. Before applying algorithm *MineIcebergsIncrementally*, only one database i.e. the database for the $(k + 1)$th year needs to be mined. But if we apply algorithm *MineIcebergs* then all the previous databases i.e. the databases for the first year to $(k + 1)$th year need to be mined. Thus, the algorithm *MineIcebergsIncrementally* practically saves a significant amount of time.

7.8 Significant Year

In the scenario of possessing time-stamped data over several years, the organization might be interested in knowing the years when sales of a large number of items either increase or decrease as well as both increase and decrease. These years have importance in the history of the organization. Further analysis of such rare events might influence the strategies of the organization. In view of this, we define a significant year as follows.

Definition 7.7 Let there be n items transacted over the years. Year i is significant if the ratio of the number of icebergs and n exceeds user-defined threshold δ.

δ is a fraction lies between 0 and 1. Such interestingness condition could also be stated with respect to either upward iceberg or downward iceberg.

For the purpose of mining significant years, we use array *GN*, where all the interesting icebergs are stored. We use the attributes *item*, *year*, and *type* of *GN* to mine significant years. A significant year can be described by the following attributes: *year* (year), *type* (upward /downward icebergs). Variable *count* is used to store total number of (upward /downward icebergs) in a particular year. Array *SigYear* is used to store significant years. The variables i, j, and *index* have been used to access arrays F, *SigYear* and *GN*, respectively. The proposed algorithm finds all the significant years that are attributed by large number of icebergs. The algorithm is presented below.

Algorithm 7.3. Mine the significant years.
procedure *MineSignificantYears* (*n*, *k*, δ, *F*, *GN*)
Inputs: *GN*
n: total number of items in the database
k: number of yearly databases
δ: user-defined threshold
F: array of frequencies of items in the yearly databases
GN: array of interesting icebergs
Outputs:
Significant years in time databases
01: **let** *j* = 1;
02: **for** *i* = 2 to (*k*+1) **do**
03:     **let** *index* = 1; **let** *count* = 0;
04:     **while** not end of *GN* **do**
05:         **if** $F(1, i) = GN(index).year$ **then**
06:             increase *count* by 1;
07:         **end if**
08:         increase *index* by 1;
09:     **end while**
10:     **if** (*count* / *n* $\geq$ δ) **then**
11:         *SigYear(j).year* = $F(1, i)$; *SigYear(j).count* = *count*; increase *j* by 1;
12:     **end if**
13: **end for**
14: display years from *SigYear*;
end procedure

Algorithm *MineSignificantYears* finds significant years with respect to both upward and downward icebergs. One might be interested in finding significant years with respect to upward or downward iceberg. The algorithms finds a number of icebergs year-wise; see lines 5–8. Then the interestingness of the year is checked at line 10. Let the number of icebergs in *GN* be *p*. Then the complexity of the algorithm is O($k \times p$).

7.9 Experimental Studies

We present experimental results using four real databases and two synthetic databases. The databases *mushroom, retail* (Frequent itemset mining dataset repository) *ecoli* and *BMS-WebView-1* are real-world databases. The real databases *BMS-Web-Wiew-1* can be found from KDD CUP 2000 (Frequent itemset mining dataset repository). Database *ecoli* is a subset of *ecoli database* (UCI ML repository). The synthetic dataset *T10I4D100K* was generated using the generator from

Table 7.1 Database characteristics

Database	NT	ALT	AFI	NI
mushroom (M)	8,124	24.000	1,624.800	120
ecoli (E)	336	7.000	25.835	91
random-68 (R)	3,000	5.460	280.985	68
retail (Rt)	88,162	11.306	99.674	10,000
BMS-WebView-1	149,639	2.000	44.575	6,714
T10I4D100K	100,000	11.102	1,276.124	870

the IBM Almaden Quest research group. *Random-68* is also a synthetic database and has been generated for the purpose of conducting experiments. The characteristics of these databases are given in Table 7.1.

The symbols used in Tables 7.1 and 7.2 come with the following meaning: *D, NT, ALT, AFI*, and *NI* denote database, the number of transactions, average length of a transaction, average frequency of an item, and number of items, respectively. The databases *mushroom, ecoli, random-68* and *retail* have been divided into 10 sub-databases, called yearly databases, for the purpose of conducting experiments. The databases *BMS-WebView*-1 and *T10I4D100K* have been divided into 20 databases. The databases obtained from *mushroom, ecoli, random-68* and *retail* are named as M_i, E_i, R_i, and Rt_i, $i = 0, 1, \ldots, 9$. The databases obtained from *BMS-WebView*-1 and *T10I4D100K* are named as B_i and T_i, $i = 0, 1, \ldots, 19$. We present some characteristics of the input databases in Table 7.2.

In Tables 7.3, 7.4, 7.5 and 7.6 we have represented upward generalized notch as '*u*' and downward generalized notch as '*d*'. In *mushroom*, there are many items having high frequency as it is relatively dense. Also, we get many generalized notches having relatively large height as shown in Table 7.3. On the other hand, the items in *retail* is somewhat skewed in the sense that some generalized notches for few items have large height. But, the items in *random-68* and *ecoli* have got more or less uniform distribution. Many generalized notches in these two databases have similar height. Unlike *mushroom* and *retail*, *ecoli* and *random-68* are smaller in size and contain items with lesser variations. In these two databases the maximum heights are 13 and 30, respectively. With respect to width of a generalized notch, we have obtained similar characteristics. In *mushroom* and *retail*, the generalized notches are wider than that of other two databases. These facts are quite natural, since *mushroom* and *retail* are bigger than *random-68* and *ecoli*. The variation of sales over the years for an item in *mushroom* and *retail* is higher. Also, we observe that many upward generalized notch is followed by a downward generalized notch and vice versa. This is because of the fact that two consecutive different types (a '*u*' type followed by a '*d*' type or a '*d*' type followed by an '*u*' type) generalized

Table 7.2 Characteristics of the time databases

D	NT	ALT	AFI	NI	D	NT	ALT	AFI	NI
M_0	812	24.000	295.273	66	M_5	812	24.000	221.454	88
M_1	812	24.000	286.588	68	M_6	812	24.000	216.533	90
M_2	812	24.000	249.846	78	M_7	812	24.000	191.059	102
M_3	812	24.000	282.435	69	M_8	812	24.000	229.271	85
M_4	812	24.000	259.840	75	M_9	816	24.000	227.721	86
E_0	33	7.000	4.620	50	E_5	33	7.000	3.915	59
E_1	33	7.000	5.133	45	E_6	33	7.000	3.500	66
E_2	33	7.000	5.500	42	E_7	33	7.000	3.915	59
E_3	33	7.000	4.812	48	E_8	33	7.000	3.397	68
E_4	33	7.000	3.397	68	E_9	39	7.000	4.550	60
R_0	300	5.590	28.677	68	R_5	300	5.140	26.677	68
R_1	300	5.417	28.000	68	R_6	300	5.510	28.353	68
R_2	300	5.360	27.647	68	R_7	300	5.497	28.338	68
R_3	300	5.543	28.456	68	R_8	300	5.537	28.471	68
R_4	300	5.533	28.382	68	R_9	300	5.477	28.235	68
Rt_0	9,000	11.244	12.070	8,384	Rt_5	9,000	10.856	16.710	5,847
Rt_1	9,000	11.209	12.265	8,225	Rt_6	9,000	11.200	17.416	5,788
Rt_2	9,000	11.337	14.597	6,990	Rt_7	9,000	11.155	17.346	5,788
Rt_3	9,000	11.490	16.663	6,206	Rt_8	9,000	11.997	18.690	5,777
Rt_4	9,000	10.957	16.039	6,148	Rt_9	7,162	11.692	15.348	5,456
B_0	7,482	2.000	5.016	2,983	B_{10}	7,482	2.000	4.573	3,272
B_1	7,482	2.000	4.494	3,330	B_{11}	7,482	2.000	4.895	3,057
B_2	7,482	2.000	5.782	2,588	B_{12}	7,482	2.000	4.636	3,228
B_3	7,482	2.000	4.359	3,433	B_{13}	7,482	2.000	4.805	3,114
B_4	7,482	2.000	4.228	3,539	B_{14}	7,482	2.000	4.192	3,570
B_5	7,482	2.000	4.194	3,568	B_{15}	7,482	2.000	4.656	3,214
B_6	7,482	2.000	3.786	3,952	B_{16}	7,482	2.000	5.379	2,782
B_7	7,482	2.000	3.477	4,304	B_{17}	7,482	2.000	4.863	3,077
B_8	7,482	2.000	4.168	3,590	B_{18}	7,482	2.000	4.654	3,215
B_9	7,482	2.000	4.365	3,428	B_{19}	7,481	2.000	4.953	3,021
T_0	5,000	11.123	64.968	856	T_{10}	5,000	11.113	64.913	856
T_1	5,000	10.987	63.880	860	T_{11}	5,000	11.165	64.988	859
T_2	5,000	11.189	65.128	859	T_{12}	5,000	11.127	64.617	861
T_3	5,000	11.078	64.330	861	T_{13}	5,000	11.089	64.694	857
T_4	5,000	11.003	63.895	861	T_{14}	5,000	11.169	65.088	858
T_5	5,000	11.131	64.867	858	T_{15}	5,000	11.028	64.338	857
T_6	5,000	11.171	64.645	864	T_{16}	5,000	11.132	64.795	859
T_7	5,000	11.075	64.764	855	T_{17}	5,000	11.031	64.661	853
T_8	5,000	11.123	65.121	854	T_{18}	5,000	11.072	64.374	860
T_9	5,000	11.151	64.755	861	T_{19}	5,000	11.090	64.856	855

Table 7.3 Top 10 generalized notches in *M*, *E* and *R* databases (according to height)

$M(\alpha)$ at $\beta = 4$				$E(\alpha)$ at $\beta = 2$				$R(\alpha)$ at $\beta = 3$			
Item	Year (sales)	Type	Height	Item	Year (sales)	Type	Height	Item	Year (sales)	Type	Height
116	4(1,489)	u	1,344	42	2 (14)	u	13	48	4(48)	u	30
116	7(353)	d	1,136	42	6 (1)	d	13	48	5(18)	d	30
114	2(1,131)	u	1,062	35	6 (0)	d	12	27	5(19)	d	26
114	4(69)	d	1,062	0	6 (10)	d	10	27	8 (45)	u	26
56	5(810)	u	796	0	7 (20)	u	10	36	6(39)	u	26
56	9(14)	d	796	39	4 (11)	d	10	36	8(13)	d	26
67	6(103)	d	704	39	6 (1)	d	10	18	8(21)	d	25
94	5(2)	d	650	41	3 (11)	u	10	3	5(41)	u	24
94	9(652)	u	650	41	6 (1)	d	10	3	7(17)	d	24
1	3(69)	d	644	34	2 (2)	d	9	36	2(41)	u	23

Table 7.4 Top 10 generalized notches in Rt, B and T databases (according to height)

Rt (α) at $\beta = 2$				B (α) at $\beta = 3$				T (α) at $\beta = 2$			
Item	Year (sales)	Type	Height	Item	Year (sales)	Type	Height	Item	Year (sales)	Type	Height
41	4(2,617)	u	2,617	333,469	9(582)	u	506	966	4(297)	d	122
41	9(2,355)	u	2,355	333,469	12(76)	d	506	966	6(419)	u	122
0	2(2,331)	d	1,315	333,469	4(151)	d	388	998	2(346)	u	103
48	9(4,544)	u	932	333,469	6(539)	u	388	998	6(243)	d	103
48	4(4,704)	u	711	333,449	9(474)	u	379	966	8(395)	u	93
0	3(2,403)	u	609	333,449	11(95)	d	379	966	10(302)	d	93
0	4(1,794)	d	609	333,449	4(148)	d	372	829	16(431)	u	90
0	7(1,619)	u	571	333,449	6(520)	u	372	966	11(389)	u	87
8,978	2(556)	u	556	110,877	5(353)	u	344	419	4(179)	d	85
0	5(2,089)	u	512	110,877	10(9)	d	344	419	6(264)	u	85

Table 7.5 Top 10 generalized notches in different databases (according to width) at a given α

M (β) at $\alpha = 300$				E (β) at $\alpha = 2$				R (β) at $\alpha = 5$			
Item	Year (sales)	Type	Width	Item	Year (sales)	Type	Width	Item	Year (sales)	Type	Width
56	5(810)	u	8	35	6(0)	d	6	1	5(19)	d	6
11	4(427)	u	7	42	6(1)	d	6	11	8(21)	d	6
13	6(39)	d	7	33	7(1)	d	5	43	5(36)	u	6
16	5(361)	u	7	41	3(11)	u	5	6	3(32)	u	5
67	6(103)	d	7	42	2(14)	u	5	11	4(32)	u	5
94	5(2)	d	7	49	6(1)	d	5	18	8(21)	d	5
98	2(404)	u	7	52	4(1)	d	5	47	5(31)	u	5
11	8(32)	d	6	40	4(8)	u	4	50	4(18)	d	5
52	6(703)	u	6	45	2(8)	u	4	50	8(35)	u	5
53	6(109)	d	6	45	5(1)	d	4	53	6(40)	u	5

notches share a common leg. For example, a 'u' type generalized notch at year 4 is followed by a 'd' type generalized notch at year 7, for item 116 in *mushroom*. Similarly, a 'd' type generalized notch at year 6 is followed by a 'u' type generalized notch at year 7, for item 0 in *ecoli*. Also, we observe that some items have both long height and long width. For example, item 56 has height 796 and width 8. These values are significantly high as compare to other items in the time databases. Also, this is true for item 67. In *BMS-WebView*-1 database items 333,469 and 110,877 have maximum variation. Therefore, only these two items are appearing among top ten generalized notches and their heights vary from 344 to 506. Similarly, items 966, 998 and 419 in *T10I4D100K* share common legs and they have more variations. From Table 7.6 one could conclude that generalized notches are sharper in *BMS-WebView*-1 as compared to *T10I4D100K* (Table 7.7).

We have also reported an execution time with respect to the number of data sources. We observe in Figs. 7.2, 7.3, 7.4, 7.5, 7.6 and 7.7 that the execution time increases as the number of databases increases. The size of each input database generated from *mushroom*, *retail*, *BMS-WebView*-1 and *T10I4D100K* are significantly larger than that of *ecoli*. As a result, we observe a steeper graph in Figs. 7.3 and 7.6. We have fixed α for *ecoli* and *random-68* at lower level, since variation of frequencies of an item is lesser. We observe that the execution time of *retail* is significantly larger than for other databases, since each of the time databases is comparatively larger. In Figs. 7.7 and 7.8 we have considered same α and β for *BMS-WebView-1* and *T10I4D100K*, respectively. But execution time of *BMS-WebView-1* is significantly larger than the one reported for *T10I4D100K*.

In Figs. 7.8, 7.9, 7.10, 7.11, 7.12, 7.13, 7.14, 7.15, 7.16, 7.17, 7.18 and 7.19 we have presented how the number of interesting icebergs decreases with respect to the increase of the values of α and β. In *mushroom*, *ALT* and *AFI* are higher as compared to other databases. We start changing the values α proceeding from 100 (Fig. 7.8). Initially, the number of icebergs decreases significantly. Afterwards, the decrease is not so significant. As shown In Fig. 7.9, the number of icebergs decreases significantly when the width remains lower. Afterwards, the number of icebergs decreases slowly.

ALT is smaller for *ecoli* and *random-68*. As a result, the height of an iceberg remains smaller. We start α from 2 and 5 for *ecoli* and *random-68*, respectively (Figs. 7.10 and 7.12). We have not discovered any interesting icebergs for the width greater than or equal to 7.

Table 7.6 Top 10 generalized notches in different databases (according to width) at a given α

Rt (β) at $\alpha = 20$				B (β) at $\alpha = 50$				T (β) at $\alpha = 5$			
Item	Year(sales)	Type	Width	Item	Year (sales)	Type	Width	Item	Year (sales)	Type	Width
9,823	7(30)	u	9	112,551	9(6)	d	10	673	8(71)	d	8
2,046	4(133)	u	8	335,213	10(4)	d	10	651	11(82)	u	8
3,321	7(24)	u	8	112,339	5(224)	u	9	524	4(29)	u	8
411	4(32)	u	8	335,185	4(130)	u	9	283	15(229)	u	7
3,121	4(0)	d	8	112,407	5(107)	u	9	487	15(139)	d	7
2,919	6(32)	u	8	335,181	5(54)	u	9	336	13(42)	d	7
103	7(267)	u	7	335,213	5(54)	u	9	523	11(117)	u	7
855	3(118)	u	7	335,177	5(51)	u	9	658	14(88)	d	7
1,659	3(116)	u	7	110,877	5(353)	u	8	733	16(63)	u	7
976	6(55)	d	7	110,315	11(310)	u	8	807	10(29)	u	7

Table 7.7 Significant years in different databases at a given δ

M	E	R	Rt	B	T
$\delta = 0.04$	$\delta = 0.05$	$\delta = 0.04$	$\delta = 0.0004$	$\delta = 0.0006$	$\delta = 0.003$
Significant years	Significant years	Significant years	Significant years	Significant years	Significant years
6	6	8	7	4	6
4	4	7	6	2	
3		5	4		

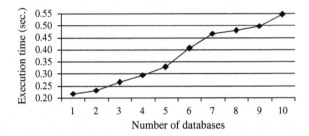

Fig. 7.2 Execution time versus the number of databases (*mushroom* at $\alpha = 50$, $\beta = 3$)

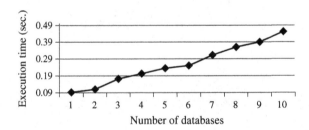

Fig. 7.3 Execution time versus the number of databases (*ecoli* at $\alpha = 3$, $\beta = 2$)

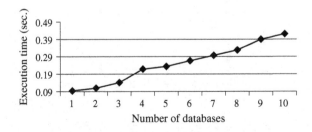

Fig. 7.4 Execution time versus the number of databases (*random-68* at $\alpha = 3$, $\beta = 2$)

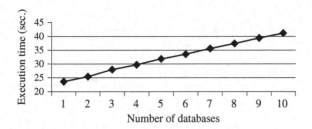

Fig. 7.5 Execution time versus the number of databases (*retail* at $\alpha = 20$, $\beta = 2$)

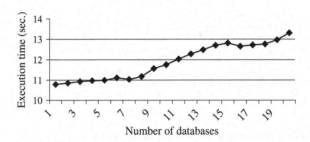

Fig. 7.6 Execution time versus the number of databases (*BMS-WebView-1* at $\alpha = 30$, $\beta = 4$)

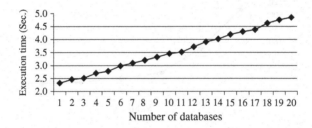

Fig. 7.7 Execution time versus the number of databases (*T10I4D100K* at $\alpha = 30$, $\beta = 4$)

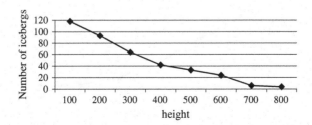

Fig. 7.8 Number of interesting icebergs versus height (α) for *mushroom* ($\beta = 3$)

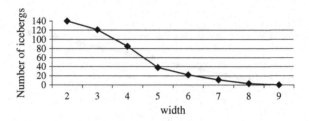

Fig. 7.9 Number of interesting icebergs versus width (β) for *mushroom* ($\alpha = 50$)

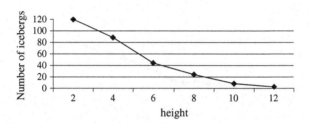

Fig. 7.10 Number of interesting icebergs versus height (α) for *ecoli* ($\beta = 2$)

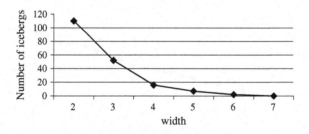

Fig. 7.11 Number of interesting icebergs versus width (β) for *ecoli* ($\alpha = 3$)

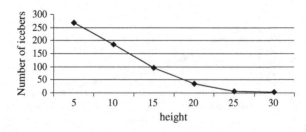

Fig. 7.12 Number of interesting icebergs versus height (α) for *random-68* ($\beta = 2$)

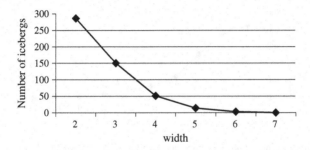

Fig. 7.13 Number of interesting icebergs versus width (β) for *random-68* ($\alpha = 3$)

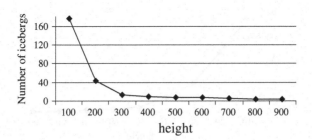

Fig. 7.14 Number of interesting icebergs versus height (α) for *retail* ($\beta = 2$)

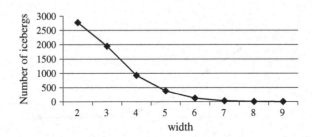

Fig. 7.15 Number of interesting icebergs versus width (β) for *retail* ($\alpha = 20$)

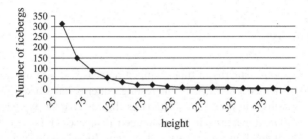

Fig. 7.16 Number of interesting icebergs versus height (α) for *BMS-WebView-1* ($\beta = 4$)

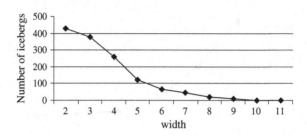

Fig. 7.17 Number of interesting icebergs versus width (β) for *BMS-WebView-1* (α = 30)

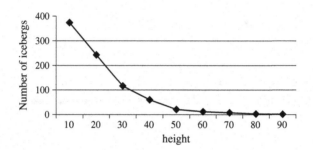

Fig. 7.18 Number of interesting icebergs versus height (α) for *T10I4D100K* (β = 5)

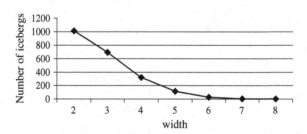

Fig. 7.19 Number of interesting icebergs versus width (β) for *T10I4D100K* (α = 30)

7.10 Conclusions

The study of temporal patterns in time-stamped databases is an important research and application-oriented issue. Many interesting patterns have been discovered in transactional as well as in time series databases. In this chapter, we have proposed definitions of different patterns in time-stamped databases. First, we have introduced the notion of notch in a sales series of an item. Based on this pattern, we have introduced two more patterns viz., generalized notch and iceberg notch, present in sales series of an item. Iceberg notch represents a special sales pattern of an item over time. It could be considered as an exceptional pattern in time-stamped databases. The investigations of such patterns could be important to understand the purchasing behaviour of customers. They also help identifying the reasons for such a behaviour. We have designed an algorithm to extract icebergs in time-stamped databases and presented experimental results for real-world and synthetic time-stamped databases.

References

Adhikari A, Rao PR (2008) Mining conditional patterns in a database. Pattern Recogn Lett 29 (10):1515–1523

Adhikari J, Rao PR, Pedrycz W (2011) Mining icebergs in time-stamped databases. In: Proceedings of Indian international conferences on artificial intelligence, pp 639–658

Adhikari J, Rao PR, Adhikari A (2009) Clustering items in different data sources induced by stability. Int Arab J Inf Technol 6(4):394–402

Adhikari J, Rao PR (2010) Measuring influence of an item in a database over time. Pattern Recogn Lett 31(3):179–187

Agrawal R, Imielinski T, Swami A (1993) Mining association rules between sets of items in large databases. In: Proceedings of ACM SIGMOD conference on management of data, pp 207–216

Agrawal R, Srikant R (1994) Fast algorithms for mining association rules. In: Proceedings of the international conference on very large databases, pp 487–499

Agrawal R, Srikant R (1995) Mining sequential patterns. In: Proceedings of international conference on data engineering, pp 3–14

Agrawal R, Psaila G, Wimmers EL, Zaït M (1995) Querying shapes of histories. In: Proceedings of 21st VLDB Conference, pp 502–514

Böttcher M, Hoppner F, Spiliopoulou (2008) On exploiting the power of time in data mining. SIGKDD Explor 10(2):3–11

Box G. Jenkins G, Reinsel G (2003) Time series analysis, 3rd edn. Pearson Education, Upper Saddle River

Brockwell J, Richard D (2002) Introduction to time series and forecasting. Springer, Berlin

Castellana L, D'Addabbo A, Pasquariello G (2007) A composed supervised/unsupervised approach to improve change detection from remote sensing. Pattern Recogn Lett 28 (4):405–413

Fink E, Gandhi HS (2007) Important extrema of time series. In: Proceedings of the IEEE international conference on systems, man and cybernetics, pp 366–372

Frequent itemset mining dataset repository, http://fimi.cs.helsinki.fi/data

Han J, Pei J, Yiwen Y (2000) Mining frequent patterns without candidate generation. In: Proceedings of ACM SIGMOD international conference management of data, pp 1–12

Han J, Kamber M (2001) Data mining: concepts and techniques. Morgan Kauffmann, Burlington

Hong TP, Wu YY, Wang SL (2009) An effective mining approach for up-to-date patterns. Expert Syst Appl (36):9747–9752

Keogh E (1997) A fast and robust method for pattern matching in time series databases. In: Proceedings of 9th international conference on tools with AI (ICTAI), pp 578–584

Keogh E, Lonardi S, Chiu B (2002) Finding surprising patterns in a time series database in linear time and space, In: Proceedings of KDD international conference, pp 23–26

Keogh E, Lin J, Lee SH, Herle HV (2006) Finding the most unusual time series subsequence: algorithms and applications. Knowl Inf Syst 11(1):1–27

Leonard M, Wolfe B (2005) Mining transactional and time series data. Proceedings of the SUGI 30:18–24

Li D, Deogun JS (2005) Discovering partial periodic sequential association rules with time lag in multiple sequences for prediction. Foundations of Intelligent Systems LNCS, vol 3488, pp 1–24

Li Y, Ning P, Wang XS, Jajodia S (2003) Discovering calendar-based temporal association rules. Data Knowl Eng 44(2):193–218

Lomet DB, Hong M, Nehme RV, Zhang R (2008) Transaction time indexing with version compression. Proc VLDB Endowment 1(1):870–881

Mahanta AK, Mazarbhuiya FA, Baruah HK (2008) Finding calendar-based periodic patterns. Pattern Recogn Lett 29(9):1274–1284

Perng CS, Wang H, Zhang SR, Parker DS (2000) Landmarks: a new model for similarity-based pattern querying in time-series databases. In: Proceedings of 16th international conference on data engineering, pp 33–42

Pratt K, Fink E (2002) Search for patterns in compressed time series. Int J Image Graph 2 (1):89–106

Roddick JF, Spillopoulou M (1999) A Bibliography of temporal, spatial and spatio-temporal data mining research. ACM SIGKDD Explor 1(1):34–38

Singh S, Stuart E (1998) A pattern matching tool for time series forecasting. In: Proceedings of 14th international conference on pattern recognition, pp 103–105

Tsay RS (1989) Identifying multivariate time series models. J Time Ser Anal 10(4):357–372

UCI ML repository, http://www.ics.uci.edu/~mlearn/MLSummary.html

Wang L, Chang ES, Li H (2010) A tree-construction search approach for multivariate time series motifs discovery. Pattern Recogn Lett 31(9):869–875

Wu X, Zhang C, Zhang S (2004) Efficient mining of both positive and negative association rules. ACM Trans Inf Syst 22(3):381–405

Chapter 8
Synthesizing Exceptional Patterns in Different Data Sources

Many multi-branch organizations transact from different branches, and the transactions are stored locally. The number of multi-branch companies as well as the number of branches of a multi-branch company is increasing over time. Thus, it is important to study data mining on related data sources. A global exceptional pattern describes interesting individuality of few branches. Therefore, it is interesting to identify such patterns. The gist of the chapter is given as follows: (i) Type I and type II global exceptional frequent itemsets in multiple data sources are presented. (ii) The notion of exceptional sources for a type II global exceptional frequent itemset is discussed. (iii) Also the type I and type II global exceptional association rules in multiple data sources are discussed. (iv) An algorithm for synthesizing type II global exceptional frequent itemsets is designed. Experimental results are presented on both artificial and real datasets. We also compare this algorithm with the existing algorithm theoretically and experimentally. The experimental results show that the proposed algorithm is effective.

8.1 Introduction

Many multi-branch companies transact from different locations. Many of them collect a huge amount of data continuously through their different branches. Due to a growth-oriented and liberal economic policy adopted by many countries across the globe, the number of multi-branch companies as well as the number of branches of a multi-branch a company is increasing over time. Moreover, the most of the previous data mining work are based on a single database. Thus, it is important to study data mining on multiple databases. Analysis and synthesis of patterns in multiple databases is an important as well as interesting issue.

Based on the number of data sources, the patterns in multiple databases could be classified as follows: local patterns, global patterns and patterns that are neither local nor global. A pattern based on a single database is called a local pattern. Local patterns are useful for local data analysis. But global patterns are based on all the data sources under consideration. They are useful in global data analyses (Adhikari

© Springer International Publishing Switzerland 2015   127
A. Adhikari and J. Adhikari, *Advances in Knowledge Discovery in Databases*,
Intelligent Systems Reference Library 79, DOI 10.1007/978-3-319-13212-9_8

and Rao 2008c; Wu and Zhang 2003) and global decision making problems (Adhikari and Rao 2008b; Wu et al. 2005a). In many data mining applications we deal with various types of patterns. Some examples are frequent itemsets (Agrawal et al. 1993), positive associative rules (Agrawal et al. 1993) and conditional patterns (Adhikari and Rao 2008a). There is no fixed set of attributes to describe these patterns, since patterns are of different types. Each type of pattern could be described by a specific set of attributes. In general, it might be difficult to define a pattern in a database by using certain attributes.

Itemset patterns influence KDD research heavily in following ways: Firstly, many interesting algorithms have been reported on mining itemset patterns in a database (Agrawal and Srikant 1994; Han et al. 2000; Savasere et al. 1995). Secondly, an itemset could be considered as a basic type of pattern in a transactional database, since many patterns are derived from the itemset patterns in a database. Some examples of derived patterns are positive association rule (Agrawal et al. 1993), negative association rule (Wu et al. 2004), high frequency association rule (Wu and Zhang 2003) and heavy association rule (Adhikari and Rao 2008c). Considerable amount of work have been reported on mining/synthesizing such derived patterns in databases. Thirdly, solutions of many problems are based on the analysis of patterns in a database. Such applications process patterns in a database for the purpose of making some decisions (Wu et al. 2005a; Wang et al. 2001; Adhikari and Rao 2008b). Thus, the mining and analysis of itemset patterns in a database is an interesting as well as important issue.

In the context of itemset pattern we discuss a few concepts and notations that are used in this chapter. *Support* (Agrawal et al. 1993) of an itemset X in database D, denoted by $supp(X, D)$, could be defined as the fraction of transactions in D containing all the items of X. In most of the cases, the importance of an itemset is judged by its support. The itemset X is *frequent* in D if $supp(X, D) \geq \alpha$, where α is user defined level of *minimum support*. Let $SFIS(D)$ be the set of frequent itemsets in database D. Frequent itemsets determine major characteristics of a database. Wu et al. (2005b) have proposed a solution of inverse frequent itemset mining. Authors argued that one could efficiently generate a synthetic market basket dataset from the frequent itemsets and their supports. Let X and Y be two itemsets in D. The characteristics of D are revealed more by the pair $(X, supp(X, D))$ than that of $(Y, supp(Y, D))$, if $supp(X, D) > supp(Y, D)$. Thus, it is important to study frequent itemsets more than infrequent itemsets. In this chapter, we introduce two types of exceptional patterns in multiple databases based on local frequency itemsets. We discuss algorithms for synthesizing such patterns. There are useful applications of exceptional frequency itemsets. For example, a company might plan to collect the feedback of customers for the exceptional products and implement strategies to increase their sales. Also, the company could identify the branches having high sales of the exceptional items. It might plan to manufacture and/or procure such items locally to reduce the transportation cost. The exceptional frequency items would affect many decisions of a multi-branch company.

The first question comes to our mind whether a traditional data mining technique could deal with the multiple large databases. To apply a traditional data mining technique, we need to amass all the databases together. A single computer may take an unreasonable amount of time to process the entire database. Sometimes it might not be feasible to carry out the mining task. Another solution would be to employ parallel machines. It requires high investment on hardware and software. Moreover, it is difficult to identify local patterns when a mining technique is applied to the entire database. Thus, the traditional data mining techniques are not suitable in this situation. It seems it is a different problem. Hence, it is required to be dealt with it in a different way. One could employ the model of local pattern analysis (Zhang et al. 2003) for mining multiple databases. Under this model of mining multiple databases, each branch requires mining the local database using a traditional data mining technique. Afterwards, each branch forwards the pattern base to the central office. Then the central office processes the pattern bases collected from different branches for synthesizing the global patterns, or making decisions related to different problems.

The rest of the chapter is organized as follows. In Sect. 8.2, we discuss different exceptional patterns in multiple databases. Section 8.3 presents the problem formally. We discuss related work in Sect. 8.4. In Sect. 8.5 a technique for synthesizing support of an itemset is presented. Section 8.6 presents an algorithm for synthesizing type II global exceptional itemsets in multiple databases. Two types of error are defined in Sect. 8.7. Experimental results are presented in Sect. 8.8. Finally, we conclude the chapter in Sect. 8.9.

8.2 Exceptional Patterns in Multiple Data Sources

In this section we discuss type I and type II global exceptional frequent itemsets as well as type I and type II global exceptional association rules in multiple databases (Adhikari 2012). We also discuss other exceptional patterns in multiple databases.

Consider a multi-branch company that has n branches, and with the restriction that all the local transactions are stored locally. Let D_i be the database corresponding to the ith branch, $i = 1, 2, ..., n$. Also, let D be the union of these databases. In the context of multiple databases, one could define global exceptional patterns in two ways: (i) A pattern that has been extracted from the most of the databases, but has a low support in D (Type I exceptional pattern). (ii) A pattern that has been extracted from a few databases, but has a high support in D (Type II exceptional pattern). Both the types of exceptional patterns are global in nature, since all the branch databases are considered. In this chapter, we are interested in mining type II exceptional patterns, we will define here formally both types of global exceptional patterns. Zhang et al. (2004b) have proposed a strategy for mining local exceptions in multiple databases. The authors have defined an exceptional pattern as follows:

A pattern p in local instances is an exceptional pattern if $EPI(p) \geq minEP$, where $EPI(p)$ is an interestingness measure of p and has been defined as follows:

$$EPI(p) = \frac{nExtrn(p) - avgNoExtrn}{-avgNoExtrn},$$ (8.1)

where $nExtrn(p)$ and $avgNoExtrn$ are the number of times p gets extracted and the average number of times a pattern gets extracted from different data sources, respectively. $minEP$ is the user-defined threshold of minimum interest degree. Also, the authors have defined interestingness of a pattern in a branch as follows:

A pattern p in ith branch is of interest if $RI_i(p) \geq minEPsup$,

where $RI_i(p)$ is interestingness degree of p in the ith branch and has been defined as follows:

$$RI_i(p) = (supp(p, D_i) - \alpha_i)/\alpha_i,$$ (8.2)

where α_i is the minimum support given for mining D_i, $i = 1, 2, …, n$; $minEPsup$ is the user-defined threshold for minimum interest degree.

From the above two definitions, we observe the following points: (i) The definition of exceptional pattern is considered with respect to all the databases. The definition of interestingness of a pattern is considered with respect to a local database. Thus, an exceptional pattern in multiple databases and interestingness of the pattern in a local branch are of two different issues. (ii) For a pattern p in local instances, the authors have shown that $0 < EPI(p) \leq 1$. We take the following example to show that the above property does not hold always. Let there be only four patterns p_1, p_2, p_3, and p_4 in 15 databases. The number of extractions of these patterns are given as follows: $nExtrn(p_1) = 2$, $nExtrn(p_2) = 15$, $nExtrn(p_3) = 4$, $nExtrn(p_4) = 5$. Thus, $avgNoExtrn = 26/4 = 6.5$. $EPI(p_1) = (2 - 6.5)/(-6.5) = 0.69$, $EPI(p_2) = (15 - 6.5)/(-6.5) = -1.31$, $EPI(p_3) = (4 - 6.5)/(-6.5) = 0.38$, $EPI(p_4) = (5 - 6.5)/(-6.5) = 0.23$. Thus, $EPI(p_2) \notin (0,1]$. (iii) An interesting exceptional pattern might not emerge as a global exceptional pattern, since the support of the pattern is not considered in the union of all databases. It is reasonable that an exceptional frequency itemset should be constrained on the number of times it gets extracted, and its support in the union of all databases. Thus, none of the above two definitions, nor the both the definitions together does serve as a definition of exceptional frequency itemset in multiple databases.

Zhang et al. (2004a) have proposed a technique for identifying global exceptional patterns in multiple databases. The authors have described global exceptional pattern as follows:

Global exceptional patterns are highly supported by only a few branches, that is to say, these patterns have very high support in these branches and zero support in other branches.

From the above descriptions, we observe the following points:

(i)   Let there be ten branches of a multi-branch company. A pattern p has very high support in first two databases that have small sizes. Also, p does not get extracted from the remaining databases. According to the above description, p is a global exceptional pattern. We observe that pattern p might not have high support in the union of all databases. Thus, such description does not serve the purpose. Also, it is not necessarily true that a type II exceptional pattern will have zero supports in the remaining databases. Thus, the above description does not describe type II global exceptional patterns in true sense. Also, we observe the following points in *IdentifyExPattern* algorithm (Zhang et al. 2004a) for identifying exceptional patterns in multiple databases. We believe that the size (i.e., the number of transactions) of a database and support of an itemset in the database are two important parameters for determining the presence of an itemset in a database, since the number of transactions containing the itemset X in a database D_1 is equal to $supp(X, D_1) \times size(D_1)$. The algorithm does not consider size of a database to synthesize the global support of a pattern. Global support of a pattern has been synthesized using only supports of the pattern in the individual databases. We take following example to illustrate this issue. Let there be two databases D_1 and D_2, where $size(D_1)$ is significantly greater than $size(D_2)$. At a given α, we assume that pattern p does not get extracted from D_2, and pattern q does not get extracted from D_1. Thus, $supp(p, D_2)$ and $supp(q, D_1)$ both are assumed as 0. Then, $supp(p, D_1 \cup D_2)$ could be synthesized by $[supp(p, D_1) \times size(D_1) + 0 \times size(D_2)]/size(D_1 \cup D_2)$. If $supp(p, D_1) < supp(q, D_2)$ then it might so happen that $supp(p, D_1 \cup D_2) > supp(q, D_1 \cup D_2)$. In particular, let $size(D_1) = 10{,}000$, $size(D_2) = 100$. At $\alpha = 0.05$, let $supp(p, D_1) = 0.1$, $supp(q, D_1) = 0$, $supp(p, D_2) = 0$, and $supp(q, D_2) = 0.2$. We note that $supp(p, D_1) < supp(q, D_2)$. But, $supp(p, D_1 \cup D_2) = 0.099$, and $supp(q, D_1 \cup D_2) = 0.002$. So, $supp(p, D_1 \cup D_2) > supp(q, D_1 \cup D_2)$. Thus, the size of a database is an important parameter to synthesize the support of a pattern in the union of all databases.

(ii) The algorithm does not identify type II global exceptional patterns correctly in all the situations. For example, let there be 10 similar databases. Assume that the number of times each pattern gets extracted is either 8, or 9, or 10. Thus, these patterns are supported by most of the databases. According to the nature of type II global exceptional patterns, a high voted pattern is not a type II global exceptional pattern. But, the algorithm would report some of them as type II global exceptional patterns.

(iii) The algorithm returns patterns that have high supports among the patterns that are extracted less than average number of times. It is reasonable that a type II global exceptional pattern should have the following properties: (a) the support of a type II global exceptional pattern in the union of all databases is greater than or equal to a user-defined threshold, and (b) the number of extractions of a type II global exceptional pattern is less than a user-defined threshold.

The difficulty of synthesizing a type II global exceptional frequency itemsets is that a frequent itemset in a database may not get reported from every data source. Apart from synthesized support of an itemset in D, the number of extractions of the itemset is an important issue. An itemset may have high frequency or, low frequency, or neither high frequency nor low frequency. In the context of type II global exceptional pattern, we may need to consider only low frequency itemsets. One could arrive in such a conclusion only if there is a predefined threshold of minimum number of extractions. Thus, a type II global exceptional frequency itemset in multiple data sources could be judged against two thresholds, viz., high support and low extraction. Let γ_1 and γ_2 be the thresholds of low extraction of an itemset and high extraction of an itemset respectively, $0 < \gamma_1 < \gamma_2 \leq 1$. Low and high frequency itemsets are defined as follows.

Definition 8.1a An itemset X has been extracted from k out of n data sources. Then X has low frequency, if $k < n \times \gamma_1$, where γ_1 is the user-defined threshold of low extraction.

Definition 8.1b An itemset X has been extracted from k out of n data sources. Then X has high frequency, if $k \geq n \times \gamma_2$, where γ_2 is the user-defined threshold of high extraction.

Among low frequency itemsets, we shall search for type II global exceptional frequent itemsets . An itemset may not get extracted from all the data sources. Sometimes we need to estimate the support of an itemset in a database to synthesize the support of the itemset in D. Let $supp_a(X, D_i)$ be the actual support of an itemset X in D_i, $i = 1, 2, \ldots, n$. Let μ_1 and μ_2 be the thresholds of low support and high support for an itemset in a database respectively, $0 \leq \mu_1 < \alpha < \mu_2 \leq 1$. For a single database, we define an itemset with high or low support as follows:

Definition 8.2a Let X be an itemset in database D_i, $i = 1, 2, \ldots, n$. X possesses high support in D_i if $supp_a(X, D_i) \geq \mu_2$, where μ_2 ($>\alpha$) is the user-defined threshold of high support, $i = 1, 2, \ldots, n$.

Definition 8.2b Let X be an itemset in database D_i, $i = 1, 2, \ldots, n$. X possesses low support in D_i if $supp_a(X, D_i) < \mu_1$, where μ_1 ($<\alpha$) is the user-defined threshold of low support, $i = 1, 2, \ldots, n$.

The method of synthesizing support of an itemset is discussed in Sect. 8.6. Let $supp_s(X, D)$ be the synthesized support of the itemset X in D. For multiple databases, we define an itemset with high or low support as follows:

Definition 8.3a Let D be the union of all branch databases. An itemset X in D possesses high support if $supp_s(X, D) \geq \mu_2$, where μ_2 is the user-defined threshold of high support.

Definition 8.3b Let D be the union of all branch databases. An itemset X in D possesses low support if $supp_s(X, D) < \mu_1$, where μ_1 is the user-defined threshold of low support.

Based on the concepts stated above, we propose type II global exceptional frequent itemset in D as follows:

Definition 8.4 Let D be the union of all branch databases. Let X be a frequent itemset in some branch databases. Then X is a type II global exceptional itemset in D if it has low frequency but high support in D.

In a similar manner, we propose type I global exceptional frequent itemset in D as follows:

Definition 8.5 Let D be the union of all branch databases. Let X be a frequent itemset in some branch databases. Then X is a type I global exceptional itemset in D if it has high frequency but low support in D.

Association rule mining has received a lot of attention in KDD community (Zhang and Wu 2011; Zhang and Zhang 2002; Adamo 2001; Wu et al. 2004). It is based on support-confidence framework established by Agrawal et al. (1993). Let I be set of items in database DB. An association rule r has been expressed symbolically as $X \rightarrow Y$, where $X = \{x_1, x_2, \ldots, x_p\}$, and $Y = \{y_1, y_2, \ldots, y_q\}$; $x_i, y_j \in I$, $i = 1, 2, \ldots, p, j = 1, 2, \ldots, q$. It expresses association between the itemsets X and Y, called the antecedent and consequent of r respectively. The meaning attached to this implication could be expressed as follows. If the items in the itemset X are purchased by a customer then the items in the itemset Y are likely to be purchased by the same customer at the same time. The interestingness of an association rule could be expressed by its support and confidence. The support and confidence of association rule r in DB could be expressed as follows: $supp_a(r, DB) = supp_a(XY, DB)$, and $conf_a(r, DB) = supp_a(XY, DB)/supp_a(X, DB)$. One needs to deal with synthesized support ($supp_s$) and synthesized confidence ($conf_s$) of an association rule, since actual support and actual confidence of an association rule in multiple databases might not be available. There exist many techniques by which one could synthesize support of an itemset. Using pipelined feedback technique (Adhikari et al. 2010b) one could obtain a good estimate of support of an itemset in multiple databases. An association rule r in database DB is *interesting* if $supp_a(r, DB) \geq$ *minimum support* (α), and $conf_a(r, DB) \geq$ *minimum confidence* (β). Here the parameters α and β are user-defined. We define *size* of database DB as the number of transactions in DB, denoted by $size(DB)$. In the following we present the concept of heavy association rule in multiple databases. First, we present the concept of heavy association rule in a single database. Afterward, we shall present the concept of heavy association rule in multiple databases.

Definition 8.6a An association rule r in database DB is heavy if $supp_a(r, DB) \geq \mu$, and $conf_a(r, DB) \geq v$, where μ $(>\alpha)$ and v $(>\beta)$ are the user-defined thresholds of high support and high confidence for identifying heavy association rules in DB respectively (Adhikari and Rao 2008c).

If an association rule is heavy in a local database then it might not be heavy in the union of all databases D. An association rule in D might have different statuses in different local databases. For example, it might be a heavy association rule, or an association rule, or absent in a local database. Thus, one needs to synthesize an association rule for determining its overall status in D. A method of synthesizing an association rule is discussed by Adhikari and Rao (2008c). After synthesizing an association rule, we get synthesized support and synthesized confidence of the association rule in D. Now, we present the concept of heavy association rule in D as follows.

Definition 8.6b Let D be the union of local databases. An association rule r in D is heavy if $supp_s(r, D) \geq \mu$, and $conf_s(r, D) \geq v$, where μ and v are the user-defined thresholds of high support and high confidence for identifying heavy association rules in D respectively (Adhikari and Rao 2008c).

Apart from synthesized support and synthesized confidence of an association rule, the frequency of an association rule is an important issue in multi-database mining. We define *frequency* of an association rule as the number of extractions of the association rule from different data sources. If an association rule is extracted from k out of n databases then the frequency of the association rule is k, $0 \leq k \leq n$. An association rule may have low frequency or, high frequency or, neither high frequency nor low frequency in multiple data sources. We could arrive in such a conclusion only if we have user-defined thresholds of low frequency (γ_1), and high frequency (γ_2) of an association rule, $0 < \gamma_1 \leq \gamma_2 \leq 1$. Earlier we have discussed with γ_1 and γ_2 while defining type I and type II global exceptional frequent itemsets . In multi-database mining using local pattern analysis, the concepts of high frequency association rule (Wu and Zhang 2003) and low frequency association rule (Adhikari and Rao 2008c) are presented as follows.

Definition 8.7a Let an association rule be extracted from k out of n data sources. Then the association rule has low frequency if $k < n \times \gamma_1$, where γ_1 is the user-defined threshold of low frequency.

Definition 8.7b Let an association rule be extracted from k out of n data sources. Then the association rule has high frequency if $k \geq n \times \gamma_2$, where γ_2 is the user-defined threshold of high frequency.

While synthesizing heavy association rules in multiple databases, it may be worth noting the other attributes of a synthesized association rule. For example, high frequency, low frequency, and exceptionality are interesting as well as important attributes of a synthesized association rule. In similar to Definitions 8.4 and 8.5, we propose type I and type II global exceptional association rules in multiple data sources as follows.

Definition 8.8 A high frequency association rule in multiple databases is type I global exceptional if it has low support (Adhikari 2012).

Definition 8.9 A heavy association rule in multiple databases is type II global exceptional if it has low frequency (Adhikari 2012).

Adhikari et al. (2011) have proposed iceberg pattern in multiple time-stamped databases. The authors have proposed an iceberg as a generalized notch that satisfies the following conditions: (i) The height of the generalized notch is greater than or equal to α, and (ii) The width of the generalized notch is greater than or equal to β. Both α and β are user-defined thresholds. Authors have proposed an algorithm for mining icebergs in multiple time-stamped data sources.

8.3 Problem Statement

Let X be a type II global exceptional frequent itemset in D. Without any loss of generality, let X be extracted from $D_1, D_2, ..., D_k$, $0 < k < n$. Also, let the support of X in D_i be $supp_a(X, D_i)$, $i = 1, 2, ..., k$. Then the average of these supports is obtained by the following formula:

$$avg(supp(X), D_1, D_2, ..., D_k) = \left(\sum_{i=1}^{k} supp_a(X, D_i) \right) / k \qquad (8.3)$$

Database D_i is called an *exceptional source* with respect to the type II global exceptional frequent itemset X, if $supp_a(X, D_i) \geq avg(supp(X), D_1, D_2, ..., D_k)$, $i = 1, 2, ..., k$. We take an example to explain this issue. Let X be a global exceptional frequent itemset in D, and it has been extracted from $D_1, D_4,$ and D_7 out of ten data sources $D_1, D_2, ..., D_{10}$. Let $supp_a(X, D_1) = 0.09$, $supp_a(X, D_4) = 0.17$, and $supp_a(X, D_7) = 0.21$. Then, $avg(supp(X), D_1, D_4, D_7) = (0.09 + 0.17 + 0.21)/3 = 0.157$. The databases D_4 and D_7 are exceptional sources for type II global exceptional frequent itemset X, since $0.17 > 0.157$, and $0.21 > 0.157$. We state the problem as follows:

Let there be n data sources D_1, D_2, …, D_n, and D be the union of these data sources. Let SFIS(D_i) be the set of frequent itemsets in D_i, i = 1, 2, …, n. Find the type II global exceptional frequent itemsets in D using SFIS(D_i), i = 1, 2, …, n. Also, report the exceptional sources for type II global exceptional frequent itemsets in D.

8.4 Related Work

Multi-database mining has been recently recognized as an important research topic in data mining (Zhang et al. 2004c; Adhikari et al. 2010a). Zhang et al. (2003) have proposed local pattern analysis for mining multiple large databases. Afterwards, we have proposed an extended model of local pattern analysis (Adhikari and Rao 2008c). The extended model provides a way to mine and analyze multiple large databases approximately. Later, we have proposed pipelined feedback technique (Adhikari et al. 2010b) that improves the accuracy of multi-database mining significantly.

Distributed data mining (DDM) algorithms deal with mining multiple databases distributed over different geographical regions. In the last few years, researchers have started addressing problems where the databases stored at different places cannot be moved to a central storage area for variety of reasons. In multi-database mining, there are no such restrictions. DDM environment often comes with different distributed sources of computation. The advent of ubiquitous computing (Green-field 2006), sensor networks (Zhao and Guibas 2004), grid computing (Wilkinson 2009), and privacy-sensitive multiparty data (Kargupta et al. 2008) present examples where centralization of data is either not possible, or at least not always desirable.

In the context of support estimation of frequent itemsets, Jaroszewicz and Simovici (2002) have proposed a method for estimating supports of frequent itemsets using Bonferroni-type inequalities (Galambos and Simonelli 1996). Also, the maximum-entropy approach to support estimation of a general Boolean expression is proposed by Pavlov et al. (2000). But, these support estimation techniques are suitable for problems that deal with single database.

Existing parallel mining techniques could also be used to deal with multiple databases (Agrawal and Shafer 1999; Chattratichat et al. 1997; Cheung et al. 1996).

8.5 Synthesizing Support of an Itemset

Synthesizing support of an itemset in multiple databases is an important issue. First we present here the support estimation technique proposed in Adhikari and Rao (2008c). In real databases, the trend of the customers' behaviour exhibited in one database is usually present in other databases. In particular, a frequent itemset in a database is usually present in some transactions of other databases even if it does not get reported there. The estimation procedure captures such trend and estimates the support of an itemset that fails to get reported in a database. The estimated support of a missing itemset usually reduces the error of synthesizing a frequent itemset in multiple data sources. If an itemset X fails to get reported from database D_1, then we assume that D_1 contributes some amount of support of X. The support of X in D_1 satisfies the following inequality: $0 \leq supp_a(X, D_1) < \alpha$. The procedure of finding an estimated support of an itemset, called *average low-support* (*als*), is discussed below.

Let the itemset X be reported from m databases, $1 \leq m < n$. Without any loss of generality, we assume that X has been extracted from the first m databases. We shall use the average behaviour of the customers of the first m branches to estimate the average behaviour of the customers in remaining branches. Let $D_{1,m}$ be the union of $D_1, D_2, ...,$ and D_m. Then $supp_a(X, D_{1,m})$ could be viewed as the average behaviour of the customers of the first m branches with respect to X. Thus, $supp_a(X, D_{1,m})$ could be obtained by the following formula.

$$supp_a(X, D_{1,m}) = \sum_{i=1}^{m} supp_a(X, D_i) \times size(D_i) / \sum_{i=1}^{m} size(D_i) \qquad (8.4)$$

One could estimate the support of X for each of the remaining databases as follows.

$$als(X, D_i) = \alpha \times supp_a(X, D_{1,m}), \quad i = m+1, m+2, ..., n \qquad (8.5)$$

The technique discussed above might not be suitable for synthesizing type II global exceptional frequent itemsets . The reason is given as follows. A type II global exceptional frequent itemset X gets extracted from a few databases. During the process of synthesis, we need to estimate the supports of X for the remaining databases. So, the number of actual supports of X is much less than the number of estimated supports of X. Thus, the error of synthesizing the support of X in D might be high. Thus, we shall follow a different strategy for synthesizing support of an itemset in D. The strategy is explained as follows. We shall mine databases at a reasonably low value of α. If the itemset X fails to get reported from D_i then we assume that $supp_a(X, D_i) = 0$, for some i. Here the itemset X is present in D_i, $i = 1, 2,$ ..., m. Then the number of the transactions containing X in D_i is $supp_a(X, D_i) \times size$ (D_j), $i = 1, 2, ..., m$. We assume that the estimated number of the transactions containing X in D_i is 0, for $i = m + 1, m + 2, ..., n$. Thus, the estimated support of X in a database is given as follows:

$$supp_e(X, D_i) = \begin{cases} supp_a(X, D_i), & i = 1, 2, \ldots, m \\ 0, & i = m + 1, m + 2, \ldots, n \end{cases} \qquad (8.6)$$

The synthesized support of X in D could be obtained by the following formula.

$$supp_s(X, D) = \left(\sum_{i=1}^{n} supp_e(X, D_i) \times size(D_i) \right) / \sum_{i=1}^{n} size(D_i) \qquad (8.7)$$

8.6 Synthesizing Type II Global Exceptional Itemsets

In this section, we present an algorithm for synthesizing type II global exceptional frequent itemsets in D. We discuss here various data structures required to implement the algorithm. Let N be the number of frequent itemsets in $D_1, D_2, ...,$ and D_n. The frequent itemsets are kept in a two dimensional array *SFIS*. The (i, j)th element of *SFIS* stores the jth frequent itemset extracted from D_i, $j = 1, 2, ..., |SFIS(i)|$, $i = 1,$ 2, ..., n. An itemset could be described by the following attributes: *itemset, supp* and *did*. Here the attributes *itemset, supp* and *did* represent itemset, support and database identification of the frequent itemset, respectively. Synthesized type II global exceptional frequent itemsets are kept in array *synFIS*. Each type II global exceptional itemset has been described by the following attributes: *itemset, ssupp, nSources, databases, nExSources,* and *exDbases*. The attributes *itemset* and *ssupp* represent the itemset and synthesized support of the type II global exceptional frequent itemset in D, respectively. The attributes *nSources* and *databases* store the number of sources of exceptional frequent itemsets and the list of identifications of source databases for a type II global exceptional frequent itemset, respectively. The attributes *nExSources* and *exDbases* store the number of exceptional sources and the list of identifications of exceptional sources for a type II global exceptional frequent itemset, respectively. The algorithm is presented below:

Algorithm 8.1. Synthesize type II global exceptional frequent itemsets in multiple databases.

procedure *Type-II-Exceptional-Frequent-Itemset-Synthesis* (n, *SFIS*, μ_2, *size*, γ_1)
Input:
n: number of data sources
SFIS: array of frequent itemsets
μ_2: threshold of high support
size: array of total number of transactions in different databases
γ_1: threshold of low extraction
Output:
Type II global exceptional frequent itemsets in D
01: collect all the frequent itemsets into array *FIS*;
02: sort frequent itemsets in *FIS* in non-decreasing order on itemset attribute;
03: calculate the total number of transactions into *totTrans*;
04: *nSynFIS* = 0; *curPos* = 1;
05: **while** (*curPos* ≤ N) **do**
06:    i = *curPos*; *count* = 0;
07:    *nTransCurFIS* = 0; *totSupp* = 0;
08:    **while** (i ≤ *curPos* + n) **do**
09:      **if** (*FIS*(i).*itemset* = *FIS*(*curPos*).*itemset*) **then**
10:        update *count*, *sources*(*count*), *totalSupp*, *nTransCurFIS*, *supports*(*count*);
11:      **else** exit from the *while*-loop;
12:      **end if**
13:     increase i by 1;
14:   **end while**
15:   **if** ((*count* / n) < γ_1) **and** (*nTransCurFIS* / *totTrans* ≥ μ_2)) **then**
16:    increase *nSynFIS* by 1;
17:    update attributes *supp*, *itemset* and *nSources* of *synFIS*(*nSynFIS*);
18:    *avgSupp* = *totalSupp* / *count*; *exCount* = 0;
19:    **for** j = 1 to *count* **do**
20:      *synFIS*(*nSynFIS*).*databases*(j) = *source*(j);
21:      **if** (*supports*(*exCount*) ≥ *avgSupp*) **then**
22   increase *exCount* by 1;
23:        *synFIS*(*nSynFIS*).*exDbases*(*exCount*) = *sources*(j) ;
24:      **end if**
25:    **end for**
26:    *synFIS*(*nSynFIS*).*nExSources* = *exCount*;
27:   **end if**
28:   *curPos* = i;
29: **end while**
30: sort itemsets in *synFIS*;
31: **for** i = 1 to *nSynFIS* **do**
32:   display details of *synFIS*(i);
33: **end for**
34: **end procedure**

The main steps of the algorithm is explained here. The frequent itemsets with the same *itemset* attribute are kept consecutive in *FIS*. It helps processing one itemset at a time. A type II global exceptional frequent itemset is synthesized using the lines 5–29. The array *sources* is used to store the database identifications of all the databases that report the current frequent itemset. Also, the array *supports* is used to store the supports of the current frequent itemset in different databases. The information about the current itemset is obtained by the *while*-loop in lines 8–14. The information includes the number of extractions, database identifications of the source databases, supports in different databases, total supports, number of trans-actions containing current frequent itemset in different databases. We explain the update-statement at line 10 as follows. The number of extraction of current frequent itemset, *count*, is increased by 1. The database identification, *did*, of the current database is copied into cell *sources(count)*. Variable *nCurFIS* is added by expression $FIS(i).supp \times size(FIS(i).did)$. Variable totalSupp is also added by expression $FIS(i).supp$. The support of frequent itemset in the current database is copied into *supports(count)*. We also explain the update-statement at line 17 as follows. The synthesized support of current type II global exceptional frequent itemset is obtained by expression *nCurFIS/totTrans*. The *itemset* attribute of current type II global exceptional frequent itemset is the same as the *itemset* attribute of current frequent itemset. The variable *count* is copied into *synFIS(nSynFIS). nSources*. The *if*-statement in lines 15–27 checks whether the current itemsets is a type II global exceptional frequent itemset in multiple data sources, and it syn-thesizes each type II global exceptional frequent itemset. All the frequent itemsets are processed using the lines 4–29. We sort global exceptional itemsets for better presentation at line 30. All the type II global exceptional itemsets are kept in non-decreasing order on the length of the itemset. Again, the itemsets of the same length are sorted on non-increasing order on support of the itemset. Finally, type II global exceptional itemsets and supports are displayed using lines 31–33. For every type II global exceptional frequent itemset, we display the data sources from which it has been extracted. Also, for every type II global exceptional frequent itemset, we also display the exceptional data sources from which it has been highly supported.

Theorem 8.1 *The time complexity of the procedure Type-II-Exceptional-Frequent-Itemset-Synthesis is maximum{$O(N \times \log(N))$, $O(n \times N)$}, where N is the number of frequent itemsets extracted from n databases.*

Proof The time complexity of line 1 is $O(N)$, since there are N frequent itemsets in all the databases. Line 2 sorts N frequent itemsets in $O(N \times \log(N))$ time. The time complexity of line 3 is $O(n)$, since there are n databases. The program segment lines 5–29 repeats maximum N times. Within this program segment, there is a *while*-loop and a *for*-loop. The *while*-loop in lines 8–14 takes $O(n)$ time. Also, the *for*-loop in lines 19–25 takes $O(n)$ time. Thus, the time complexity of the program segment lines 5–29 is $O(n \times N)$. Line 30 takes $O(N \times \log(N))$ time for sorting maximum N synthesized type II global exceptional itemsets. To display the details of a type II global exceptional itemset it takes $O(n)$ time, since there are maximum n sources of the itemset. Thus, the program segment in lines 31–33 take $O(n \times N)$ time.

Table 8.1 Sorted frequent itemsets in different databases

Itemset	A	A	A	B	B	C	C	C	D	E	F	AB	CD
supp	0.12	0.10	0.11	0.14	0.20	0.20	0.25	0.60	0.16	0.16	0.77	0.11	0.12
did	1	2	3	1	2	1	2	3	2	2	3	1	2

Table 8.2 Synthesized frequent itemsets in multi-databases

Itemset	A	B	C	D	E	F	AB	CD
Synthesized supp	0.11	0.07	0.44	0.03	0.03	0.45	0.03	0.02

Therefore, the time complexity of the procedure *Type-II-Exceptional-Frequent-Itemset-Synthesis* is *maximum*$\{O(N \times \log(N)), O(n \times N)\}$. $\quad\square$

In the following example, we manually execute the above algorithm and verify that it works correctly.

Example 8.1 Let D_1, D_2 and D_3 be three databases of sizes 4,000 transactions, 3,290 transactions, and 10,200 transactions respectively. Let *DB* be the union of D_1, D_2, and D_3. Assume that $\alpha = 0.1$, $\gamma = 0.4$, $\mu = 0.25$. The sets of frequent itemsets extracted from these databases are given as follows. $SFIS(D_1) = \{A(0.12), B(0.14), AB(0.11), C(0.20)\}$, $SFIS(D_2) = \{A(0.10), B(0.20), C(0.25), D(0.16), CD(0.12), E (0.16)\}$, $SFIS(D_3) = \{A(0.11), C(0.60), F(0.77)\}$. The symbol $X(\eta)$ is used to indicate that X is a frequent itemset in the local database with support η. We keep frequent itemsets in array *FIS* and sort them on *itemset* attribute. The sorted frequent itemsets are given in Table 8.1.

Here, *totTrans* is 17,490. We synthesize the frequent itemsets in *FIS*. Synthesized frequent itemsets are given in Table 8.2.

In Algorithm 8.1, we maintain synthesized global exceptional frequent itemset in array *synFIS*. For the purpose of explanation, Table 8.2 has been introduced here. From Table 8.2, we find that the synthesized supports of itemsets C and F are high, since $supp_s(C, DB) \geq \mu$ and $supp_s(F, DB) \geq \mu$. Itemset F has been extracted from one out of three databases. Thus, F is low-voted, since $1/3 < \gamma$. Thus, the itemset F is a type II global exceptional frequent itemset in *DB*.

In the following theorem, we determine time complexity of algorithm *Identify-ExPattern* (Zhang et al. 2004a) for comparing algorithm *IdentifyExPattern* with algorithm *Type-II-Exceptional-FrequentItemset-Synthesis* theoretically.

Theorem 8.2 The algorithm *IdentifyExPattern* takes $O(n^2 \times N \times \log(N))$ time, where N is the number of frequent itemsets extracted from n databases.

Proof We refer to algorithm *IdentifyExPattern*. Step 5 of the algorithm ranks candidate exceptional patterns based on their global supports. Step 4 calculates global support of a candidate exceptional pattern based on the number of databases that support the pattern. Step 1 counts the number databases that support a specific pattern. Thus, step 5 is based on step 4, and step 4 is based on step 1. Step 1 takes O (n) time for a specific pattern. This implies that step 4 takes $O(n \times n)$ time for each candidate exceptional pattern. Thus, step 5 takes $O(n^2 \times N \times \log(N))$ time, and hence the theorem follows.   □

From Theorems 8.1 and 8.2, we conclude that the proposed algorithm executes faster than algorithm *IdentifyExPattern*. We also compare these two algorithms experimentally in Sect. 8.8.

8.7 Error Calculation

To evaluate the proposed technique of synthesizing type II global exceptional frequent itemsets, we have measured amount of error occurred in an experiment. Error of the experiment is relative to the number of transactions (i.e., the size of the database), number of items, and length of a transaction in a database. Thus, the error of the experiment needs to be expressed along with the *ANT*, *ALT*, and *ANI* in a database, where *ANT*, *ALT*, and *ANI* denote the average number of transactions, average length of a transaction and average number of items in a database respectively. The error of the experiment is based on the global exceptional frequent itemsets in D. Let $\{X_1, X_2, \ldots, X_m\}$ be the set of type II global exceptional frequent itemsets in D. There are several ways one could define the error of an experiment. We have defined following two types of errors of an experiment.

1. Average Error (*AE*)

$$AE(D, \alpha, \mu, \gamma) = \frac{1}{m} \sum_{i=1}^{m} |supp_a(X_i, D) - supp_s(X_i, D)| \qquad (8.8)$$

2. Maximum Error (*ME*)

$$AE(D, \alpha, \mu, \gamma) = \text{maximum}\{|supp_a(X_i, D) - supp_s(X_i, D)|, \quad i = 1, 2, \ldots, m\} \qquad (8.9)$$

Actual support of X_i in D, $supp_a(X_i, D)$ is obtained by mining D using a traditional data mining technique, $i = 1, 2, \ldots, m$. Synthesized support of X_i in D, $supp_s(X_i, D)$ is obtained by the technique presented in Sect. 8.6, $i = 1, 2, \ldots, m$. Then we compute error of synthesizing support of X_i in D as $|supp_a(X_i, D) - supp_s(X_i, D)|$, $i = 1, 2, \ldots, m$.

Example 8.2 With reference to Example 8.1, the itemset F is the only type II global exceptional frequent itemset present in D. Thus, $AE(D, 0.1, 0.25, 0.4) = ME(D, 0.1, 0.25, 0.4) = |supp_a(F, D) - supp_s(F, D)|$.

Table 8.3 Dataset
characteristics

Dataset	NT	ALT	AFI	NI
check(C)	40	3.03	3.10	39
retail(R)	88,162	11.31	99.71	10,000
SD100000(S)	1,00,000	15.11	151.10	10,000

8.8 Experiments

The experimental results are presented on both artificial and real datasets. An artificial database *check* is constructed to verify that the proposed algorithm works correctly. Each item is represented by an integer number to perform experiments more conveniently. Thus, a transaction in *check* is a collection of integer numbers separated by commas. The dataset *retail* is obtained from an anonymous Belgian retail supermarket store.[1] Also, we have generated a synthetic dataset *SD100000*. We present some characteristics of these datasets in Table 8.3.

The symbols *NT*, *AFI*, *ALT* and *NI* denote the number of transactions, average frequency of an item, average length of a transaction and number of items in the dataset respectively. Each of the above datasets has been divided into 10 databases for the purpose of carrying out experiments. These databases are called input databases. The proposed work is based on the frequent itemsets in the input databases. There are many algorithms (Agrawal and Srikant 1994; Han et al. 2000; Savasere et al. 1995) for mining frequent itemsets in a database. Thus, there exist many implementations of mining frequent itemsets in a database (FIMI 2004).

The dataset *check* consists of 40 transactions. The input databases obtained from *check* are given as follows: $C_0 = \{\{1, 4, 9, 31\}, \{1, 4, 7, 10, 50\}, \{1, 4, 10, 20, 24\}, \{1, 4, 10, 23\}$; $C_1 = \{\{1, 4, 10, 34\}, \{1, 3, 44\}, \{1, 2, 3, 10, 20, 44\}, \{2, 3, 20, 39\}\}$; $C_2 = \{\{2, 3, 20, 44\}, \{2, 3, 45\}, \{2, 3, 44, 50\}, \{2, 3, 20, 44, 50\}\}$; $C_3 = \{\{3, 44\}, \{3, 19, 50\}, \{5, 7, 21\}, \{5, 8\}\}$; $C_4 = \{\{5, 41, 45\}, \{5, 49\}, \{5, 7, 21\}, \{5, 11, 21\}\}$; $C_5 = \{\{6, 41\}, \{6, 15, 19\}, \{11, 12, 13\}, \{11, 21, 49\}\}$; $C_6 = \{\{11, 19\}, \{21\}, \{21, 24, 39\}, \{22, 26, 38\}\}$; $C_7 = \{\{22, 30, 31\}, \{24, 35\}, \{25, 39, 49\}, \{26, 41, 46\}\}$; $C_8 = \{\{30, 32, 42\}, \{32, 49\}, \{41, 45, 59\}, \{42, 45\}\}$; $C_9 = \{\{42, 47\}, \{45, 49\}, \{47, 48, 49\}, \{49\}\}$. The input databases obtained from *retail* and *DS100000* are named as R_i and D_i, $i = 0, 1, ..., 9$. For the purpose of mining input databases, we have implemented apriori algorithm (Agrawal and Srikant 1994), since it is simple and popular. Some characteristics of these input databases (*DB*) are presented in the Tables 8.4 and 8.5.

The global exceptional frequent itemsets corresponding to *check*, *retail* and *SD100000* are presented in Tables 8.6, 8.7 and 8.8 respectively.

A global exceptional frequent itemset might not be supported with equal degree from the source databases. For example, the global exceptional frequent itemset {5}

---

[1] Frequent itemset mining dataset repository, http://fimi.cs.helsinki.fi/data/.

Table 8.4 The characteristics of databases obtained from *retail*

DB	NT	ALT	AFI	NI	DB	NT	ALT	AFI	NI
R_0	9,000	11.24	12.07	8,384	R_5	9,000	10.86	16.71	5,847
R_1	9,000	11.21	12.26	8,225	R_6	9,000	11.20	17.42	5,788
R_2	9,000	11.34	14.60	6,990	R_7	9,000	11.16	17.35	5,788
R_3	9,000	11.49	16.66	6,206	R_8	9,000	11.99	18.69	5,777
R_4	9,000	10.96	16.04	6,148	R_9	7,162	11.69	15.35	5,456

Table 8.5 The characteristics of databases obtained from *SD100000*

DB	NT	ALT	AFI	NI	DB	NT	ALT	AFI	NI
S_0	10,000	16.01	17.08	9,371	S_5	10,000	15.77	15.95	9,886
S_1	10,000	15.23	16.16	9,426	S_6	10,000	16.69	19.58	8,524
S_2	10,000	13.93	17.35	8,031	S_7	10,000	14.93	15.77	9,468
S_3	10,000	12.55	14.69	8,541	S_8	10,000	15.01	16.95	8,856
S_4	10,000	17.56	19.05	9,003	S_9	10,000	13.97	15.31	9,119

Table 8.6 Global exceptional frequent itemsets in $\{C_0, C_1, ..., C_9\}$ at $\alpha = 0.05$, $\gamma_1 = 0.4$ and $\mu_2 = 0.09$ (top 10)

Itemset	ssupp	Sources	Exceptional sources
{1}	0.175	C_0, C_1	C_0
{2}	0.175	C_1, C_2	C_2
{3}	0.200	C_1, C_2, C_3	C_1, C_2
{5}	0.150	C_3, C_4	C_4
{44}	0.150	C_1, C_2, C_3	C_1, C_2
{2,3}	0.150	C_1, C_2	C_2
{3,44}	0.150	C_1, C_2, C_3	C_2, C_3
{4}	0.125	C_0, C_1	C_0
{10}	0.125	C_0, C_1	C_0
{20}	0.125	C_0, C_1, C_2	C_1, C_2

has been extracted from databases C_3 and C_4. But, the database C_4 reports itemset {5} exceptionally more.

Some databases do no report any global exceptional frequent itemsets. On the contrary, some other databases are the source of many global exceptional frequent itemsets. In Tables 8.9, 8.10 and 8.11, we present the distributions of global exceptional frequent itemsets in $\{C_0, C_1, ..., C_9\}$, $\{R_0, R_1, ..., R_9\}$ and $\{S_0, S_1, ..., S_9\}$ respectively. In Table 8.9, we notice that three out of ten databases are not source of any global exceptional frequent itemset.

The distribution of global exceptional frequent itemsets in $\{R_0, R_1, ..., R_9\}$ is different from that in $\{C_0, C_1, ..., C_9\}$. In Table 8.10, we notice that seven out of ten databases are not the source of any exceptional global frequent itemset. In Table 8.11, we observe that the most of the databases are the source of some exceptional global frequent itemsets.

Table 8.7 Global exceptional frequent itemsets in $\{R_0, R_1, \ldots, R_9\}$ at $\alpha = 0.02$, $\gamma_1 = 0.4$ and $\mu_2 = 0.1$

Itemset	*ssupp*	Sources	Exceptional sources
{2, 6, 9}	0.102	R_0, R_1	R_0
{2, 9, 41}	0.103	R_0, R_1, R_7	R_0, R_7
{6, 9, 41}	0.107	R_0, R_1, R_3	R_0, R_1
{8, 9, 271}	0.102	R_0	R_0
{9, 41, 48}	0.102	R_1	R_1

Table 8.8 Global exceptional frequent itemsets in $\{S_0, S_1, \ldots, S_9\}$ at $\alpha = 0.02$, $\gamma_1 = 0.4$ and $\mu_2 = 0.05$ (top 10)

Itemset	*ssupp*	Sources	Exceptional sources
{22}	0.247	S_0, S_5, S_6	S_0
{32}	0.208	S_2, S_4, S_5	S_4, S_5
{45}	0.185	S_3, S_4, S_5	S_3
{4}	0.167	S_2, S_5	S_5
{49}	0.161	S_4, S_9	S_4
{22, 32}	0.158	S_0, S_2, S_5	S_0, S_2
{22, 45}	0.152	S_0, S_4, S_5	S_0, S_4
{22, 4}	0.132	S_0, S_2, S_5	S_0, S_5
{22, 49}	0.130	S_0, S_4	S_4
{32, 45}	0.121	S_2, S_4	S_2

Table 8.9 Distribution of exceptional source for global exceptional frequent itemsets in $\{C_0, C_1, \ldots, C_9\}$ at $\alpha = 0.05$, $\gamma_1 = 0.4$ and $\mu_2 = 0.09$

Database	C_0	C_1	C_2	C_3	C_4	C_5	C_6	C_7	C_8	C_9
Number of sources of global exceptional frequent itemsets	7	8	10	4	2	1	1	0	0	0

Table 8.10 Distribution of global exceptional frequent itemsets in $\{R_0, R_1, \ldots, R_9\}$ at $\alpha = 0.02$, $\gamma_1 = 0.4$ and $\mu_2 = 0.1$

Database	R_0	R_1	R_2	R_3	R_4	R_5	R_6	R_7	R_8	R_9
Number of sources of global exceptional frequent itemsets	4	2	0	0	0	0	0	1	0	0

Also, we have studied the number of global exceptional frequent itemsets in multi-databases at different γ_1 s. As we increase γ_1, we allow more frequent itemsets to be global exceptional. In Figs. 8.1, 8.2 and 8.3, we study the relationship between γ_1 and the number of global exceptional frequent itemsets in multiple databases.

Table 8.11 Distribution of global exceptional frequent itemsets in $\{S_0, S_1, …, S_9\}$ at $\alpha = 0.02$, $\gamma_1 = 0.4$ and $\mu_2 = 0.05$

Database	S_0	S_1	S_2	S_3	S_4	S_5	S_6	S_7	S_8	S_9
Number of sources of global exceptional frequent itemsets	11	0	6	9	1	5	7	3	3	5

We have observed that the number of global exceptional frequent itemsets do not vary much at different γ_1s. In fact, there is only one change in both the graphs of Figs. 8.1 and 8.2.

Also, we have studied the number of global exceptional frequent itemsets in multiple databases at different αs. We present experimental results in Figs. 8.4, 8.5 and 8.6. The number of global exceptional frequents in $\{C_0, C_1, …, C_9\}$ remains fixed at 21 over different αs.

But, we find a different trend with respect to the number of global exceptional frequent itemsets in $\{R_0, R_1, …, R_9\}$. At lower and upper values of α, the number of global exceptional frequent itemsets is 0. Again, we get some global exceptional frequent itemsets for some middle values of α. Thus, there is no fixed relationship between the number of global exceptional frequent itemsets and α.

Also, we have calculated the error of the experiments. In Table 8.12, we present the error of the experiments at a given value of tuple (α, γ, μ).

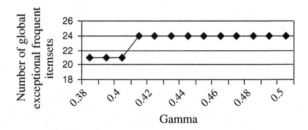

Fig. 8.1 Number of global exceptional frequent itemsets in $\{C_0, C_1, …, C_9\}$ at $\alpha = 0.05$, and $\mu_2 = 0.1$

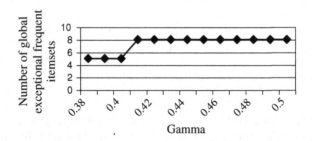

Fig. 8.2 Number of global exceptional frequent itemsets in $\{R_0, R_1, …, R_9\}$ at $\alpha = 0.02$, and $\mu_2 = 0.1$

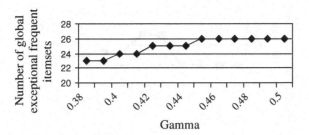

Fig. 8.3 Number of global exceptional frequent itemsets in $\{S_0, S_1, ..., S_9\}$ at $\alpha = 0.02$, and $\mu_2 = 0.05$

8.8.1 Comparison with the Existing Algorithm

In this section we make comparison between algorithms *IdentifyExPattern* (Zhang et al. 2004a) and *Exceptional-FrequentItemset-Synthesis*. We analyze and compare these two algorithms on the basis of experiments conducted on the following two issues: (i) average error versus α, and (ii) synthesizing time versus number of databases.

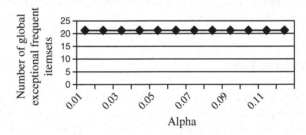

Fig. 8.4 Number of global exceptional frequent itemsets in $\{C_0, C_1, ..., C_9\}$ at $\gamma_1 = 0.4$, and $\mu_2 = 0.1$

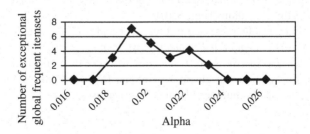

Fig. 8.5 Number of global exceptional frequent itemsets in $\{R_0, R_1, ..., R_9\}$ at $\gamma_1 = 0.4$, and $\mu_2 = 0.1$

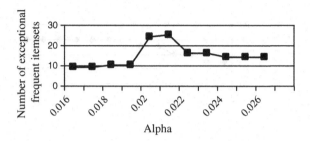

Fig. 8.6 Number of global exceptional frequent itemsets in $\{S_0, S_1, \ldots, S_9\}$ at $\gamma_1 = 0.4$, and $\mu_2 = 0.05$

8.8.1.1 Average Error

We have calculated AEs at different αs to study the relationship between them. Experimental results are presented in Figs. 8.5 and 8.6. We observe that there is no fixed relationship between AE and α.

In both the cases, algorithm *Type-II-Exceptional-FrequentItemset-Synthesis* performs better than algorithm *IdentifyExPattern*. In dataset *check*, the global exceptional frequent itemsets are not uniformly distributed. The global exceptional frequent itemsets appear only in few databases, while they remain absent in the remaining databases. The proposed technique finds average error 0 at different αs, since the error of synthesizing each global exceptional frequent itemset in $\{C_0, C_1, \ldots, C_9\}$ is 0 (Figs. 8.7, 8.8 and 8.9).

8.8.1.2 Synthesizing Time

Also, we have calculated the time for synthesizing global exceptional frequent itemsets by varying the number of databases. In Figs. 8.11 and 8.12, we show time (in ms.) required for synthesizing global exceptional frequent itemsets in multiple databases. In case of the experiment conducted on $\{C_0, C_1, \ldots, C_9\}$, we observe that the synthesizing time does not increase as the number of databases increases. This is due to the fact that the size of each of the databases is very small. In fact, the time required for synthesizing global exceptional frequent itemsets in $\{C_0, C_1, \ldots, C_9\}$ is 0 ms., for both the algorithms (Fig. 8.10).

Considering the results presented in Figs. 8.11 and 8.12, we could conclude that algorithm *Type-II-Exceptional-FrequentItemset-Synthesis* executes faster than algorithm *IdentifyExPattern*. Also, it matches with the theoretical results. In general, the time for synthesizing global exceptional frequent itemsets either remains same or, increases as the number of databases increases.

Table 8.12 Errors of the experiments at a given value of tuple (α, γ, μ)

Databases	α	γ_1	μ_2	(AE, ANT, ALT, ANI)	(ME, ANT, ALT, ANI)
$C_0, C_1, \ldots, C_9$	0.05	0.4	0.1	(0, 4, 3.025, 8.4)	(0, 4, 3.025, 8.4)
$R_0, R_1, \ldots, R_9$	0.02	0.4	0.1	(0.064, 8,816.2, 11.306, 5,882.1)	(0.085, 8,816.2, 11.306, 5,882.1)
$S_0, S_1, \ldots, S_9$	0.02	0.4	0.05	(0.052, 10,000.0, 15.11, 9,022.5)	(0.075, 10,000.0, 15.11, 9,022.5)

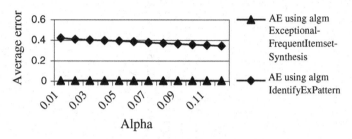

Fig. 8.7 Average error versus α for *check* at $\gamma_1 = 0.4$, and $\mu_2 = 0.1$

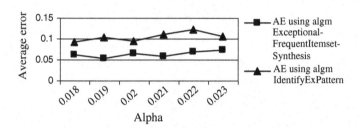

Fig. 8.8 Average error versus α for *retail* at $\gamma_1 = 0.4$, and $\mu_2 = 0.1$

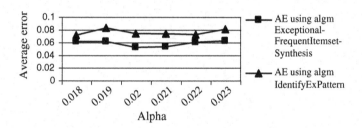

Fig. 8.9 Average error versus α for *retail* at $\gamma_1 = 0.4$, and $\mu_2 = 0.05$

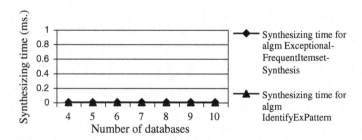

Fig. 8.10 Synthesizing time versus number of databases obtained from *check* at $\gamma_1 = 0.4$, and $\mu_2 = 0.1$

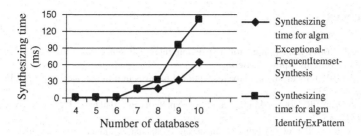

Fig. 8.11 Synthesizing time versus number of databases obtained from *retail* at $\alpha = 0.02$, $\gamma_1 = 0.4$, and $\mu_2 = 0.1$

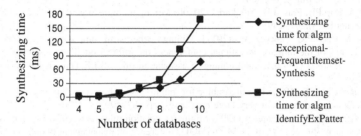

Fig. 8.12 Synthesizing time versus number of databases obtained from *SD10000l* at $\alpha = 0.02$, $\gamma_1 = 0.4$, and $\mu_2 = 0.05$

8.9 Conclusions

Synthesis of global exceptional patterns is an important component of a multi-database mining system. Many corporate decisions of a multi-branch company would depend on global exceptional patterns in branch databases. We have identified the short-comings of the existing concepts and the algorithm to identify global exceptional patterns, and a definition of a global exceptional frequent itemset. Also, the notion of exceptional sources for a global exceptional frequent itemset is presented. The proposed algorithm identifies global exceptional frequent itemsets and their exceptional sources in multiple databases. We have also compared proposed algorithm with the existing algorithm. The performance of the proposed algorithm has been observed on the following issues: (i) error of the experiment and (ii) execution time, and a comparison is made with the existing algorithm. We have observed that the proposed algorithm executes faster when the number of databases increases. Also, we have shown theoretically that the proposed algorithm executes faster than the existing algorithm. The proposed solution is simple and effective in synthesizing global exceptional frequent itemsets in multi-databases.

References

Adamo J-M (2001) Data mining for association rules and sequential patterns: sequential and parallel algorithms. Springer, Berlin

Adhikari A (2012) Synthesizing global exceptional patterns in different data sources. J Intell Syst 21(3):293–323

Adhikari A, Rao PR (2008a) Mining conditional patterns in a database. Pattern Recogn Lett 29 (10):1515–1523

Adhikari A, Rao PR (2008b) Efficient clustering of databases induced by local patterns. Decis Support Syst 44(4):925–943

Adhikari A, Rao PR (2008c) Synthesizing heavy association rules from different real data sources. Pattern Recogn Lett 29(1):59–71

Adhikari A, Ramachandrarao P, Pedrycz W (2010a) Developing multi-database mining applications. Springer, London

Adhikari A, Rao PR, Prasad B, Adhikari J (2010b) Mining multiple large data sources. Int Arab J Inf Technol 7(3):241–249

Adhikari J, Rao PR, Pedrycz W (2011) Mining icebergs in time-stamped databases. In: Proceedings of the Indian international conferences on artificial intelligence, pp 639–658

Agrawal R, Shafer J (1999) Parallel mining of association rules. IEEE Trans Knowl Data Eng 8 (6):962–969

Agrawal R, Srikant R (1994) Fast algorithms for mining association rules. In: Proceedings of the international conference on very large data bases, pp 487–499

Agrawal R, Imielinski T, Swami A (1993) Mining association rules between sets of items in large databases. In: Proceedings of ACM SIGMOD conference on management of data, pp 207–216

Chattratichat J, Darlington J, Ghanem M, Guo Y, Hüning H, Köhler M, Sutiwaraphun J, To HW, Yang D (1997) Large scale data mining: challenges, and responses. In: Proceedings of the 3rd international conference on knowledge discovery and data mining, pp 143–146

Cheung D, Ng V, Fu A, Fu Y (1996) Efficient mining of association rules in distributed databases. IEEE Trans Knowl Data Eng 8(6):911–922

FIMI (2004) http://fimi.cs.helsinki.fi/src/

Galambos J, Simonelli I (1996) Bonferroni-type inequalities with applications. Springer, New York

Greenfield A (2006) Everyware: the dawning age of ubiquitous computing, 1st edn. New Riders Publishing, San Francisco

Han J, Pei J, Yiwen Y (2000) Mining frequent patterns without candidate generation. In: Proceedings ACM SIGMOD conference on management of data, pp 1–12

Jaroszewicz S, Simovici DA (2002) Support approximations using Bonferroni-type inequalities. In: Proceedings of the 6th European conference on principles of data mining and knowledge discovery, pp 212–223

Kargupta H, Han J, Yu PS, Motwani R, Kumar V (2008) Next generation of data mining. Springer, Berlin

Pavlov D, Mannila H, Smyth P (2000) Probabilistic models for query approximation with large sparse binary data sets. In: Proceedings of the 16th conference on uncertainty in artificial intelligence, pp 465–472

Savasere A, Omiecinski E, Navathe S (1995) An efficient algorithm for mining association rules in large databases. In: Proceedings of the 21st international conference on very large data bases, pp 432–443

Wang K, Zhou S, He Y (2001) Hierarchical classification of real life documents. In: Proceedings of the 1st (SIAM) international conference on data mining, pp 1–16

Wilkinson B (2009) Grid computing: techniques and applications. CRC Press, Boca Raton

Wu X, Zhang S (2003) Synthesizing high-frequency rules from different data sources. IEEE Trans Knowl Data Eng 14(2):353–367

Wu X, Zhang C, Zhang S (2004) Efficient mining of both positive and negative association rules. ACM Trans Inf Syst 22(3):381–405

Wu X, Zhang C, Zhang S (2005a) Database classification for multi-database mining. Inf Syst 30 (1):71–88

Wu X, Wu Y, Wang Y, Li Y (2005b) Privacy-aware market basket data set generation: a feasible approach for inverse frequent set mining. In: Proceedings of SIAM international conference on data mining, pp 103–114

Zhang S, Wu X (2011) Fundamentals of association rules in data mining and knowledge discovery. Wiley Interdisc Rev: Data Min Knowl Disc 1(2):97–116

Zhang C, Zhang S (2002) Association rule mining: models and algorithms. Springer, Heidelberg

Zhang S, Wu X, Zhang C (2003) Multi-database mining. IEEE Comput Intell Bull 2(1):5–13

Zhang C, Liu M, Nie W, Zhang S (2004a) Identifying global exceptional patterns in multi-database mining. IEEE Comput Intell Bull 3(1):19–24

Zhang S, Zhang C, Yu JX (2004b) An efficient strategy for mining exceptions in multi-databases. Inf Sci 165(1–2):1–20

Zhang S, Zhang C, Wu X (2004c) Knowledge discovery in multiple databases. Springer, Berlin

Zhao F, Guibas L (2004) Wireless sensor networks: an information processing approach. Morgan Kaufmann, Massachusetts

Chapter 9
Clustering Items in Time-Stamped Databases

Many multi-branch companies transact from different branches. Each branch of the company maintains a separate database over time. The variation of sales of an item over time is an important issue, and therefore, we present the notion of stability of an item. Stable items are useful in making numerous strategic decisions of the company. We have discussed two measures of stability of an item. Based on the degree of stability of an item, an algorithm is designed for finding partition among items in different data sources. Then the notion of the best cluster is introduced by considering average degree of variation of a class, and designed an algorithm to find clusters among items in different data sources. The best cluster is determined by average degree of variation in a cluster. Experimental results are provided for three transactional databases.

9.1 Introduction

Knowledge discovery in multiple databases has been recently recognized as an important area of research in data mining (Zhang et al. 2004; Adhikari et al. 2010). Many multi-branch companies deal with transactional time-stamped data. Transactional data collected over time at no particular frequency are often termed as transactional time-stamped data (Leonard and Wolfe 2005). Some examples of transactional time-stamped data are point of sales data, inventory data, and trading data. Little work has been reported on the area of mining multiple transactional time-stamped databases. In this chapter, we discuss a useful data mining application on multiple transactional time-stamped data.

Consider a multi-branch company that transacts from multiple branches, and all the transactions in a branch are stored locally. A transaction could be viewed as a collection of items with a unique identifier. Then it becomes interesting to study the characteristics of items in multiple databases. A useful characteristic of item is its variation of sales over time. The items having lower variation of sales over time are

© Springer International Publishing Switzerland 2015

A. Adhikari and J. Adhikari, *Advances in Knowledge Discovery in Databases*,
Intelligent Systems Reference Library 79, DOI 10.1007/978-3-319-13212-9_9

useful in devising strategies of the company. Thus it is important to identify such items. Consider a situation as presented in Example 9.1.

Example 9.1 Let there be ten branch databases. Suppose we are interested in studying variations of five items. We take here the support (Agrawal et al. 1993) series of the five items. Let i-th series be the series of supports in ten databases corresponds to item x_i, $i = 1, 2, 3, 4, 5$.

(1)   0.03, 0.20, 0.31, 0.11, 0.07, 0.35, 0.82, 0.62, 0.44, 0.13
(2)   0.19, 0.20, 0.18, 0.21, 0.20, 0.20, 0.19, 0.18, 0.21, 0.20
(3)   0.05, 0.11, 0.07, 0.20, 0.16, 0.12, 0.13, 0.08, 0.17, 0.10
(4)   0.03, 0.04, 0.03, 0.07, 0.08, 0.12, 0.09, 0.15, 0.17, 0.12
(5)   0.04, 0.04, 0.03, 0.05, 0.04, 0.06, 0.04, 0.05, 0.06, 0.05

Among the support series corresponding to different items, it is observed that the variation of sales corresponding to item x_5 is the least. Thus, the strategies based on item x_5 could be deemed more reliable.

The chapter is organized as follows. Work related to the problem is presented in Sect. 9.2. In Sect. 9.3, we propose a model of mining multiple transactional time-stamped databases. The problem is formulated in Sect. 9.4. In Sect. 9.5, we design an algorithm for clustering of items in multiple databases. Experimental results are provided in Sect. 9.6.

9.2 Related Work

Liu et al. (2001) have proposed stable association rules based on testing of hypothesis. The distribution of test statistic under null hypothesis follows normal distribution for large sample size. Thus, the stable association rules are determined based on some assumptions. Due to this reason, we present a construction of stable items based on the concept of stationary time series data (Brockwell and Richard 2002).

In the context of interestingness measures, Tan et al. (2002) have described several key properties of twenty one interestingness measures proposed in statistics, machine learning and data mining literature. Wu et al. (2005) have proposed two item-based similarity measures for clustering a set of databases. Based on transaction similarity, Adhikari and Rao (2008b) have proposed two similarity measures for clustering databases.

Zhang et al. (1997) have proposed an efficient and scalable data clustering method BIRCH based on a new in-memory data structure called CF-tree. Estivill-Castro and Yang (2004) have proposed an algorithm that remains efficient, generally applicable, multi-dimensional but is more robust to noise and outliers. Jain et al. (1999) have presented an overview of pattern clustering methods from a statistical pattern rec-ognition perspective, with a goal of providing useful advice and references to fun-damental concepts accessible to the broad community of clustering practitioners. In this chapter, we cluster items in multiple databases based on supports of items. Thus, the above algorithms might not be suitable under this framework.

Yang and Shahabi (2005) have proposed an algorithm to determine the sta-tionarity of multivariate time series data for improving the efficiency of many correlation-based data schemes. Matsubara et al. (2012) have proposed *TriMine* algorithm for mining meaningful patterns and forecasting complex time-stamped events. Guil and Marín (2012) have presented a tree-based structure and an algo-rithm, called *TSET—Miner*, for frequent temporal pattern mining from time-stamped datasets. The algorithm is based on mining inter-transaction association, and is characterized by the use of a single tree-based data structure for generation and storage of all frequent sequences discovered through mining. Albanese et al. (2014) have proposed to start with a known set A of activities (both innocuous and dangerous) that authors wish to monitor. In addition, authors wish to identify "unexplained" subsequences in a sequence of observations that are poorly explained by A (e.g., because they may contain occurrences of activities that have never been seen or anticipated before, i.e. they are not in A). Authors formally defined the probability that a sequence of observations was unexplained totally or partially with respect to (w.r.t.) A, and developed algorithms to identify the top-k totally and partially unexplained sequences (w.r.t.) A.

In multi-database environment we have proposed the notion of high frequency itemsets, and an algorithm for synthesizing supports of such itemsets is designed (Adhikari 2013). The existing clustering technique might cluster local frequency items at a low level, since it estimates association among items in an itemset with a low accuracy. Therefore, in this work a new algorithm for clustering local fre-quency items is proposed.

Zhang et al. (2003) designed a local pattern analysis for mining multiple dat-abases. Zhang et al. (2004) studied various issues related to multi-database mining. We have presented some strategies on developing multi-database mining applica-tions (Adhikari et al. 2010).

9.3 A Model of Mining Multiple Transactional Time-Stamped Databases

Consider a multi-branch company that has n branches. All the local transactions are stored locally. Let D_i be the transactional time-stamped database corresponding to i-th branch, $i = 1, 2, ..., n$. Web sites and transactional databases contain a large amount of time-stamped data related to suppliers and/or customers of the organization over time.

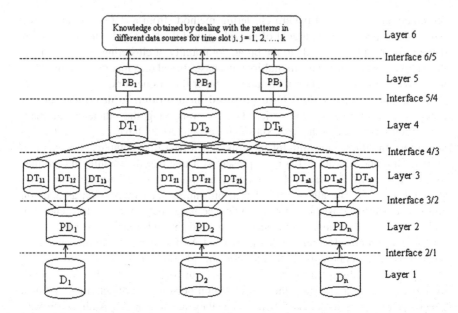

Fig. 9.1 A model of mining global patterns in multiple time-stamped databases

Mining time-stamped data could help business leaders make better decisions by listening to their suppliers or customers via their transactions collected over time (Leonard and Wolfe 2005). We present here a model of mining global patterns in multi-databases over time.

An extended model of local pattern analysis for mining multiple large databases is reported in (Adhikari and Rao 2008a). The limitation of this model is that it may return approximate global patterns. We present here a new model of mining global patterns in multiple time-stamped databases as shown in Fig. 9.1 (Adhikari et al. 2009). It has a set of interfaces and a set of layers. Each interface is a set of operations that produces dataset(s) (or, knowledge) based on the lower layer dataset(s). There are five distinct interfaces of the proposed model of synthesizing global patterns from local patterns. The function of each interface is described below. Interface 2/1 cleans/transforms/integrates/reduces data at the lowest layer. By applying these procedures we get processed database from the original database. In addition, interface 2/1 applies a filtering algorithm on each database for separating relevant data from outlier data. Also, it loads data into the respective data warehouse. At interface 3/2, each processed database PD_i is partitioned into k time databases DT_{ij}, where DT_{ij} is the processed database (if available) for the j-th time slot at the i-th branch, $j = 1, 2, ..., k$; $i = 1, 2, ..., n$. At interface 4/3 the j-th time databases of all branches are merged into a single time database $DT_j, j = 1, 2, ..., k$. A traditional data mining technique could be applied on database DT_j at the interface 5/4, $j = 1, 2, ..., k$. Let PB_j be pattern base corresponding to the time database $DT_j, j = 1, 2, ..., k$. Finally, all the pattern bases are processed for synthesizing knowledge or, making decision at the interface 6/5. Undirected lines in Fig. 9.1 are assumed as directed from bottom to top. The model of mining global patterns over time is efficient, since we form the exact global patterns in multiple databases over time.

At layer 4, we have a collection of all time databases. If any one of these databases is too large to apply a traditional data mining technique then this data mining model would fail. In this situation, we could apply an appropriate sampling technique to reduce the size of a database. Thus, we might develop approximate patterns over time.

9.4 Problem Statement

With reference to Fig. 9.1, let DT_j be the database corresponding to the j-th year, $j = 1$, $2, \ldots, k$. Each of these databases corresponds to a specific period of time. Thus, we could call them as time databases. Each of these time databases is mined using a traditional data mining technique (Han et al. 2000; Agrawal and Srikant 1994). For the specific requirement of this problem, we need to mine only items in the time databases. Let I be the set of all items in these databases. Each itemset X in a database D is associated with a statistical measure, called *support* (Agrawal et al. 1993), denoted by $supp(X, D)$. The support of an itemset is defined as the fraction of transactions containing the itemset. The variation of sales of an item over the time is an important aspect in determining stability of the item. Stable items are useful in many applications. Stable items could be useful to promote sales of other items. Modelling with stable items is more justified than modelling with unstable items.

Let $\mu_{s(x)}(t)$ be the mean support of item x in the database $DT_1, DT_2, \ldots, DT_t$. The value of $\mu_{s(x)}(t)$ is obtained by the following formula:

$$\mu_{s(x)}(t) = \left(\sum_{i=1}^{t} supp(x, DT_i) \times size(DT_i) \right) / \sum_{i=1}^{t} size(DT_i), \ t = 1, 2, \ldots, k. \quad (9.1)$$

Let $\sigma(\mu_{s(x)})$ be the standard deviation of $\mu_{s(x)}(t)$, $t = 1, 2, \ldots, k$. We call $\sigma(\mu_{s(x)})$ the *variation of means* corresponding to support of x. Let $\gamma_{s(x)}(t, t + h)$ be the auto-covariance of $supp(x, DT_t)$ at lag h, $t = 1, 2, \ldots, k - 1$. Thus, $\gamma_{s(x)}(t, t + h)$ is expressed as follows:

$$\gamma_{s(x)}(t, \ t + h) = \frac{1}{k} \sum_{t=1}^{k-h} \left(supp(x, DT_t) - \mu_{s(x)}(k) \right) \left(supp(x, DT_{t+h}) - \mu_{s(x)}(k) \right)$$

$$(9.2)$$

$\sigma(\gamma_{s(x)}(t, t + h))$ be the standard deviation of $\gamma_{s(x)}(t, t + h)$, $h = 1, 2, \ldots, k - 1$. We call this $\sigma(\gamma_{s(x)}(t, \ t + h))$ a *variation of autocovariances* corresponding to support of x. We have chosen standard deviation as a measure of dispersion (Bluman 2006). Standard deviation and mean deviation about mean are relevant measures of dispersion. Unlike the range measure of dispersion these measures take into account a variation due to each support. Skewness, being a descriptive measure of dispersion, is not suitable in this context. Before we define stability of an item, we study the following time series of supports corresponding to an item. In Example 9.2, we compute $\sigma(\mu)$ and $\sigma(\gamma)$ of support series corresponding to different items.

Example 9.2 We continue with Example 9.1. The variations of means and autocovariances of above series are given as follows: (1) $\sigma(\mu) = 0.09342$, $\sigma(\gamma) = 0.01234$, (2) $\sigma(\mu) = 0.00230$, $\sigma(\gamma) = 0.00002$, (3) $\sigma(\mu) = 0.02351$, $\sigma(\gamma) = 0.00039$, (4) $\sigma(\mu) = 0.02114$, $\sigma(\gamma) = 0.00076$, (5) $\sigma(\mu) = 0.0027986$, $\sigma(\gamma) = 0.0000124$. After adding $\sigma(\mu) + \sigma(\gamma)$ for each series, we observe that the value of total variation, $\sigma(\mu) + \sigma(\gamma)$, is the least corresponding to (5), i.e., item x_5.

We define stable items based on the concept of stationary time series (Brockwell and Richard 2002). In finding $\sigma(\mu)$, we first compute a set of means of support values. Then we compute standard deviation of these mean values. Thus, we find standard deviation of a set of fractions. In finding $\sigma(\gamma)$, we first compute a set of autocovariances of support values. Then we compute standard deviation of these autocovariances. The autocovariance of supports is an average of a set of squared fractions. Thus, we find standard deviation of a set of squared fractions. So, $\sigma(\mu) \geq \sigma(\gamma)$. In fact, $\sigma(\gamma)$ is close to 0. Thus, we define our first measure of stability *stable*$_1$ (Adhikari et al. 2009) as follows.

Definition 9.1 An item x is stable if $\sigma(\mu_{s(x)}) \leq \delta$, where δ is a user-defined maximum threshold.

More specifically, we may wish to impose restrictions on both $\sigma(\mu)$ and $\sigma(\gamma)$. We define the second measure of stability *stable*$_2$ (Adhikari et al. 2009) as follows.

Definition 9.2 An item x is stable if $\sigma(\mu_{s(x)}) + \sigma(\gamma_{s(x)}) \leq \delta$, where δ is user-defined maximum threshold.

In Definition 9.2, the expression $\sigma(\mu_{s(x)}) + \sigma(\gamma_{s(x)})$ is the determining factor of stability of an item. We define a degree of variation of an item x as follows.

$$degOfVar(x) = \sigma(\mu_{s(x)}) + \sigma(\gamma_{s(x)}) \qquad (9.3)$$

Higher values of *degOfVar* imply lower degrees of stability of the item. Based on above discussion, we state our problem as follows.

Let D_i and DT_j be the databases corresponding to i-th branch and j-th year of a multi-branch company as depicted in Fig. 9.1 respectively, $i = 1, 2, ..., n$; $j = 1, 2, ..., k$. Each of the time (year) databases has been mined using a traditional data mining technique. Based on the mining results, degree of variation of each item has *been computed as discussed above. Find the best non-trivial partition (if it exists) of the items in D_1, D_2, ..., D_n based on degree of variation of an item.*

A partition (Liu 1985) is linked with a certain type of clustering. A formal definition of a non-trivial partition is given in the following section.

9.5 Clustering Items

The proposed clustering technique is based on the notion of degree of stability of an item. Again, the degree of stability is based on the variations of means and auto-covariances. Thus, the clustering technique requires computing the degree of variation for each item in the databases. Let I be the set of all items in the databases. Given a set of yearly databases, the difference in variations between every pair of items could be expressed by a square matrix, called *difference in variation (diffInVar)*. We define the difference in variation between items x_i and x_j as follows.

$$diffInVar(i,j) = \left| degOfVar(x_i) - degOfVar(x_j) \right|, \text{ for } x_i, x_j \in I. \quad (9.4)$$

In the following example, we compute below the *diffInVar* matrix corresponding to Example 9.1.

Example 9.3 We continue here with Example 9.1. Matrix *diffInVar* is given as follows.

$$diffInVar = \begin{bmatrix} 0 & 0.103 & 0.082 & 0.084 & 0.102 \\ 0.103 & 0 & 0.021 & 0.019 & 0.001 \\ 0.082 & 0.021 & 0 & 0.002 & 0.020 \\ 0.084 & 0.019 & 0.002 & 0 & 0.018 \\ 0.102 & 0.001 & 0.020 & 0.018 & 0 \end{bmatrix}$$

Matrix *diffInVar* is symmetric. We use this matrix for clustering items in multiple databases.

Intuitively, if the difference in variations between two items is close to zero then they may be assigned to the same class. We would like to cluster the items based on this idea. Before clustering the items, we define a class as follows.

Definition 9.3 Let $I = \{i_1, i_2, ..., i_p\}$ be the set of items. A class formed at the level of difference in variation of α is expressed as follows.

$$class(I, \alpha) = \begin{cases} X : X \subseteq I, \ |X| \geq 2, \ and \ degOfVar(x_1, x_2) \leq \alpha, \ for \ x_1, x_2 \in X \\ X : X \subseteq I, \ |X| = 1 \end{cases}$$

Based on the above definition of a class, we introduce a concept of clustering as follows.

Definition 9.4 Let $I = \{i_1, i_2, \ldots, i_p\}$ be the set of items. Let $\pi(I, \alpha)$ be a clustering of items in I at the level of difference in variation α. Then, $\pi(I, \alpha) = \{X : X \in \rho(I),$ and X is a class$(I, \alpha)\}$, where $\rho(I)$ is the power set of I.

During the clustering process we may like to impose a restriction that each item belongs to at least one class. This restriction makes a clustering complete. We define a complete clustering as follows.

Definition 9.5 Let $I = \{i_1, i_2, \ldots, i_p\}$ be the set of items. Let $\pi(I, \alpha) = \{C_1(I, \alpha), C_2(I, \alpha), \ldots, C_m(I, \alpha)\}$, where $C_k(I, \alpha)$ is the k-th class of the clustering π, for $k = 1, 2, \ldots, m$. π is complete, if $\cup_{k=1}^{m} C_k(I, \alpha) = I$.

In a complete clustering, two classes may have common items. We may be interested in finding out a cluster containing mutually exclusive classes. A mutually exclusive cluster is defined as follows.

Definition 9.6 Let $I = \{i_1, i_2, \ldots, i_p\}$ be the set of items. Let $\pi(I, \alpha) = \{C_1(I, \alpha), C_2(I, \alpha), \ldots, C_m(I, \alpha)\}$, where $C_k(I, \alpha)$ is the k-th class of the clustering π, $k = 1, 2, \ldots, m$. π is mutually exclusive if $C_i(I, \alpha) \cap C_j(I, \alpha) = \phi$, for i $\neq$ j, and $1 \leq i, j \leq m$.

We may be interested in finding such a mutually exclusive and complete cluster. A partition of a set of items I is defined as follows.

Definition 9.7 Let $\pi(I, \alpha)$ be a mutually exclusive and complete clustering of a set of items I at the level of difference in variation α. $\pi(I, \alpha)$ is called a non-trivial partition if $1 < |\pi| < m$.

A partition is a cluster. But a cluster is not necessarily a partition. In the next section, we find the best non-trivial partition (if it exists) of a set of items.

The items in a class are similar with respect to their variations. We are interested in the classes of a partition where the variations of items are less. The items in these classes are useful in devising strategies for the company. Thus, we define an average degree of variation adv, of a class, as follows.

Definition 9.8 Let C be a class of partition π. Then, $adv(C|\pi) = \frac{1}{|C|} \Sigma_{x \in C} degOfVar(x)$.

9.5.1 Finding the Best Non-trivial Partition

With reference to Example 9.1, we arrange all non-zero and distinct values of *diffInVar* in non-decreasing order for finding all the non-trivial partitions, $1 \le i < j \le 5$. The arranged values of *diffInVar* are given as follows: 0.001, 0.002, 0.018, 0.019, 0.020, 0.021, 0.082, 0.084, 0.102, 0.103. We get two non-trivial partitions at $\alpha = 0.001$, and 0.002. The partitions are given as follows: $\pi^{0.001} = \{\{x_1\}, \{x_2, x_5\}, \{x_3\}, \{x_4\}\}$, and $\pi^{0.002} = \{\{x_1\}, \{x_2, x_5\}, \{x_3, x_4\}\}$. We observe that at different levels of α we come up with different partitions. We would like to determine the best partition among these partitions. The best partition is the one which maximizes the intra-class variation and minimizes the inter-class similarity. Intra-class variation and inter-class similarity are defined as follows.

Definition 9.9 The intra-class variation *intra-var*, of a partition π at the level α is defined as follows:

$$intra-var(\pi^{\alpha}) = \sum_{k=1}^{|\pi|} \Sigma_{x_i, x_j \in C_k; \, x_i < x_j} \left| degOfVar(x_i) - degOfVar(x_j) \right|$$

Definition 9.10 The inter-class similarity *inter-sim*, of a partition π at the level α is defined as follows:

$$inter-sim(\pi^{\alpha}) = \Sigma_{c_p, c_q \in \pi; \, p < q} \Sigma_{x_i \in C_p, \, x_j \in C_q} minimum\{degOfVar(x_i), \, degOfVar(x_j)\}.$$

The best partition among a set of partitions is selected on the basis of goodness of a partition. Goodness measure, *goodness*, of a partition is defined as follows.

Definition 9.11 The goodness of a partition π at level α is defined as follows: $goodness(\pi^{\alpha}) = intra-var(\pi^{\alpha}) + inter-sim(\pi^{\alpha}) - |\pi^{\alpha}|$, where $|\pi^{\alpha}|$ is the number of classes of π.

We have subtracted the term $|\pi^{\alpha}|$ from the sum of intra-class variation and inter-class similarity to remove the bias of goodness value of a partition. Better partition is obtained at higher goodness value. We would like to partition the set of items in Example 9.1 using above goodness measure.

Example 9.4 With reference to Example 9.2, we calculate goodness value of each of the non-trivial partitions.

$intra\text{-}var(\pi^{0.001}) = 0.001$, $inter\text{-}sim(\pi^{0.001}) = 0.081$, and $|\pi^{0.001}| = 4$. Thus, $goodness(\pi^{0.001}) = -3.916$.
$intra\text{-}var(\pi^{0.002}) = 0.003$, $inter\text{-}sim(\pi^{0.002}) = 0.06$, and $|\pi^{0.002}| = 3$. Thus, $goodness(\pi^{0.002}) = -2.937$.

The goodness value corresponding to the partition $\pi^{0.002}$ is the largest one. Thus, the partition $\pi^{0.002}$ is the best among the non-trivial partitions. Let us return back to Example 9.1. There are five series of supports corresponding to five items. Based on variation among the supports in a series, we could partition the series as follows: {series 1}, {series 2, series 5}, {series 3, series 4}. Hence, we get the following partition: $\{x_1\}$, $\{x_2, x_5\}$, $\{x_3, x_4\}$. The proposed clustering technique also identifies the same partition as the best partition. Thus, it verifies the correctness of the proposed clustering technique. $adv\ (\{x_1\}|\pi^{0.002}) = 0.105$, $adv\ (\{x_2, x_5\}|\pi^{0.002}) = 0.0025$ and adv $(\{x_3, x_4\}|\pi^{0.002}) = 0.022$. We find that the average degree of variation of $\{x_2, x_5\}$ is the least among the classes of $\pi^{0.002}$. Thus, the items x_2 and x_5 are most suitable among all the items in the given databases for making strategies of the company.

We design an algorithm for finding best non-trivial partition of items in multiple databases. First we describe different data structures used in designing an algorithm for determining the best partition of items. For each item there are k supports corresponding to k different years. We maintain $m \times k$ supports for m items in array *supports*. The i-th row of *supports* stores supports corresponding to i-th item for k years, $i = 1, 2, ..., m$. Let *means* be a two dimensional array such that the i-th row stores means of supports corresponding to different years for i-th item, $i = 1, 2, ..., m$. Let *autocovariances* be a two dimensional array such that the i-th row stores autocovariances of supports corresponding to different lags for i-th item, $i = 1, 2, ..., m$. For year j, we compute mean value of supports for year 1 to j. Thus, we get different mean values for different years. Let *stdDevMeans* be the standard deviation of these mean values. For year j, we also compute autocovariances of supports for year 1 to j at different lags. Thus, we get different autocovariances for different lags corresponding to a year. Let *stdDevAutocovars* be the standard deviation of these autocovariances. The degrees of variation of different items are stored in array *degInVar*. Variable S is a one dimensional array containing $^{m}C_2$ difference in variations. *adv* is a one dimensional array which stores the average degree of variation for the items in each class. The algorithm is presented below (Adhikari et al. 2009).

Algorithm 9.1 Find best non-trivial partition (if it exists) of items in multiple databases.
procedure B*estPartition* (*m*, *supports*)
Inputs:
m: number of items
supports: array of supports of different items corresponding to different years
Outputs:
Best non-trivial partition (if it exists) of items in multiple databases
01: **for** $i = 1$ to m **do**
02:    compute *means*(*i*) using formula (9.1) at different years;
03:    **let** *stdDevMeans* = standard deviation of mean values for different years;
04:    compute *autocovariance*(*i*) using formula (3.2) at different time lags;
05:    **let** *stdDevAutocovars* = standard deviation of autocovariances;
06:    compute *degOfVar*(*i*) = *stdDevMeans* + *stdDevAutocovar*;
07: **end for**
08: **for** *row* = 1 to *m* **do**
09:    **for** *col* = (*row* + 1) to *m* **do**
10:       compute *diffInVar*(*row*, *col*) using formula (9.3);
11:    **end for**
12: **end for**
13: sort distinct elements in the upper triangle of *diffInVar* in non-decreasing order into *S*;
14: **let** $k = 1$; **let** *maxGoodness* = -9999; $\pi = \phi$;
15: **while** ($k \leq |S|$) **do**
16:    **let** *curRow* = 1; **let** *curClass* = 1;
17:    **for** $i = 2$ to m **do**
18:     *classLabel*(*i*) = 0;
19:    **end for**
20:    **let** *classLabel*(1) = 1;
21:    **let** *curDiffVar* = *S*(*k*);
22:    **for** *col* = *curRow* + 1 to *m* **do**
23:       **if** (*diffInVar*(*curRow*, *col*) ≤ *curDiffVar*) **then**
24:          **if** (*classLabel* (*col*) = 0) **then**
25:           *classLabel*(*col*) = *curClass*;
26:          **else if** (*classLabel* (*col*) ≠ *curClass*) **then**
27:             partition does not exist at this level;
28:              go to line 49;
29:          **end if**
30:       **end if**
31:    **end for**
32:    increased *curRow* by 1;
33:    **if** (*classLabel*(*curRow*) = 0) **then**
34:       increased *curClass* by 1;
35:       *classLabel*(*curRow*) = *curClass*;
36:    **else** *curclass* = *classLabel*(*curRow*);
37:    **end if**
38:    **if** (*curRow* ≤ *m*) go to line 22; **end if**
39:    **let** $j = 0$;
40:    **while** ((*classLabel*(*j*) ≠ 0) **and** ($j < m$)) **do**
41:       increase *j* by 1;
42:    **end while**
43:    **if** ($j = m + 1$) **then**
44:       **if** (*maxGoodness* < goodness value of current partition) **then**
45:          *maxGoodness* = goodness value of current partition;
46:          store current partition into π;
47:       **end if**
48:    **end if**
49:    increase *k* by1;
50: **end while**
51: return π;
end procedure

In this paragraph, we explain the Algorithm 9.1. It computes *degreeOfVar* for all items using lines 1–7. Matrix *diffInVar* is constructed using lines 8–12. We check the existence of partition at every value in S. We start checking partition by assigning the first item to first class i.e., *curClass* = 1. Also, clustering process is performed row by row, starting from row number 1. At the i-th row, all the items greater than i are classified. During this process, if a labelled item gets another label then we conclude that partition does not exist at the current level. After increasing the current row by 1, we check the class label corresponding to current row. Each row corresponds to an item in the database. If the current row is not labelled yet then the class label is increased by 1. If the goodness value of the current partition is less than the *goodness* value of another partition then the current partition is ignored.

Lemma 9.1 *Algorithm 9.1 executes in $O(m^4)$ time.*

Proof Line 2 takes $O(k)$ time to compute *means(i)*, for some $i = 1, 2, ..., m$. Also, line 3 takes $O(k)$ time to compute standard deviation of mean values. To compute formula (9.2), we require $O(k)$ time. Thus, line 4 takes $O(k^2)$ time. In line 5, we compute standard deviation of $k - 1$ autocovariance values. Thus, line 5 takes $O(k^2)$ time. The *for-loop* in lines 1–7 repeat m times. Thus, the *for-loop* in lines 1–7 take $O(m \times k^2)$ time. For computing *diffInVar* at a given row and column, it takes $O(1)$ time. Thus, lines 8–12 take $O(m^2)$ time. There are maximum $^{m-1}C_2$ elements in the upper triangle of *diffInVar*. Thus, line 13 takes $O(m^2 \times \log(m))$ time. The *while-loop* at line 15 repeats maximum $^{m-1}C_2$ times. Each of the loops at lines 17, 22, and 40 takes $O(m)$ time. To store a partition it takes $O(m)$ time. To compute goodness value for a particular partition, it takes $O(m^2)$ time. Thus, the lines 15–50 take $O(m^4)$ time. The time complexity of *bestPartition* algorithm is $O(m^4)$.   □

In finding stable items in multiple databases, a class having minimal average degree of variation in the best partition might not be a best class at a given degree of stability. In many applications, we may need to find stable items at a given degree of stability. In this case, it might not be a requirement that the stable items need to form a class of a non trivial partition. Thus, the question of finding a partition might not always arise. To find such a class we follow a different approach.

9.5.2 Finding a Best Class

Before finding a best class, we first define the concept of the best class.

Definition 9.12 Let C be a class of items. C is called a best class at the level of difference in variation α if (i) $|degOfVar(x) - degOfVar(y)| \leq \alpha$, for $x, y \in C$, (ii) *adv* (C) is the minimum among all classes of maximal size, and (iii) C has a maximal size.

In Lemma 9.2, we show that it might not be possible to find two classes of maximal size having the same average degree of variation.

Lemma 9.2 *Best class is unique.*

Proof Let $x_1, x_2, \ldots, x_m$ be the items sorted in a non-decreasing degree of variation. Thus the item x_1 has the maximum stability, and the item x_m has the minimum stability. At level α, let the stabilities of items $x_1, x_2, \ldots, x_k$ be less than or equal to α, and the stabilities of items $x_{k+1}, x_{k+2}, \ldots, x_m$ be greater than α, for $1 \leq k \leq m$. The best class has least average degree of variation. Also, the difference in variation of two items in the class is less than or equal to α. Thus, $\{x_1, x_2, \ldots, x_k\}$ forms the best class. We are not concerned whether it becomes a member of a partition. $adv(\{x_1, x_2, \ldots, x_k\})$ is the minimum, and hence the best class is unique. $\square$

We might be interested in finding best class of items in multiple databases. We use array *class* to hold the best class of items. In the following, we provide an algorithm in finding best class of items in multiple databases.

Algorithm 9.2. Find the best class of items in multiple databases induced by stability.
procedure *BestClass* (*m*, α, *supports*)
Inputs:
m: number of items
α: level of degree of variation
supports: array of supports of different items corresponding to different years
Outputs:
Best class of items in multiple databases
01: perform lines 01 – 07 of Algorithm 9.1;
02: sort array *degOfVar* in non-decreasing order;
03: **let** *class* (1) = *degOfVar*(1); **let** *count* = 1; **let** *avgVar* = 0;
04: **for** $i = 2$ to *m* **do**
05:    *class* (*i*) = -1;
06: **end for**
07: **for** i = 2 to *m* **do**
08:    **if** ((*degOfVar*(*i*) – *degOfVar*(1)) $\leq \alpha$)
09:        *class* (*i*) = *degOfVar*(*i*);
10:        increase *count* by 1;
11:        *avgVar* = *avgVar* + *degOfVar*(*i*);
12: **end for**
13: *avgVar* = *avgVar* / *count*;
14: return (*class*, *count*, *avgVar*);
end procedure

Let us elaborate on the individual lines of the above algorithm. We compute degree of variations for all items using lines 1–7 and store them in array *degreeOfVar* in a non-decreasing order. The best class would contain the first item of *degreeOfVar*. The item with least *degreeOfVar* is assigned to *class* 1. An item *i* is included in the best class if (*degOfVar*(*i*) − *degOfVar*(1)) $\leq \alpha$. Algorithm 9.2 returns best class *class*, the number of items in the best class *count*, and the average degree of variation (*avgVar*) of the best class.

Lemma 9.3 *Algorithm 9.2 executes in maximum* $\{O(m \times k^2), O(m \times \log(m))\}$ *time.*

Proof Line 1 executes in $O(m \times k^2)$ time [Lemma 9.2], where k is the number of years. There are two for loops in Algorithm 9.2 apart from loops placed in line 1. Each of these loops executes in $O(m)$ time. Line 2 takes $O(m \times \log(m))$ time. Thus, the lemma follows. $\square$

9.6 Experiments

We present the experimental results using two real datasets *mushroom* (Frequent Itemset Mining Dataset Repository 2004), and *ecoli* (UCI ML repository). Dataset *ecoli* is a subset of *ecoli database* and it has been processed for the purpose of conducting experiments. *Random-68* is a synthetic database, which has been generated for the purpose of conducting experiments.

Let *DB*, *NT*, *ALT*, *AFI*, and *NI* denote database, the number of transactions, average length of a transaction, average frequency of an item, and number of items respectively. We present some characteristics of these datasets in Table 9.1. Each dataset has been divided into ten databases, called input databases, for the purpose of conducting experiments. The input databases obtained from *mushroom* and *ecoli* are named as M_i, and E_i, $i = 0, 1, \ldots, 9$. We present some characteristics of the input databases in Table 9.2. In Table 9.3, we present top 10 stable items encountered in multiple databases.

The best partition of items in *mushroom* dataset is obtained at level 0.25. The intra variation, inter similarity, and goodness value are 413.60, 377.58, and 789.18, respectively. The best partition contains two classes. The best class is given as follows: {1, 2, 3, 4, 5, 6, 7, 8, 9, 10, 11, 12, 13, 14, 15, 16, 17, 18, 19, 20, 21, 22, 23, 24, 25, 26, 27, 28, 29, 30, 31, 32, 33, 34, 35, 36, 37, 38, 39, 40, 41, 42, 43, 44, 45, 46, 47, 48, 49, 50, 51, 52, 53, 54, 55, 57, 58, 59, 60, 61, 62, 63, 64, 65, 66, 67, 68, 69, 70, 71, 72, 73, 74, 75, 76, 77, 78, 79, 80, 81, 82, 83, 84, 85, 86, 87, 88, 89, 90, 91, 92, 93, 94, 95, 96, 97, 98, 99, 100, 101, 102, 103, 104, 105, 106, 107, 108, 109, 110, 111, 112, 113, 114, 115, 116, 117, 118, 119}. It has average degree of variation 0.06.

The best partition of items in *ecoli* dataset is obtained at level 0.06. The amount of intra variation, inter similarity, and goodness value are 44.18, 185.61, and 227.79 respectively. The best partition contains two classes. The best class is given as follows: {0, 1, 3, 4, 5, 6, 7, 8, 10, 11, 12, 14, 15, 16, 17, 18, 19, 20, 21, 22, 23, 24, 25, 26, 27, 28, 29, 30, 31, 32, 33, 34, 36, 37, 38, 39, 40, 41, 43, 45, 46, 47, 48, 49,

Table 9.1 Dataset characteristics

Dataset	*NT*	*ALT*	*AFI*	*NI*
mushroom	8,124	24.00	1,624.80	120
ecoli	336	7.00	25.84	91
random-68	3,000	5.46	280.99	68

Table 9.2 Time database characteristics

DB	NT	ALT	AFI	NI	DB	NT	ALT	AFI	NI
M_0	812	24.00	295.27	66	M_5	812	24.00	221.45	88
M_1	812	24.00	286.59	68	M_6	812	24.00	216.53	90
M_2	812	24.00	249.85	78	M_7	812	24.00	191.06	102
M_3	812	24.00	282.43	69	M_8	812	24.00	229.27	85
M_4	812	24.00	259.84	75	M_9	816	24.00	227.72	86
E_0	33	7.00	4.62	50	E_5	33	7.00	3.92	59
E_1	33	7.00	5.13	45	E_6	33	7.00	3.50	66
E_2	33	7.00	5.50	42	E_7	33	7.00	3.92	59
E_3	33	7.00	4.81	48	E_8	33	7.00	3.40	68
E_4	33	7.00	3.40	68	E_9	39	7.00	4.55	60
R_0	300	5.59	28.68	68	R_5	300	5.14	26.68	68
R_1	300	5.42	28.00	68	R_6	300	5.51	28.35	68
R_2	300	5.36	27.65	68	R_7	300	5.50	28.34	68
R_3	300	5.54	28.46	68	R_8	300	5.54	28.47	68
R_4	300	5.53	28.38	68	R_9	300	5.48	28.24	68

Table 9.3 Top 10 stable items in multiple databases

mushroom		ecoli		random-68	
Item	degOfVar	Item	degOfVar	Item	degOfVar
85	0.0000	1	0.0013	42	0.0012
8	0.0002	99	0.0018	41	0.0018
12	0.0003	91	0.0023	37	0.0023
75	0.0003	4	0.0025	67	0.0027
89	0.0003	94	0.0036	11	0.0028
62	0.0009	15	0.0039	45	0.0029
22	0.0010	12	0.0039	18	0.0030
20	0.0011	19	0.0040	56	0.0031
82	0.0012	3	0.0040	3	0.0031
33	0.0014	10	0.0044	28	0.0031

50, 51, 52, 53, 54, 55, 56, 57, 58, 59, 60, 61, 62, 63, 64, 65, 66, 67, 69, 70, 71, 72, 73, 74, 75, 76, 77, 78, 79, 80, 81, 82, 83, 84, 85, 86, 87, 88, 89, 90, 91, 92, 94, 99, 100}. It has average degree of variation 0.02.

The best partition of items in *random-68* dataset is obtained at the level 0.02. The amount of intra variation, inter similarity, and goodness value are 4.63, 18.61, and 21.24, respectively. The best partition contains two classes. The best class is given as follows: {1, 2, 3, 5, 6, 7, 8, 9, 10, 11, 12, 13, 14, 15, 16, 17, 18, 19, 20, 21, 22, 23, 24, 25, 26, 27, 28, 29, 30, 31, 32, 33, 34, 35, 36, 37, 38, 39, 40, 41, 42, 43, 44, 45, 46, 47, 48, 49, 50, 51, 52, 53, 54, 55, 56, 57, 58, 59, 60, 61, 62, 63, 64, 65, 66, 67, 68}. It has average degree of variation 0.01.

Table 9.4 Five best classes in multiple databases (*mushroom*)

α	Items	adv
0.0002	{85, 8}	0.00009
0.0003	{85, 8, 12, 75, 89}	0.00021
0.0010	{85, 8, 12, 75, 89, 62}	0.00033
0.0012	{85, 8, 12, 75, 89, 62, 22, 20}	0.00051
0.0014	{85, 8, 12, 75, 89, 62, 22, 20, 82}	0.00059

Table 9.5 Five best classes in multiple databases (*ecoli*)

α	Items	adv
0.0020	{1, 99, 91, 4}	0.00196
0.0025	{1, 99, 91, 4, 94}	0.00230
0.0030	{1, 99, 91, 4, 94,15, 12, 19, 3}	0.00303
0.0035	{1, 99, 91, 4, 94, 15, 12, 19, 3, 10, 6}	0.00331
0.0050	{1, 99, 91, 4, 94, 15, 12, 19, 3, 10, 6, 18}	0.00354

In Tables 9.4, 9.5 and 9.6 we present five best classes and their average degree of variations for a given value of α for each database.

We have analyzed execution time with respect to number of data sources. We observe in Figs. 9.2, 9.3 and 9.4 that the execution time increases as the number of data sources increases.

Table 9.6 Five best classes in multiple databases (*random-68*)

α	Items	adv
0.0010	{42, 41}	0.00152
0.0015	{42, 41, 37}	0.00178
0.0017	{42, 41, 37, 67, 11, 45}	0.00229
0.0020	{42, 41, 37, 67, 11, 45, 18, 56, 3, 28}	0.00261
0.0022	{42, 41, 37, 67, 11, 45, 18, 56, 3, 28, 7, 53}	0.00271

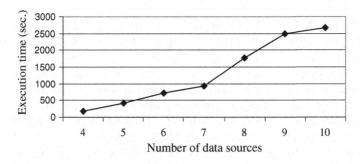

Fig. 9.2 Execution time versus number of data sources for *mushroom*

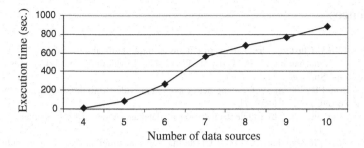

Fig. 9.3   Execution time versus number of data sources for *ecoli*

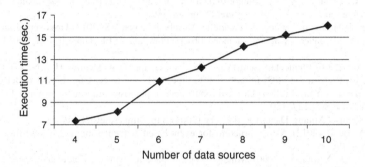

Fig. 9.4   Execution time versus number of data sources for *random-68*

9.7   Conclusions

Many organizations deal with a large number of items collected for a long period of time. In this scenario, data analyses are carried out for the data items dealt with. Studying stability of an item is an important issue, since there are a series of time databases. For the purpose of conducting experiments, we have considered yearly databases. Thus, the notion of stability of an item is introduced here. Stable items are useful for modelling various strategies of an organization. We design an algorithm for clustering items in multiple databases based on degree of stability. Afterwards we have introduced the notion of best class in a cluster. We have designed an algorithm to find the best class in a cluster. The best class is obtained by considering an average degree of variation in a class. A class having the least average degree variation is the best one. We have conducted experiments on three datasets, and provided detailed data analyses related to stability of an item.

References

Adhikari A (2013) Clustering local frequency items in multiple databases. Inf Sci 237:221–241

Adhikari A, Ramachandrarao P, Pedrycz W (2010) Developing multi-databases mining applications. Springer, London

Adhikari A, Rao PR (2008a) Synthesizing heavy association rules from different real data sources. Pattern Recogn Lett 29(1):59–71

Adhikari A, Rao PR (2008b) Efficient clustering of databases induced by local patterns. Decis Support Syst 44(4):925–943

Adhikari J, Rao PR, Adhikari A (2009) Clustering items in different data sources induced by stability. Int Arab J Inf Technol 6(4):394–402

Agrawal R, Imielinski T, Swami A (1993) Mining association rules between sets of items in large databases. In: Proceedings of ACM SIGMOD conference management of data, pp 207–216

Agrawal R, Srikant R (1994) Fast algorithms for mining association rules. In: Proceedings of 20th very large databases (VLDB) conference, pp 487–499

Albanese M, Molinaro C, Persia F, Picariello A, Subrahmanian VS (2014) Discovering the top-k unexplained sequences in time-stamped observation data, IEEE Trans Knowl Data Eng 26 (3):577–594

Bluman AG (2006) Elementary statistics: a step by step approach. Mcgraw Hill, New York

Brockwell J, Richard D (2002) Introduction to time series and forecasting. Springer, Berlin

Estivill-Castro V, Yang J (2004) Fast and robust general purpose clustering algorithms. Data Min Knowl Disc 8(2):127–150

Frequent Itemset Mining Dataset Repository (2004) http://fimi.cs.helsinki.fi/data

Guil F, Marín R (2012) A tree structure for event-based sequence mining. Knowl-Based Syst 35:186–200

Han J, Pei J, Yiwen Y (2000) Mining frequent patterns without candidate generation. In: Proceedings of ACM SIGMOD conference management of data, pp 1–12

Jain AK, Murty MN, Flynn PJ (1999) Data clustering: a review. ACM Comput Surv 31 (3):264–323

Leonard M, Wolfe B (2005) Mining transactional and time series data. In: Proceedings of SUGI 30, paper 080–30

Liu B, Ma Y, Lee R (2001) Analyzing the interestingness of association rules from the temporal dimension. In: Proceedings of IEEE international conference on data mining, pp 377–384

Liu CL (1985) Elements of discrete mathematics. McGraw-Hill, New York

Matsubara Y, Sakurai Y, Faloutsos C, Iwata T, Yoshikawa M (2012) Fast mining and forecasting of complex time-stamped events. In: Proceedings of KDD conference, pp 271–279

Tan PN, Kumar V, Srivastava J (2002) Selecting the right interestingness measure for association patterns. In: Proceedings of SIGKDD conference, pp 32–41

UCI ML Repository Content Summary. http://www.ics.uci.edu/~mlearn/MLSummary.html

Wu X, Zhang C, Zhang S (2005) Database classification for multi-database mining. Inf Syst 30 (1):71–88

Yang K, Shahabi C (2005) On the stationarity of multivariate time series for correlation-based data. In: Proceedings of ICDM, pp 805–808

Zhang T, Ramakrishnan R, Livny M (1997) BIRCH: a new data clustering algorithm and its applications. Data Min Knowl Disc 1(2):141–182

Zhang S, Wu X, Zhang C (2003) Multi-database mining. IEEE Comput Intell Bull 2(1):5–13

Zhang S, Zhang C, Wu X (2004) Knowledge discovery in multiple databases. Springer, Berlin

Chapter 10
Synthesizing Some Extreme Association Rules from Multiple Databases

The model of local pattern analysis provides sound solutions to many multi-database mining problems. In this chapter, we discuss different types of extreme association rules in multiple databases viz., heavy association rule, high-frequency association rule, low-frequency association rule, and exceptional association rule. Also, we show how one can apply the model of local pattern analysis systematically and effectively. For this purpose, an extended model of local pattern analysis is presented. The extended model has been applied to mine heavy association rules in multiple databases. Also, we justify why the extended model works more effectively. An algorithm for synthesizing heavy association rule in multiple databases is given. Furthermore, we show that the algorithm identifies whether a heavy association rule is high-frequency rule or exceptional rule. Experimental results are provided for both synthetic and real-world datasets and a detailed error analysis is carried out. Furthermore, we present a comparative analysis by contrasting the proposed algorithm with some of those reported in the literature. This analysis is completed by taking into consideration the criteria of execution time and average error.

10.1 Introduction

In many decision support applications, an approximate knowledge stemming from multiple large databases might result in significant savings when being used in decision-making. Hence the model of local pattern analysis (Zhang et al. 2003) used for mining multiple large databases can constitute a viable solution. In this chapter, we show how one could apply the model of local pattern analysis in a systematic and efficient manner for mining different types of extreme association rules in multiple databases.

The analysis of relationships existing among variables is a fundamental task positioned at the heart of many data mining problems. Mining association rules has received a lot of attention to the data mining community. For instance, an association rule expresses how the purchase of a group of items, called an *itemset*, affects

© Springer International Publishing Switzerland 2015   173
A. Adhikari and J. Adhikari, *Advances in Knowledge Discovery in Databases*,
Intelligent Systems Reference Library 79, DOI 10.1007/978-3-319-13212-9_10

the purchase of another group of items. Association rule mining is based on two measures quantifying the quality of the rules, that is support (*supp*) and confidence (*conf*); see Agrawal et al. (1993). An association rule r in database *DB* can be expressed symbolically as $X \rightarrow Y$, where X and Y are two itemsets in database *DB*. It expresses an association between the itemsets X and Y, called the antecedent and consequent of r, respectively. The meaning attached to this type of implication could be clarified as follows. If the items in X are purchased by a customer then the items in Y are likely to be purchased by the same customer at the same time. The interestingness of an association rule could be expressed by its support and confidence. Let E be a Boolean expression defined on the items in *DB*. Support of E in *DB* is defined as the fraction of transactions in *DB* such that the Boolean expression E is true for each of these transactions. We denote the support of E in *DB* as $supp_a(E, DB)$. The support and confidence of association rule r is expressed as follows:

$$supp_a(r, DB) = supp_a(X \cap Y, \ DB), \text{ and}$$
$$conf_a(r, DB) = supp_a(X \cap Y, DB)/supp_a(X, DB)$$

Later, we shall be dealing with synthesized support and synthesized confidence of an association rule. Thus, it is required to differentiate between actual support/confidence with synthesized support/confidence of an association rule. The subscript a used in the notation of support/confidence for referring the actual support/confidence of an association rule. On the other hand, the subscript s in the notation of support/confidence is used to refer synthesized support/confidence of an association rule. A synthesized support/confidence of an association rule might depend on the technique applied to synthesizing support/confidence. We present here a technique for synthesizing support and confidence of an association rule in multiple databases. We say that an association rule r in database *DB* is *interesting* if the following relationships hold

$$supp_a(r, DB) \geq minimum\ support\ (\alpha), \text{ and}$$
$$conf_a(r, DB) \geq minimum\ confidence\ (\beta)$$

The values of the parameters α and β are user-defined. The collection of association rules extracted from a database for the given values of α and β is called a *rulebase*.

In a multi-database mining environment, often one needs to handle multiple large databases. As a result, one may come across various types of patterns. Association rule mining (Agrawal et al. 1993) is an important and popular data mining task. It has many applications to different areas of computing (Zhang and Wu 2011). In this chapter, we are interested in mining association rules in multiple databases that are extreme in some sense. These association rules are induced by different data sources, and thus, these rules are specific to multi-database mining environment. An association rule in multiple databases becomes more interesting if it possesses higher support and higher confidence. This type of association rules is called heavy association rules (Adhikari and Rao 2008). Sometimes the number of times an association rule gets reported from local databases becomes an interesting issue. In the context of multiple databases, an association rule is called high-frequency rule (Wu and Zhang 2003) if it is extracted from many databases. In this context an association rule is called low-frequency rule if it is extracted from a few databases (Adhikari and Rao 2008). Some association rules possess high support but have been extracted from a few databases only. These association rules are called exceptional association rules (Adhikari and Rao 2008). Many corporate decisions could be influenced by these types of extreme association rules in multiple databases. Thus, it is important to mine them. In the next section, we present different extreme association rules, and then we present a model of mining such association rules.

The chapter is organized as follows. We discuss some "extreme" types of association rules (Sect. 10.2). In Sect. 10.3, we present the problem formally. An extended model of local pattern analysis is presented in Sect. 10.4. We discuss related work in Sect. 10.5. We present an algorithm for synthesizing different extreme association rules in Sect. 10.6. In this section, we have also defined error of the experiment. We present experimental result in Sect. 10.7. Finally, some conclusions are provided in Sect. 10.8.

10.2 Some Extreme Types of Association Rule in Multiple Databases

Consider a large company with transactions originating from n branches. Let D_i be the database corresponding to the i-th branch of this multi-branch company, $i = 1, 2, \ldots, n$. Furthermore let D be the union of all branch databases. First, we define a heavy association rule in a single database. Afterwards, we define a heavy association rule in multiple databases.

Definition 10.1 An association rule r in database DB is heavy if $supp_a(r, DB) \geq \mu$, and $conf_a(r, DB) \geq v$, where μ ($>\alpha$) and v ($>\beta$) are the user-defined thresholds of high-support and high-confidence for identifying heavy association rules in DB, respectively.

If an association rule is heavy in a local database then it might not be heavy in D. An association rule in D might have different statuses in different local databases. For example, it might be a heavy association rule, or an association rule, or a suggested association rule (defined later), or absent in a local database. Thus, we need to synthesize an association rule for determining its overall status in D. The method of synthesizing an association rule is discussed in Sect. 10.6. After synthesizing an association rule, we get its synthesized support and synthesized confidence in D. Let $supp_s(r, DB)$ and $conf_s(r, DB)$ denote synthesized support and synthesized confidence of association rule r in DB, respectively. A heavy association rule in multiple databases is defined as follows:

Definition 10.2 Let D be the union of all local databases. An association rule r in D is heavy if $supp_s(r, D) \geq \mu$, and $conf_s(r, D) \geq v$, where μ and v are the user-defined thresholds of high-support and high-confidence used for identifying heavy association rules in D, respectively.

Apart from synthesized support and synthesized confidence of an association rule, the frequency of an association rule is an important issue in multi-database mining. We define *frequency* of an association rule as the number of extractions of the association rule from different databases. If an association rule is extracted from k out of n databases then the frequency of the association rule is k, $0 \leq k \leq n$. An association rule may be high-frequency rule or, low-frequency rule, or neither high-frequency rule nor low-frequency rule in multiple databases. We could arrive in such a conclusion only if we have user-defined thresholds of low-frequency (γ_1) and high-frequency (γ_2) of an association rule, for $0 < \gamma_1 < \gamma_2 \leq 1$. A low-frequency association rule is extracted from less than $n \times \gamma_1$ databases. On the other hand, a high-frequency association rule is extracted from at least $n \times \gamma_2$ databases. In the context of multi-database mining using local pattern analysis, we define a high-frequency association rule and a low-frequency association rule as follows:

Definition 10.3 Let an association rule be extracted from k out of n databases. Then the association rule is low-frequency rule if $k < n \times \gamma_1$, where γ_1 is the user-defined threshold of low-frequency.

Definition 10.4 Let an association rule be extracted from k out of n databases. Then the association rule is high-frequency rule if $k \geq n \times \gamma_2$, where γ_2 is the user-defined threshold of high-frequency.

While synthesizing heavy association rules in multiple databases, it may be worth noting some other attributes of a synthesized association rule. For example, high-frequency, low-frequency, and exceptionality are interesting as well as important attributes of a synthesized association rule. We have already defined high-frequency association rule and low-frequency association rule in multiple databases. We now define an exceptional association rule in multiple databases:

Definition 10.5 A heavy association rule in multiple databases is exceptional if it is a low-frequency rule.

It may be worth contrasting a heavy association rule, a high-frequency association rule with an exceptional association rule in multiple databases.

- An exceptional association rule is also a heavy association rule.
- A high-frequency association rule is not an exceptional association rule, and vice versa.
- A high-frequency association rule is not necessarily be a heavy association rule.
- There may exist heavy association rules that are neither high-frequency rule nor exceptional rule.

The goal of this chapter is to extract these extreme association rules from multiple databases. For this purpose, we present an extended model of local pattern analysis.

10.3 Problem Statement

In the previous section, we learnt different types of extreme association rules. We have observed some difficulties in extracting different extreme association rules in the union of all branch databases by employing a traditional data mining technique (Adhikari et al. 2010). Therefore, we synthesize different extreme association rules by using patterns in branch databases. Let D be the union of all branch databases. Also, let RB_i and SB_i be the rulebase and suggested rulebase corresponding to database D_i, respectively. An association rule $r \in RB_i$, if $supp_a(r, D_i) \geq \alpha$, and $conf_a(r, D_i) \geq \beta$, $i = 1, 2, ..., n$. An association rule $r \in SB_i$, if $supp_a(r, D_i) \geq \alpha$, and $conf_a(r, D_i) < \beta$. There is a tendency of a suggested association rule in a database to become an association rule in another database. Apart from the association rules, we also consider the suggested association rules for synthesizing heavy association rules in D. The reasons for considering suggested association rules are given as follows. Firstly, we could synthesize support and confidence of an association rule in D more accurately. Secondly, we could synthesize high-frequency association rules in D more accurately. Thirdly, some experimental results have shown that the number of suggested association rules could be significant for some databases. In general, the accuracy of synthesizing an association rule increases as the number of extractions of the association rule increases. Thus, we consider suggested association rules also in synthesizing heavy association rules in D. In addition, the number of transactions in a database would be required in synthesizing an association rule. We define *size* of database DB as the number of transactions in DB, denoted by $size(DB)$. We state the problem as follows.

Let there be n databases $D_1, D_2, ..., D_n$. Let RB_i and SB_i be the set of association rules and suggested association rules in D_i, respectively, $i = 1, 2, ..., n$. Synthesize heavy association rules in the union of all databases (D) based on RB_i and SB_i, $i = 1, 2, ..., n$. Also, notify whether each heavy association rule is high-frequency rule or exceptional rule in D.

10.4 An Extended Model of Local Pattern Analysis
for Synthesizing Global Patterns

Let D_i be the database corresponding to i-th branch of the organization, $i = 1, 2, ...,$
n. Patterns in multiple databases could be grouped into the following categories
based on the number of databases: local patterns, global patterns, and patterns that
are neither local nor global. A pattern based on a branch database is called a *local
pattern*. On the other hand, a *global pattern* is based on all databases under consideration. An essence of the extended model of local pattern analysis (Adhikari and
Rao 2008) is illustrated in Fig. 10.1. The extended model comes with a set of
interfaces and a set of layers. Each interface realizes a set of operations and produces dataset(s) (or, knowledge) based on the dataset(s) available at the next lower
layer. There are four interfaces of the proposed model of synthesizing global patterns from local patterns.

Interface 2/1 is concerned with different operations on data realized at the lowest
layer. By applying these operations, we come up with a processed database
resulting from a local (original) database. These operations are performed on each
branch database. Interface 3/2 applies a filtering algorithm to each processed
database to separate relevant data from possible outliers. In particular, if we are
interested in studying durable items then the transactions containing only non-
durable items could be treated as outlier transactions. Different interesting criteria
could be set to filter data. This interface supports loading data into the respective
data warehouse. Interface 4/3 mines (local) patterns in each local data warehouse.

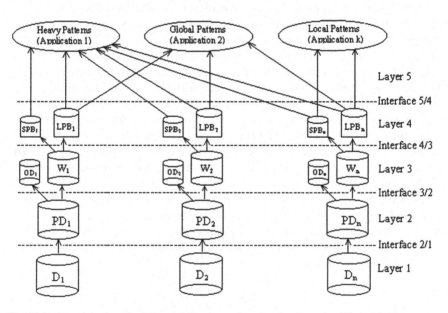

Fig. 10.1 A model of synthesizing global patterns from local patterns in different databases

There are two types of local patterns: local patterns and suggested local patterns. A suggested local pattern is close but fails to fully satisfy the requisite interestingness criteria. The reasons for considering suggested patterns are given as follows. Firstly, by admitting these patterns, we could synthesize patterns more accurately. Secondly, due to the stochastic nature of the transactions, the number of suggested patterns could be significant in some databases. Thirdly, there is a tendency that a suggested pattern of one database could become a local pattern in some other databases. Thus, the correctness of synthesizing global patterns would increase as the number of local patterns increases. Therefore, the extended model becomes effective in synthesizing non-local patterns. Consider a multi-branch company having n databases. Let LPB_i and SPB_i be the local pattern base and suggested local pattern base corresponding to i-th branch of the organization, respectively, $i = 1, 2, ..., n$. Interface 5/4 synthesizes global patterns, or analyses local patterns in order to find solutions to many problems.

At the lowest layer, all the local databases are retained. We may need to process these databases for the purpose of data mining task. Various data preparation techniques (Pyle 1999)—data preprocessing like data cleaning, data transformation, data integration, and data reduction are applied to data in the local databases. We get the processed database PD_i corresponding to the original database D_i, $i = 1, 2, ..., n$. Then we retain all the data that are relevant to the data mining applications. Using a relevance analysis, one could detect outlier data (Last and Kandel 2001) from processed database. A relevance analysis is dependent on the context and varies from one application to another. Let OD_i be the outlier database corresponding to the i-th branch, $i = 1, 2, ..., n$. Sometimes these databases are also used in some other applications. After removing outliers from the processed database we form data warehouse, and the data present there become ready for data mining task. Let W_i be the data warehouse corresponding to i-th branch. Local patterns for the i-th branch are extracted from W_i, $i = 1, 2, ..., n$. Finally, the local patterns are forwarded to the central office for synthesizing global patterns, or completing analysis of local patterns. Many data mining applications could be developed based on the local patterns in different databases. In particular, if we are interested in synthesizing global frequent itemsets then an itemset may not be extracted from all the databases. It might be required to estimate the support of a frequent itemset in a database that fails to report it. Thus, in essence, a global frequent itemset synthesized from local frequent itemsets is approximate. If any one of the local databases is too large to apply a traditional data mining technique then this model would fail. In this situation, one could apply an appropriate sampling technique to reduce the size of the corresponding local database. Otherwise, the database could be partitioned into sub-databases. As a result, the error associated with the results produced in data analysis would increase.

Though the above model introduces many layers and interfaces for synthesizing global patterns, in a real life application, some of these layers might not be fully exploited. In this chapter, we discuss a problem of multi-database mining that uses the above model.

10.5 Related Work

Association rule mining gives rise to interesting association between two itemsets in a database. The notion of association rule was introduced by Agrawal et al. (1993). The authors have proposed an algorithm to mine frequent itemsets in a database. Many algorithms to extract association rules have been reported in the literature. In what follows, we present a few interesting algorithms for extracting association rules in a database. Agrawal and Srikant (1994) have proposed apriori algorithm that uses breadth-first search strategy to count the supports of itemsets. The algorithm uses an improved candidate generation function, which exploits the downward closure property of support and makes it more efficient than earlier algorithm. Han et al. (2000) have proposed data mining method of FP-growth (frequent pattern growth) which uses an extended prefix-tree (FP-tree) structure to store the database in a compressed form. FP-growth adopts a divide-and-conquer approach to decompose both the mining tasks and databases. It uses a pattern fragment growth method to avoid the costly process of candidate generation and testing. Savasere et al. (1995) have introduced partition algorithm. The database is scanned only twice. In the first scan, the database is partitioned and in each partition support is counted. Then the counts are merged to generate potential frequent itemsets. In the second scan, the potential frequent itemsets are counted to find the actual frequent itemsets.

Existing parallel mining techniques (Agrawal and Shafer 1999; Chattratichat et al. 1997; Cheung et al. 1996) could also be used to mine different extreme association rules in multi-databases. Zeng et al. (2012) surveyed state-of-the-art algorithms and applications in distributed data mining. Zhong et al. (2003) have proposed a theoretical framework for peculiarity oriented mining in multiple data sources. Zhang et al. (2009) have proposed a nonlinear method, named KEMGP, which adopts kernel estimation method for synthesizing global patterns from local patterns. Shang et al. (2008) have proposed an extension to Piatetsky-Shapiro's minimum interestingness condition to mine association rules in multiple databases.

Yi and Zhang (2007) have proposed a privacy-preserving distributed association rule mining protocol based on a semi-trusted mixer model. Rozenberg and Gudes (2006) have presented their work on association rule mining from distributed vertically partitioned data with the goal of preserving the confidentiality of each database. The authors have presented two algorithms for discovering frequent itemsets and for calculating the confidence of the rules.

Zhu and Wu (2007) have proposed a framework DRAMA for discovering patterns from different databases with patterns' relationships satisfying the user-specified constraints. It builds a HFP-tree from multiple databases, and mine patterns from the HFP-tree by integrating users' constraints into the pattern mining process. Wu et al. (2013) have studied the problem of frequent pattern mining without user-specified gap constraints and proposed PMBC to solve the problem of finding patterns without involving user-specified gap requirements. Given a sequence and a support threshold value, PMBC intends to discover all subsequences with their support values equal to or greater than the given threshold value.

Liu et al. (2010) have presented a top-down mining of sequential patterns (TD-Seq) from high-dimensional stock sequence databases. A two-phase mining method has been proposed, in which a top-down transposition-based searching strategy as well as a new support counting method are exploited.

10.6   Synthesizing an Association Rule

The technique of synthesizing heavy association rules is suitable for the real databases, where the trend of the customers' behavior exhibited in one database is usually present in other databases. In particular, a frequent itemset in one database is usually present in some transactions of other databases even if it does not get extracted. Our estimation procedure captures such trend and estimates the support of a missing association rule in a database. Let $E_1(r, DB)$ be the amount of error in estimating support of a missing association rule r in database DB. Also, let $E_2(r, DB)$ be the level of error in assuming support as 0 for the missing association rule in DB. Then the value of $E_1(r, DB)$ is usually lower than $E_2(r, DB)$. The estimated support and confidence of a missing association rule usually reduce the error of synthesizing heavy association rules in different databases. We would like to estimate the support and confidence of a missing association rule rather assuming it as absent in a database. If an association rule fails to get extracted from database DB, then we assume that DB contributes some amount of support and confidence for the association rule. The support and confidence of an association rule r in database DB satisfy the following inequality:

$$0 \leq supp_a(r, DB) \leq conf_a(r, DB) \leq 1 \qquad (10.1)$$

At a given $\alpha = \alpha_0$, we observe that the confidence of an association rule r varies over the interval $[\alpha_0, 1]$ as explained in Example 10.1.

Example 10.1 Let $\alpha = 0.333$. Assume that database D_1 contains the following transactions: $\{a1, b1, c1\}$, $\{a1, b1, c1\}$, $\{b2, c2\}$, $\{a2, b3, c3\}$, $\{a3, b4\}$ and $\{c4\}$. The support and confidence of association rule r: $\{a1\} \rightarrow \{b1\}$ in D_1 are 0.333 and 1.0 (highest), respectively. Assume that database D_2 contains the following transactions: $\{a1, b1, c1\}$, $\{a1, b1\}$, $\{a1, c1\}$, $\{a1\}$, $\{a1, b2\}$ and $\{a1, b3\}$. The support and confidence of r in D_2 are 0.333 and 0.333 (lowest), respectively.

As the support of an association rule is expressed as the lower bound of its confidence, the confidence goes up as support increases. The support of an association rule is distributed over $[0, 1]$. If an association rule is not extracted from a database, then the support falls in $[0, \alpha)$, since the suggested association rules are also considered for synthesizing association rules. We would be interested in estimating the support of such rules. Assume that the association rule r: $\{c\} \rightarrow \{d\}$ has been extracted from m databases, $1 \leq m \leq n$. Without any loss of generality, we assume that the association rule r has been reported from the first m databases. We shall use the average behavior of the customers of the first m branches to estimate the average behavior of the customers in remaining branches. Let $D_{i,j}$ denote the union of databases D_i, D_{i+1}, ..., D_j, for $1 \leq i \leq j \leq n$. Then, $supp_a(\{c, d\}, D_{1,m})$ could be viewed as the average behavior of customers of the first m branches for purchasing items c and d together at the same time. Then, $supp_a(\{c, d\}, D_{1,m})$ is obtained by using the following formula:

$$supp_a\left(\{c,d\},D_{1,m}\right) = \left(\sum_{i=1}^{m} supp_a(r,D_i) \times size(D_i)\right) \bigg/ \sum_{i=1}^{m} size(D_i) \quad (10.2)$$

We estimate the support of association rule r for each of the remaining $(n-m)$ databases as follows:

$$supp_s\left(r, D_{m+1,n}\right) = \alpha \times supp_a\left(\{c,d\}, D_{1,m}\right) \quad (10.3)$$

The number of the transactions containing the itemset $\{c, d\}$ in D_i is $supp_a(r, D_i) \times size(D_i)$, for $i = 1, 2, ..., m$. The association rule r is not present in D_i, for $i = m + 1, m + 2, ..., n$. Then the estimated number of the transactions containing the itemset $\{c, d\}$ in D_i is $supp_s(r, D_{m+1, n}) \times size(D_i)$, for $i = m + 1, m + 2, ..., n$. The estimated support of association rule r in D_i is determined in the form:

$$supp_e(r, D_i) = \begin{cases} supp_a(r,D_i), & \text{for } i = 1, 2, ..., m \\ supp_s\left(r,D_{m+1,n}\right), & \text{for } i = m+1, m+2, ..., n \end{cases} \quad (10.4)$$

Then the synthesized support of association rule r in D is expressed as follows.

$$supp_s(r, D) = \left(\sum_{i=1}^{n} supp_e(r,D_i) \times size(D_i)\right) \bigg/ \sum_{i=1}^{n} size(D_i) \quad (10.5)$$

The confidence of the association rule r depends on the supports of the itemsets $\{c\}$ and $\{c, d\}$. The support of itemset $\{c, d\}$ has been synthesized. Now, we need to synthesize the support of itemset $\{c\}$. Without any loss of generality, let the itemset $\{c\}$ gets extracted from first p databases, for $1 \leq m \leq p \leq n$. The estimated support of frequent itemset $\{c\}$ in D_i is calculated as follows:

$$supp_e(\{c\}, D_i) = \begin{cases} supp_a(\{c\}, D_i), & \text{for } i = 1, 2, \ldots, p \\ supp_s(\{c\}, D_{p+1, n}), & \text{for } i = p + 1, p + 2, \ldots, n \end{cases} \quad (10.6)$$

Then the synthesized support of itemset $\{c\}$ in D is determined.

$$supp_s(\{c\}, D) = \left(\sum_{i=1}^{n} supp_e(\{c\}, D_i) \times size(D_i) \right) \bigg/ \sum_{i=1}^{n} size(D_i) \quad (10.7)$$

We compute the synthesized confidence of association rule r in D.

$$conf_s(r, D) = supp_s(r, D) / supp_s(\{c\}, D). \quad (10.8)$$

10.6.1 Design of Algorithm

Here we present an algorithm for synthesizing heavy association rules in D. The algorithm also indicates whether a heavy association rule is high-frequency rule or exceptional rule. Let N and M be the number of association rules and the number of suggested association rules in different local databases, respectively. The association rules and suggested association rules are kept in arrays RB and SB, respectively. An association rule could be described by following attributes: ant, con, did, $supp$ and $conf$. The attributes ant, con, did, $supp$ and $conf$ represent antecedent, consequent, database identification, support, and confidence of a rule, respectively. An attribute x of the i-th association rule of RB is denoted by $RB(i).x$, $i = 1, 2, \ldots,$ $|RB|$. All the synthesized association rules are kept in array SR. Each synthesized association rule could be described by following attributes: ant, con, did, $ssupp$ and $sconf$. The attributes $ssupp$ and $sconf$ represent synthesized support and synthesized confidence of a synthesized association rule, respectively. In the context of mining heavy association rules in D, the following additional attributes are also considered: $heavy$, $highFreq$, $lowFreq$ and $except$. The attributes $heavy$, $highFreq$, $lowFreq$ and $except$ are used to indicate whether an association rule is a heavy rule, high-frequency rule, low-frequency rule and exceptional rule in D, respectively. An attribute y of the i-th synthesized association rule of SR is denoted by $SR(i).y$, $i = 1, 2, \ldots, |SR|$.

Algorithm 10.1 Synthesize heavy association rules in D. Also, it indicates whether a heavy association rule is high-frequency rule or exceptional rule.

procedure *Association-Rule-Synthesis* (n, *RB*, *SB*, μ, v, *size*, γ_1, γ_2)
Inputs:
n: number of databases
RB: array of association rules
SB: array of suggested association rules
μ: threshold of high-support for determining heavy association rules
v: threshold of high-confidence for determining heavy association rules
size: array of the number of transactions in different databases
γ_1: threshold of low-frequency for determining low-frequency association rules
γ_2: threshold of high-frequency for determining high-frequency association rules
Outputs:
Heavy association rules along with their high-frequency and exceptionality statuses
01: copy rules of *RB* and *SB* into array *R*;
02: sort rules of *R* based on attributes *ant* and *con*;
03: calculate total number of transactions in all the databases and store it in *totalTrans*;
04: **let** *nSynRules* = 1;
05: **let** *curPos* = 1;
06: **while** (*curPos* ≤ |R|) **do**
07:    calculate number of occurrences of current rule *R(curPos)* and store it in *nExtractions*;
08:    **let** *SR(nSynRules).highFreq* = false;
09:    **if** ((*nExtractions* / *n*) ≥ γ_2) **then**
10:       *SR(nSynRules). highFreq* = true;
11:    **end if**
12:    **let** *SR(nSynRules).lowFreq* = false;
13:    **if** ((*nExtractions* / *n*) < γ_1) **then**
14:       *SR(nSynRules).lowFreq* = true;
15:    **end if**
16:    calculate $supp_s(R(curPos), D)$ using formula (10.5);
17:    calculate $conf_s(R(curPos), D)$ using formula (10.8);
18:    **let** *SR(nSynRules).heavy* = false;
19:    **if** (($supp_s(SR(nSynRules), D) \geq \mu$) **and** ($conf_s(SR(nSynRules), D) \geq v$)) **then**
20:       *SR(nSynRules).heavy* = true;
21:    **end if**
22:    **let** *SR(nSynRules).except* = false;
23:    **if** ((*SR(nSynRules)* is a low-frequency rule) **and** (*SR(nSynRules)* is a heavy rule)) **then**
24:       *SR(nSynRules).except* = true;
25:    **end if**
26:    update index *curPos* for processing the next association rule;
27:    increase index *nSynRules* by 1;
28: **end while**
29: **for** each synthesized association rule τ in *SR* **do**
30:    **if** τ is heavy **then**
31:       display τ along with its high-frequency and exceptionality statuses;
32:    **end if**
33: **end for**
end procedure

The above algorithm works as follows. The association rules and suggested association rules are copied into R. All the association rules in R are sorted on the pair of attributes {*ant, con*}, so that the same association rule extracted from different databases remains together after sorting. Thus, it would help synthesizing a single association rule at a time. The synthesis process is realized in the *while*-loop shown in line 6. Based on the number of extractions of an association rule, we could determine its high-frequency and low-frequency status. The number of extractions of current association rule has been determined as indicated in line 7. The high-frequency status of current association rule is determined—see lines 8–11. Also, the low-frequency status of current association rule is determined (lines 12–15). We synthesize support and confidence of current association rule based on (10.5) and (10.8), respectively. Once the synthesized support and synthesized confidence have been determined, we could identify the heavy and exceptional statues of current association rule. The heavy status of current association rule is determined using the part of the procedure covered in lines 18–21. Also, the exceptional status of current association rule is determined using lines 22–25. At line 26, we determine the next association rule in R for the synthesizing process. Heavy association rules are displayed along with their high-frequency and exceptionality statuses using lines 29–33. The shaded lines of the pseudo code have been added to report the high-frequency and exceptional statuses of heavy association rules.

Theorem 10.1 *The time complexity of procedure Association-Rule-Synthesis is maximum*$\{O((M + N) \times \log(M + N)), O(n \times (M + N))\}$*, where N and M are the number of association rules and the number of suggested association rules extracted from n databases.*

Proof The lines 1 and 2 take time in $O(M + N)$ and $O((M + N) \times \log(M + N))$ respectively, since there are $M + N$ rules in different local databases. The *while*-loop at line 6 repeats maximum $M + N$ times. Line 7 takes $O(n)$ time, since each rule is extracted maximum n number of times. Lines 8–15 take $O(1)$ time. Using formula (10.3), we could calculate the average behavior of customers of the first m databases in $O(m)$ time. Each of lines 16 and 17 takes $O(n)$ time. Lines 18–25 take $O(1)$ time. Line 26 could be executed during execution of line 7. Thus, the time complexity of *while*-loop 6–28 is $O(n \times (M + N))$. The time complexity of lines 29–33 is $O(M + N)$, since the number of synthesized association rules is less than or equal to $M + N$. Thus, time complexity of procedure *Association-Rule-Synthesis* is *maximum* $\{O((M + N) \times \log(M + N)), O(n \times (M + N)), O(M + N)\} = maximum\{O((M + N) \times \log(M + N)), O(n \times (M + N))\}$. □

Wu and Zhang (2003) have proposed *RuleSynthesizing* algorithm for synthesizing high-frequency association rules in different databases. The algorithm is based on the weights of the different databases. Again, the weight of a database would depend on the association rules extracted from the database. The proposed algorithm executes in $O(n^4 \times maxNosRules \times totalRules^2)$ time, where n, *max-NosRules*, and *totalRules* are the number of data sources, the maximum among the numbers of association rules extracted from different databases, and the total

Table 10.1 Heavy association rules in the union of databases given in Example 10.2

$r: ant \rightarrow con$	ant	con	$supp_s(r, D)$	$conf_s(r, D)$	Heavy	High freq	Except
r_2	C	G	0.31	0.66	True	False	False
r_5	A	B	0.57	0.90	True	False	True

number of association rules in different databases, respectively. Ramkumar and Srivinasan (2008) have proposed a modification of *RuleSynthesizing* algorithm. In this modified algorithm, the weight of an association rule is based on the size of a database. This assumption seems to be more logical. For synthesizing confidence of an association rule, the authors have described a method which was originally proposed by Adhikari and Rao (2008). Though the time complexity of modified *RuleSynthesizing* algorithm is the same as that of the original *RuleSynthesizing* algorithm, but it reduces the average error in synthesizing an association rule. The algorithm *Association-Rule-Synthesis* could synthesize heavy association rules, high-frequency association rules, and exceptional association rules in *maximum* $\{O(totalRules \times \log(totalRules)), O(n \times totalRules)\}$ time. Thus, algorithm *Association-Rule-Synthesis* takes much less time than the existing algorithms. Moreover, the proposed algorithm is simple and straight forward. We illustrate the performance of the proposed algorithm using the following example.

Example 10.2 Let D_1, D_2 and D_3 be three databases of sizes 4,000 transactions, 3,290 transactions, and 10,200 transactions, respectively. Let D be the union of the databases D_1, D_2, and D_3. Assume that $\alpha = 0.2$, $\beta = 0.3$, $\gamma_1 = 0.4$, $\gamma_2 = 0.7$, $\mu = 0.3$ and $v = 0.4$. The following association rules have been extracted from the given databases. $r_1: \{H\} \rightarrow \{C, G\}$, $r_2: \{C\} \rightarrow \{G\}$, $r_3: \{G\} \rightarrow \{F\}$, $r_4: \{H\} \rightarrow \{E\}$, r_5: $\{A\} \rightarrow \{B\}$. The rulebases are given as follows: $RB_1 = \{r_1, r_2\}$, $SB_1 = \{r_3\}$; $RB_2 = \{r_4\}$, $SB_2 = \{r_1\}$; $RB_3 = \{r_1, r_5\}$, $SB_3 = \{r_2\}$. The supports and confidences of the association rules are given as follows. $supp_a(r_1, D_1) = 0.22$, $conf_a(r_1, D_1) = 0.55$; $supp_a(r_1, D_2) = 0.25$, $conf_a(r_1, D_2) = 0.29$; $supp_a(r_1, D_3) = 0.20$, $conf_a(r_1, D_3) = 0.52$; $supp_a(r_2, D_1) = 0.69$, $conf_a(r_2, D_1) = 0.82$; $supp_a(r_2, D_3) = 0.23$, $conf_a(r_2, D_3) = 0.28$; $supp_a(r_3, D_1) = 0.22$, $conf_a(r_3, D_1) = 0.29$; $supp_a(r_4, D_2) = 0.40$, $conf_a(r_4, D_2) = 0.45$; $supp_a(r_5, D_3) = 0.86$, $conf_a(r_5, D_3) = 0.92$. Also, let $supp_a(\{A\}, D_3) = 0.90$, $supp_a(\{C\}, D_1) = 0.80$, $supp_a(\{C\}, D_3) = 0.40$, $supp_a(\{G\}, D_1) = 0.29$, $supp_a(\{H\}, D_1) = 0.31$, $supp_a(\{H\}, D_2) = 0.33$, and $supp_a(\{H\}, D_3) = 0.50$. Heavy association rules are presented in Table 10.1.

The association rules r_2 and r_5 have synthesized support greater than or equal to 0.3 and synthesized confidence greater than or equal to 0.4. So, r_2 and r_5 are heavy association rules in D. The association rule r_5 is a exceptional rule, since it is a heavy and low-frequency rule. But the association rule r_2 is neither a high-frequency nor exceptional rule. Though the association rule r_1 is a high-frequency rule but it is not a heavy rule, since $supp_s(r_1, D) = 0.21$ and $conf_s(r_1, D) = 0.48$.

10.6.2 Error Calculation

To evaluate the proposed technique of synthesizing heavy association rules we have determined the error which has occurred in the experiments. More specifically, the error is expressed relative to the number of transactions, number of items, and the length of a transaction in the databases. Thus the error of an experiment needs to be expressed along with *ANT*, *ALT*, and *ANI* in the given databases, where *ANT*, *ALT* and *ANI* denote the average number of transactions, the average length of a transaction and the average number of items in a database, respectively. There are several ways one could define the error. The proposed definition of error is based on the frequent itemsets generated from heavy association rules. Let r: $\{c\} \rightarrow \{d\}$ be a heavy association rule. Then the frequent itemsets generated from association rule r are $\{c\}$, $\{d\}$, and $\{c, d\}$. Let $\{X_1, X_2, \ldots, X_m\}$ be set of frequent itemsets generated from all the heavy association rules in D. We define the following two types of error.

1. *Average Error* (AE)

$$AE(D, \alpha, \mu, v) = \frac{1}{m} \sum_{i=1}^{m} |supp_a(X_i, D) - supp_s(X_i, D)| \qquad (10.9)$$

2. *Maximum Error* (ME)

$$ME(D, \alpha, \mu, v) = maximum\{ |supp_a(X_i, D) - supp_s(X_i, D)|, i = 1, 2, \ldots, m \} \qquad (10.10)$$

where $supp_a(X_i, D)$ and $supp_s(X_i, D)$ are actual support i.e., the support based on apriori algorithm and synthesized support of the itemset X_i in D, respectively. In Example 10.3, we illustrate the behavior of the measures given above.

Example 10.3 With reference to Example 10.2, r_2: $C \rightarrow G$ and r_5: $A \rightarrow B$ are heavy association rules in D. The frequent itemsets generated from r_2 and r_5 are A, B, C, G, AB and CG. For the purpose of finding the error of an experiment, we need to find the actual supports of the itemsets generated from the heavy association rules. The actual support of an itemset generated from a heavy association rule could be obtained by mining all the databases D_1, D_2, and D_3 together. Thus,

$$\begin{aligned}
AE(D, 0.2, 0.3, 0.4) = \frac{1}{6}\{&|supp_a(\{A\}, D) - supp_s(\{A\}, D)| \\
&+ |supp_a(\{B\}, D) - supp_s(\{B\}, D)| \\
&+ |supp_a(\{C\}, D) - supp_s(\{C\}, D)| \\
&+ |supp_a(\{G\}, D) - supp_s(\{G\}, D)| \\
&+ |supp_a(\{A, B\}, D) - supp_s(\{A, B\}, D)| \\
&+ |supp_a(\{C, G\}, D) - supp_s(\{C, G\}, D)|\}.
\end{aligned}$$

$$\mathrm{ME}(D,\ 0.2,\ 0.3,\ 0.4) = maximum\{|supp_a(\{A\},\ D) - supp_s(\{A\},\ D)|,$$
$$|supp_a(\{B\},\ D) - supp_s(\{B\},\ D)|,$$
$$|supp_a(\{C\},\ D) - supp_s(\{C\},\ D)|,$$
$$|supp_a(\{G\},\ D) - supp_s(\{G\},\ D)|,$$
$$|supp_a(\{A,\ B\},\ D) - supp_s(\{A,\ B\},\ D)|,$$
$$|supp_a(\{C,\ G\},\ D) - supp_s(\{C,\ G\},\ D)|\}.$$

10.7 Experiments

We have carried out several experiments to study the effectiveness of the approach presented in this chapter. We present the experimental results using three real databases. The database *retail* (Frequent itemset mining dataset repository 2004) is obtained from an anonymous Belgian retail supermarket store. The databases *BMS-Web-Wiew-1* and *BMS-Web-Wiew-2* can be found from KDD CUP 2000 (Frequent itemset mining dataset repository 2004). We present some characteristics of these databases in Table 10.2. We use notation *DB*, *NT*, *AFI*, *ALT* and *NI* to denote a database, the number of transactions, the average frequency of an item, the average length of a transaction and the number of items in the database, respectively.

Each of the above databases is divided into 10 subsets for the purpose of carrying out experiments. The databases obtained from *retail*, *BMS-Web-Wiew-1* and *BMS-Web-Wiew-2* are named as R_i, B_{1i} and B_{2i} respectively, $i = 0, 1, …, 9$. The databases R_j and B_{ij} are called branch databases, $i = 1, 2$, and $j = 0, 1, …, 9$. Some characteristics of the branch databases are presented in Table 10.3.

The results of the three experiments using Algorithm 10.1 are presented in Table 10.4. The choice of different parameters is an important issue. We have selected different values of α and β for different databases. But, they are kept the same for branch databases obtained from the same database. For example, α and β are the same for branch databases R_i, for $i = 0, 1, …, 9$.

Table 10.2 Dataset characteristics

Dataset	NT	ALT	AFI	NI
Retail	88,162	11.31	99.67	10,000
BMS-Web-Wiew-1	1,49,639	2.00	155.71	1,922
BMS-Web-Wiew-2	3,58,278	2.00	7,165.56	100

Table 10.3 Branch database characteristics

DB	NT	ALT	AFI	NI	DB	NT	ALT	AFI	NI
R_0	9,000	11.24	12.07	8,384	R_5	9,000	10.86	16.71	5,847
R_1	9,000	11.21	12.27	8,225	R_6	9,000	11.20	17.42	5,788
R_2	9,000	11.34	14.60	6,990	R_7	9,000	11.16	17.35	5,788
R_3	9,000	11.49	16.66	6,206	R_8	9,000	12.00	18.69	5,777
R_4	9,000	10.96	16.04	6,148	R_9	7,162	11.69	15.35	5,456
B_{10}	14,000	2.00	14.94	1,874	B_{15}	14,000	2.00	280.00	100
B_{11}	14,000	2.00	280.00	100	B_{16}	14,000	2.00	280.00	100
B_{12}	14,000	2.00	280.00	100	B_{17}	14,000	2.00	280.00	100
B_{13}	14,000	2.00	280.00	100	B_{18}	14,000	2.00	280.00	100
B_{14}	14,000	2.00	280.00	100	B_{19}	23,639	2.00	472.78	100
B_{20}	35,827	2.00	1,326.93	54	B_{25}	35,827	2.00	716.54	100
B_{21}	35,827	2.00	1,326.93	54	B_{26}	35,827	2.00	716.54	100
B_{22}	35,827	2.00	716.54	100	B_{27}	35,827	2.00	716.54	100
B_{23}	35,827	2.00	716.54	100	B_{28}	35,827	2.00	716.54	100
B_{24}	35,827	2.00	716.54	100	B_{29}	35,835	2.00	716.70	100

Table 10.4 First five heavy association rules reported from different databases (sorted in non-increasing order on synthesized support)

Data base	α	β	μ	ν	Heavy assoc rules	Syn supp	Syn conf	High freq	Excep tional
$\cup_{i=0}^{9} R_i$	0.05	0.2	0.1	0.5	$\{48\} \rightarrow \{39\}$	0.33	0.68	Yes	No
					$\{39\} \rightarrow \{48\}$	0.33	0.56	Yes	No
					$\{41\} \rightarrow \{39\}$	0.13	0.63	Yes	No
					$\{38\} \rightarrow \{39\}$	0.12	0.66	Yes	No
					$\{41\} \rightarrow \{48\}$	0.10	0.51	Yes	No
$\cup_{i=0}^{9} B_{1i}$	0.01	0.2	0.007	0.1	$\{1\} \rightarrow \{5\}$	0.01	0.13	No	No
					$\{5\} \rightarrow \{1\}$	0.01	0.11	No	No
					$\{7\} \rightarrow \{5\}$	0.01	0.12	No	No
					$\{5\} \rightarrow \{7\}$	0.01	0.11	No	No
					$\{3\} \rightarrow \{5\}$	0.01	0.12	No	No
$\cup_{i=0}^{9} B_{2i}$	0.006	0.01	0.01	0.1	$\{3\} \rightarrow \{1\}$	0.02	0.14	Yes	No
					$\{1\} \rightarrow \{3\}$	0.02	0.14	Yes	No
					$\{7\} \rightarrow \{1\}$	0.02	0.14	Yes	No
					$\{1\} \rightarrow \{7\}$	0.02	0.14	Yes	No
					$\{5\} \rightarrow \{1\}$	0.02	0.14	Yes	No

10.7.1 Results of Experimental Studies

After mining a branch database from a group of branch databases using a rea-
sonably low values α and β, one could fix α and β for the purpose data mining task.
If α and β are smaller, then synthesized support and synthesized confidence values
are closer to their actual values. Thus, the synthesized association rules are closer to
the true association rules in multiple databases.

The choice of the values of μ and v are context dependent. Also if μ and v are
kept fixed then some databases might not report heavy association rules, while other
databases might report many heavy association rules. While generating association
rule one could estimate the average synthesized support and confidence based on
the generated association rules. Thus, it gives an idea of thresholds for high-support
and high-confidence for synthesizing heavy association rules in different databases.
Also, the choice of γ_1 and γ_2 are also context dependent. It has been found that
"reasonable" values of γ_1 and γ_2 could lie in the interval [0.3, 0.4] and [0.6, 0.7],
respectively. Given these findings, we have taken $\gamma_1 = 0.35$, and $\gamma_2 = 0.60$ for
synthesizing heavy association rules.

The experiments conducted on the three databases have resulted in no exceptional association rule. Normally, exceptional association rules are rare. Also, we
have not found any association rule which is a heavy rule as well as high-frequency
rule in multiple databases obtained from *BMS-Web-Wiew-1*.

In many applications, the suggested association rules are significant. While
synthesizing the association rules from different databases we might need to consider the suggested association rules for the correctness of synthesizing association
rules. We have observed that the number of suggested association rules in the set of
databases $\{R_0, R_1, ..., R_9\}$ and $\{B_{10}, B_{11}, ..., B_{19}\}$ are significant. But, the set of
databases $\{B_{20}, B_{21}, ..., B_{29}\}$ do not generate any suggested association rule. We
present the number of association rules and the number of suggested association
rules for different experiments in Table 10.5.

The error of synthesizing association rules in a database is relative to the following parameters: the number of transactions, the number of items, and the length
of transactions in the given databases. If the number of transactions in database
increases, the error of synthesizing association rules also increases, provided other
two parameters remain constant. If the lengths of transactions of a database

Table 10.5 Number of association rules and suggested association rules extracted from multiple
databases

Database	α	β	Number of association rules (N)	Number of suggested association rules (M)	$M/(N + M)$
$\cup_{i=0}^{9} R_i$	0.05	0.2	821	519	0.39
$\cup_{i=0}^{9} B_{1i}$	0.01	0.2	50	96	0.66
$\cup_{i=0}^{9} B_{2i}$	0.006	0.01	792	0	0

Table 10.6 Error of synthesizing different extreme association rules

Database	α	β	μ	ν	(AE, ANT, ALT, ANI)	(ME, ANT, ALT, ANI)
$\cup_{i=0}^{9} R_i$	0.05	0.2	0.1	0.5	(0.00, 8,816.2, 11.31, 5,882.1)	(0.00, 8,816.2, 11.31, 5,882.1)
$\cup_{i=0}^{9} B_{1i}$	0.01	0.2	0.007	0.1	(0.00, 14,963.9, 2.0, 277.4)	(0.00, 14,963.9, 2.0, 277.4)
$\cup_{i=0}^{9} B_{2i}$	0.006	0.01	0.01	0.1	(0.000118, 35,827.8, 2.0, 90.8)	(0.00, 35,827.8, 2.0, 90.8)

increase, the error of synthesizing association rules is likely to increase, provided that two other parameters remain constant. Lastly, if the number of items increases, then the error of synthesizing association rules is likely to decrease, provided that two other parameters remain constant. Thus, the error needs to be reported along with the *ANT, ALT* and *ANI* for the given databases. The obtained results are presented in Table 10.6.

10.7.2 Comparison with Existing Algorithm

In this section, we make a detailed comparison among the part of the proposed algorithm that synthesizes only high-frequency association rules, *RuleSynthesizing* algorithm (Wu and Zhang 2003) and *Modified RuleSynthesizing* algorithm (Ramkumar and Srivinasan 2008). Let the part of the proposed algorithm be *High-Frequency-Rule-Synthesis* used for synthesizing (only) high-frequency association rules in different databases. We conduct experiments for comparing these algorithms. We compare them on the basis of the following two criteria, namely average error and execution time.

10.7.2.1 Analysis of Average Error

The definitions of average error and maximum error given above and those proposed by Wu and Zhang (2003) are similar and use the same set of synthesized frequent itemsets. However the methods of synthesizing frequent itemsets for these two approaches are different. Thus, the value of error incurred in these two approaches might differ. In *RuleSynthesizing* algorithm, if an itemset fails to get extracted from a database then the support of the itemset is assumed to be 0. But in *Association-Rule-Synthesis* algorithm, if an itemset fails to get extracted from a database then the support of the itemset is estimated. The synthesized support of an itemset in the union of databases in these two approaches might be different. As the number of databases increases the relative presence of a rule normally decreases. The error of synthesizing an association rule normally increases. The AE reported in the experiment is likely to increase if the number of databases increases. We observe such phenomenon in Figs. 10.2 and 10.3.

The proposed algorithm follows a direct approach in identifying high-frequency association rules as opposed to the *RuleSynthesizing* and *Modified RuleSynthesizing* algorithms. In Figs. 10.2 and 10.3, we observe that AE of an experiment conducted

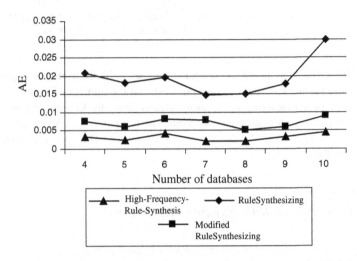

Fig. 10.2 AE versus the number of databases from *retail* at $(\alpha, \beta, \gamma) = (0.05, 0.2, 0.6)$

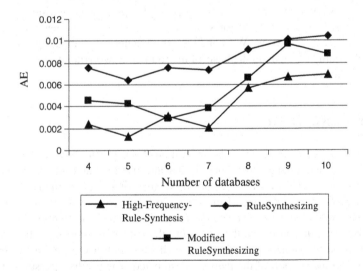

Fig. 10.3 AE versus the number of databases from *BMS-Web-Wiew-1* at $(\alpha, \beta, \gamma) = (0.005, 0.1, 0.3)$

using *High-Frequency-Rule-Synthesis* algorithm is less than that of *RuleSynthesizing* algorithm. But the *Modified RuleSynthesizing* algorithm improves the accuracy of synthesizing an association rule as compared to *RuleSynthesizing* algorithm. It remains less accurate when compared to the accuracy of the *High-Frequency-Rule-Synthesis* algorithm.

10.7.2.2 Analysis of Execution Time

We have also completed experiments to study the execution time by varying the number of databases. The number of synthesized frequent itemsets increases as the number of databases increases. The execution time increases with the increase of number of databases. We observe this phenomenon in Figs. 10.4 and 10.5. However, more significant differences are noted with the increase in the number of databases.

The time complexity of *RuleSynthesizing* and *Modified RuleSynthesizing* algorithms is the same. When the number of databases is less the *RuleSynthesizing* and *Modified RuleSynthesizing* algorithms might be faster than *High-Frequency-Rule-Synthesizing* algorithm. As the number of databases increases, *High-Frequency-Rule-Synthesizing* algorithm works faster than both *RuleSynthesizing* and *Modified RuleSynthesizing* algorithms.

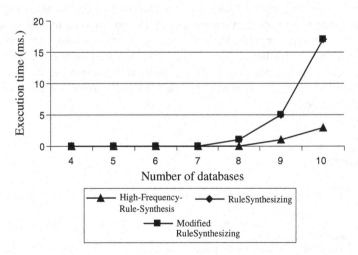

Fig. 10.4 Execution time versus the number of databases from *retail* at $(\alpha, \beta, \gamma) = (0.05, 0.2, 0.6)$

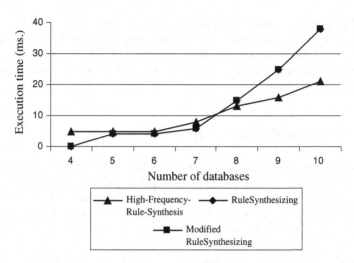

Fig. 10.5 Execution time versus the number of databases from *BMS-Web-Wiew-1* at $(\alpha, \beta, \gamma) = (0.005, 0.1, 0.3)$

10.8 Conclusions

The extended model of local pattern analysis enables us to develop useful multi-database mining applications. Although it exhibits many layers and interfaces, this general model can come with many variations. In particular, some of these layers might not be present when developing a certain application. Synthesizing heavy association rule is an important component of a multi-database mining system. In this chapter, we have presented three extreme types of association rules present in multiple databases viz., heavy association rules, high-frequency association rules, and exceptional association rules. The introduced algorithm referred to as the *Association-Rule-Synthesis* is used to synthesize these extreme association rules in multiple databases.

References

Adhikari A, Ramachandrarao P, Pedrycz W (2010) Developing multi-databases mining applications. Springer, Berlin

Adhikari A, Rao PR (2008) Synthesizing heavy association rules from different real data sources. Pattern Recogn Lett 29(1):59–71

Agrawal R, Imielinski T, Swami A (1993) Mining association rules between sets of items in large databases. In: Proceedings of ACM SIGMOD conference, pp 207–216

Agrawal R, Shafer J (1999) Parallel mining of association rules. IEEE Trans Knowl Data Eng 8 (6):962–969

Agrawal R, Srikant R (1994) Fast algorithms for mining association rules. In: Proceedings of international conference on very large data bases, pp 487–499

Chattratichat J, Darlington J, Ghanem M, Guo Y, Hüning H, Köhler M, Sutiwaraphun J, To HW, Yang D (1997) Large scale data mining: challenges, and responses. In: Proceedings of the third international conference on knowledge discovery and data mining, pp 143–146

Cheung D, Ng V, Fu A, Fu Y (1996) Efficient mining of association rules in distributed databases. IEEE Trans Knowl Data Eng 8(6):911–922

Frequent itemset mining dataset repository (2004) http://fimi.cs.helsinki.fi/data

Han J, Pei J, Yiwen Y (2000) Mining frequent patterns without candidate generation. In: Proceedings of ACM SIGMOD conference on management of data, pp 1–12

Last M, Kandel A (2001) Automated detection of outliers in real-world data. In: Proceedings of the second international conference on intelligent technologies, pp 292–301

Liu H, Lin F, He J, Cai Y (2010) New approach for the sequential pattern mining of high-dimensional sequence databases. Decis Support Syst 50(1):270–280

Pyle D (1999) Data preparation for data mining. Morgan Kufmann, San Francisco

Ramkumar T, Srivinasan R (2008) Modified algorithms for synthesizing high-frequency rules from different data sources. Knowl Inf Syst 17(3):313–334

Rozenberg B, Gudes E (2006) Association rules mining in vertically partitioned databases. Data Knowl Eng 59(2):378–396

Savasere A, Omiecinski E, Navathe S (1995) An efficient algorithm for mining association rules in large databases. In: Proceedings of the 21st international conference on very large data bases, pp 432–443

Shang S, Dong X, Li J, Zhao Y (2008) Mining positive and negative association rules in multi-database based on minimum interestingness. In: Proceedings of the 2008 international conference on intelligent computation technology and automation, pp 791–794

Wu X, Zhang S (2003) Synthesizing high-frequency rules from different data sources. IEEE Trans Knowl Data Eng 14(2):353–367

Wu X, Zhu X, He Y, Abdullah N, Arslan AN (2013) PMBC: pattern mining from biological sequences with wildcard constraints. Comp Bio Med 43(5):481–492

Yi X, Zhang Y (2007) Privacy-preserving distributed association rule mining via semi-trusted mixer. Data Knowl Eng 63(2):550–567

Zeng L, Li L, Duan L, Lü K, Shi Z, Wang M, Wu W, Luo P (2012) Distributed data mining: a survey. Inf Technol Manage 13(4):403–409

Zhang S, Wu X, Zhang C (2003) Multi-database mining. IEEE Comput Intell Bull 2(1):5–13

Zhang S, You X, Jin Z, Wu X (2009) Mining globally interesting patterns from multiple databases using kernel estimation. Expert Syst Appl Int J 36(8):10863–10869

Zhang S, Wu X (2011) Fundamentals of association rules in data mining and knowledge discovery. Wiley Interdisc Rev Data Min Knowl Discov 1(2):97–116

Zhong N, Yao YYY, Ohishima M (2003) Peculiarity oriented multidatabase mining. IEEE Trans Knowl Data Eng 15(4):952–960

Zhu X, Wu X (2007) Discovering relational patterns across multiple databases. In: Proceedings of ICDE, pp 726–735

Chapter 11
Clustering Local Frequency Items in Multiple Data Sources

Frequent items could be considered as a generic type of patterns in a database. In the context of multiple data sources, most of the global patterns are based on local frequency items. A multi-branch company transacting from different branches often needs to extract global patterns from data distributed over the branches. Global decisions could be made effectively using such patterns. Thus, it becomes important to cluster local frequency items in multiple databases. In this chapter, an overview of the existing measures of association is presented. For the purpose of selecting the suitable technique of mining multiple databases, a survey of the existing multi-database mining techniques is presented. A study on the related clustering techniques is also covered here. We present the notion of high frequency itemsets (HFISs), and an algorithm for synthesizing the supports of such itemsets is designed. It has been shown that the existing clustering technique clusters a set of items at a low level, since it estimates association among items in an itemset with low accuracy, and a new algorithm for clustering local frequency items is designed. Due to the suitability of measure of association A_2, on its basis association among items in a high frequency itemset is synthesized. The soundness of the clustering technique has been shown. Numerous experiments are conducted using five datasets, and the results concerning different aspects of the proposed problem are presented in the experimental section. The effectiveness of the proposed clustering technique is more visible in dense databases.

11.1 Introduction

Many multi-branch companies collect a huge amount of data through different branches and the local transactions are stored locally. Banks, shopping malls, and insurance companies are some examples of multi-branch companies that deal with multiple databases. In the context of multiple databases, some global patterns such as high frequency association rule (Wu and Zhang 2003), heavy association rule (Adhikari and Rao 2008d), and exceptional pattern (Adhikari 2012) are based on local frequency items in multiple databases. Thus, knowledge discovery using local

© Springer International Publishing Switzerland 2015   197
A. Adhikari and J. Adhikari, *Advances in Knowledge Discovery in Databases*,
Intelligent Systems Reference Library 79, DOI 10.1007/978-3-319-13212-9_11

frequency items becomes an important issue. The goal of this chapter is to present a technique for clustering local frequency items in multiple databases.

For the purpose of continuity of the presentation, let us start with a few definitions. An *itemset* is a collection of items in a database. Each itemset in a database is associated with a statistical measure called *support* (Agrawal et al. 1993). The support of an itemset X in database D is expressed as the fraction of transactions in D containing X, denoted by $S(X, D)$. In general, let $S(E, D)$ be the support of a Boolean expression E defined on the transactions in database D. An itemset X is called *frequent* in D if $S(X, D) \geq \alpha$, where α is user-defined level of *minimum support*. If X is frequent then $Y \subseteq X$ is also frequent, since $S(Y, D) \geq S(X, D)$, for $Y \neq \phi$. Thus, each item of a frequent itemset is also frequent. Items in a frequent itemset could be considered as a basic type of pattern in a transactional database.

The collection of frequent itemsets determines major characteristics of a database. Interesting algorithms are proposed for mining frequent itemsets in a database (Agrawal and Srikant 1994; Han et al. 2000; Savasere et al. 1995). Thus, there are many implementations for extracting frequent itemsets from a database (Frequent itemset mining implementations repository).[1] Itemset patterns influence heavily the current KDD research. We observe such influence in the following ways. Firstly, many algorithms as mentioned above, have been reported about mining frequent itemsets in a database. Secondly, many patterns are based on frequent itemsets in a database. They could be called as derived patterns in a database. For example, positive association rule (Agrawal et al. 1993), high frequency association rule (Wu and Zhang 2003) and conditional pattern (Adhikari and Rao 2008b) are examples of some derived patterns. Considerable number of studies have been reported on mining/synthesizing derived patterns in a database (Adhikari and Rao 2008d; Zhang et al. 2004a; Adhikari 2012). Finally, solutions to many problems could be based on the analysis of patterns in a database (Adhikari and Rao 2008c; Wu et al. 2005). Such applications process patterns in a database for the purpose of making some decisions. Frequent items are the components of many interesting patterns. Thus, the analysis and synthesis of local frequency items is an interesting as well as an important issue. They are used to construct the global patterns in multiple databases. Thus, clustering local frequency items in multiple databases is an important knowledge for a multi-branch company. Many corporate decisions could be taken effectively by incorporating knowledge inherent in data across the branches. An effective management of multiple large databases becomes a challenging issue (Adhikari et al. 2010a). In the next section we study the existing techniques of mining multiple large databases. Based on their suitability and performances, we choose the best among the available techniques for mining multiple databases.

This chapter is organized as follows. Section 11.2 elaborates on a study of the existing measures of association, techniques for mining multiple databases, and related clustering algorithms. The proposed problem is presented in Sect. 11.3. In Sect. 11.4, an algorithm for synthesizing supports of HFISs is designed. The

---

[1] Frequent itemset mining implementations repository. http://fimi.cs.helsinki.fi/src/.

algorithm also synthesizes association among items in a high frequency itemset of size greater than one. In Sect. 11.5, we design an algorithm for clustering local frequency items in multiple databases. Finally, experimental results are presented in Sect. 11.6 to show the effectiveness of the proposed clustering technique.

11.2 Related Work

Consider a multi-branch company that operates from n branches. Let D_i be the database corresponding to the ith branch, $i = 1, 2, \ldots, n$. Also, let D be the union of these databases. Our clustering procedure is based on itemset patterns in multiple databases. In this context, one needs a technique for mining itemset patterns in multiple databases. Afterwards, association among items in an itemset is captured using a measure of association. Finally, a clustering algorithm to cluster local frequency items in multiple databases is designed. We categorize work related to this issue into three areas viz. measures of association, techniques for mining multiple databases, and related clustering algorithms.

11.2.1 Measures of Association

The analysis of relationships among variables is a fundamental task being at the heart of many data mining problems. For instance, association rules find relationships between sets of items in a database of transactions. Such rules express buying patterns of customers e.g., finding how the presence of one item affects the presence of another, and so forth. A measure of association gives a numerical estimate of statistical dependence among a set of items. Highly associated items are likely to be purchased together. In other words, items of itemset X are highly associated, if one of the items of X is purchased then the remaining items of X are also likely to be purchased in the same transaction.

Tan et al. (2003) have described several key properties of twenty one interestingness measures proposed in statistics, machine learning and data mining literature. It might be required to examine the properties in order to select right interestingness measure for a given application domain. Hershberger and Fisher (2005) discuss some measures of association proposed in statistics. Measures of association could be categorized into two groups. Some measures deal with a set of objects, or could be generalized to deal with a set of objects. On the other hand, remaining measures could not be generalized. Confidence (Agrawal et al. 1993); conviction (Brin et al. 1997) are examples of the second category of measures. On the other hand, measures such as Jaccard (Tan et al. 2003) could be generalized to find association among a set of items in a database. Most of the existing measures are based on a 2×2 contingency table. Thus, these measures might not be suitable for measuring association among a set of items.

Agrawal et al. (1993) have proposed support measure in the context of finding association rules in a database. To find support of an itemset, it requires counting frequency of the itemset in the given database. An itemset in a transaction could be a source of association among items in the itemset. But, support of an itemset does not consider frequencies of it subsets. As a result, the support of an itemset might not be a good measure of association among items in an itemset.

Piatetsky-Shapiro (1991) has proposed leverage measure in the context of mining strong rules in a database. Aggarwal and Yu (1998) have proposed a notion of a collective strength of an itemset. Collective strength is based on the concept of violation of an itemset. An itemset X is said to be in violation of a transaction, if some items of X are present in the transaction and others are not. Collective strength of an itemset X has been defined as follows.

$$C(X) = \frac{1 - v(X)}{1 - E(v(X))} \times \frac{E(v(X))}{v(X)}, \qquad (11.1)$$

where

- $v(X)$ is the violation rate of itemset X. It is the fraction of transactions in violation of itemset X.
- $E(v(X))$ is the expected violation rate of itemset X.

The major concern regarding computation of $C(X)$ is that the computation of $E(v(X))$ is based on statistical independence of items of X.

Cosine (Han and Kamber 2011) and correlation (Han and Kamber 2011) are used to measure association between two objects. These measures might not be suitable as a measure of association among items of an itemset. Confidence and conviction are used to measure strength of association between itemsets in some sense. These measures might not be useful in the current context, since we are interested in capturing association among items of an itemset.

Zhou and Xiong (2009) have proposed to find confounding effects attributable to local associations efficiently. Authors derived an upper bound by a necessary condition of confounders, which can help us prune the search space and efficiently identify confounders. Duan and Street (2009) explored high-dimensional correlation in two ways. Initially, authors expanded the set of desirable properties for correlation measures and study the advantages and disadvantages of various measures. Then, they proposed an MFCI framework to decouple the correlation measure from the need for efficient search.

11.2.2 Multi-database Mining Techniques

Mining multiple large databases seems to be a different problem than mining a database (Adhikari et al. 2010a). One of the main challenges is that the collection of all branch databases might be very large. Traditional data mining techniques

(Agrawal and Srikant 1994; Han et al. 2000) seem to be not suitable for mining multiple large databases. In this situation, one could employ local pattern analysis (Zhang et al. 2003) to deal with multiple large databases. In this technique each branch is required to forward local patterns, instead of original databases, to the central office for synthesis and analysis of local patterns. Local pattern analysis might return approximate global patterns.

For the purpose of mining multiple databases, one could apply *partition algorithm* proposed by Savasere et al. (1995). The algorithm was designed to mine a very large database by partitioning. The algorithm works as follows. It scans the database twice. The database is divided into disjoint partitions, where each partition is small enough to fit in memory. In the first scan, the algorithm reads each partition and computes locally frequent itemsets in each partition using apriori algorithm (Agrawal and Srikant 1994). In the second scan, the algorithm counts the supports of all locally frequent itemsets toward the complete database. In this case, each local database could be considered as a partition. Though partition algorithm mines frequent itemsets accurately, it is an expensive solution to mining multiple large databases, since each database is required to be mined twice.

Zhang et al. (2004a) have proposed algorithm *IdentifyExPattern* for identifying global exceptional patterns in multi-databases. Every local database is mined separately using *random order* for synthesizing global exceptional patterns. In algorithm *IdentifyExPattern*, a pattern in a local database is assumed as zero, if it does not get reported. Let $S_a(p, DB)$ and $S_s(p, DB)$ be the actual (i.e., apriori) support and synthesized support of pattern p in database DB. Support of pattern p in D has been synthesized in the form

$$S_s(p, D) = \frac{1}{num(p)} \sum_{i=1}^{num(p)} \frac{S_a(p, D_i) - \alpha}{1 - \alpha} \tag{11.2}$$

where $num(p)$ is the number of databases that report pattern p at a given minimum support level α.

For synthesizing association rules in multiple real databases *Association-Rule-Synthesis* algorithm has been proposed in Adhikari and Rao (2008d). For real databases, the trend of the customers' behaviour exhibited in one database is usually present in other databases. In particular, a frequent itemset in one database is usually present in some transactions of other databases even if it does not get extracted. The estimation procedure captures such trend, and estimates the support of a missing association rule. Without any loss of generality, let the itemset X be extracted from first m databases, $1 \leq m \leq n$. Then trend of X in first m databases could be expressed as follows.

$$trend^{1,m}(X|\alpha) = \frac{1}{\sum_{i=1}^{m} |D_i|} \times \sum_{i=1}^{m} [S_a(X, D_i) \times |D_i|] \tag{11.3}$$

One could use the trend of X in first m databases for synthesizing support of X in D. An estimate of support of X in each of the remaining databases is obtained by $\alpha \times trend^{1,\ n}(X \mid \alpha)$. Thus, the synthesized support of X is computed as follows.

$$S_s(X, D) = \frac{trend^{1, m}(X \mid \alpha)}{\sum_{i=1}^{n} |D_i|} \times \left[(1 - \alpha) \times \sum_{i=1}^{m} |D_i| + \alpha \times \sum_{i=1}^{n} |D_i| \right] \quad (11.4)$$

For synthesizing high frequency association rules in multiple databases, Wu and Zhang (2003) have proposed *RuleSynthesizing* algorithm. Based on the association rules in different databases, the authors have estimated weights of different databases. Let w_i be the weight of ith database, $i = 1, 2, …, n$. Without any loss of generality, let the association rule r be extracted from first m databases, $1 \leq m \leq n$. Then $S_a(r, D_i)$ has been assumed as 0, for $i = m + 1, m + 2, …, n$. Then support of r in D has been synthesized in the following form.

$$S_s(r, D) = w_1 \times S_a(r, D_1) + \cdots + w_m \times S_a(r, D_m) \quad (11.5)$$

He et al. (2010) studied the problem of rule synthesizing from multiple related databases where items representing the databases may be different, and the databases may not be relevant, or similar to each other. Authors argued that, for such multi-related databases, simple rule synthesizing without a detailed understanding of the databases is not able to reveal meaningful patterns inside the data collections. Consequently, authors proposed a two-step clustering on the databases at both item and rule levels such that the databases in the final clusters contain both similar items and similar rules. A weighted rule synthesizing method then has been applied to each such cluster to generate final rules.

Adhikari et al. (2011b) have introduced a new pattern, called notch, of an item in time-stamped databases. Based on this pattern, authors have proposed two special kinds of notch, called generalized notch and iceberg notch, in time-stamped databases. Also, the authors have identified an application of generalized notch, and designed an algorithm for mining interesting icebergs in time-stamped databases.

The variation of sales of an item over time is an important issue. Thus, the notion of stability of an item is introduced in Adhikari et al. (2009). Stable items are useful in making many strategic decisions for a company. Based on the degree of stability of an item, an algorithm has been designed for clustering items in different data sources. A notion of best cluster by considering average degree of variation of a class has been proposed. Also, an alternative algorithm has been designed to find best cluster among items in different data sources.

Liu et al. (2001) have proposed multi-database mining technique that searches only the relevant databases. Identifying relevant databases is based on selecting the relevant tables (relations) that contain specific, reliable and statistically significant information pertaining to the query. Adhikari et al. (2010a), Zhang et al. (2004b) studied various strategies for mining multiple databases. Existing parallel mining techniques could also be used to deal with multiple large databases (Agrawal and Shafer 1996; Chattratichat et al. 1997; Cheung et al. 1996).

Many large organizations have multiple large databases as they transact from multiple branches. Numerous decisions are based on a set of specific items called the select items. Thus, the analysis of select items in multiple databases is an important issue. For the purpose of studying select items in multiple databases, one may need true global patterns of select items. Thus, we have proposed a model of mining global patterns of select items from multiple databases (Adhikari et al. 2011a). A measure of overall association between two items in a database is also proposed. An algorithm is designed based on the proposed measure for the purpose of grouping the frequent items in multiple databases.

Data collected for collaborative filtering (CF) purposes might be cross-distributed between two online vendors, even competing companies. Such corporations might want to integrate their data to provide more precise and reliable recommendations. However, due to privacy, legal, and financial concerns, they do not desire to disclose their private data to each other. If privacy-preserving measures are introduced, they might decide to generate predictions based on their distributed data collaboratively. Yakut and Polat (2012) have investigated how to offer hybrid CF-based referrals with decent accuracy on cross-distributed data (CDD) between two e-commerce sites while maintaining their privacy.

11.2.3 Clustering Techniques

Zhang et al. (1997) have proposed an efficient and scalable data clustering method, called BIRCH, based on a new in-memory data structure called CF-tree. Estivill-Castro and Yang (2004) have proposed an algorithm that remains efficient, generally applicable, multi-dimensional but is more robust to noise and outliers. Jain et al. (1999) have presented an overview of pattern clustering methods from a statistical pattern recognition perspective, with a goal of providing useful advice and references to fundamental concepts accessible to the broad community of clustering practitioners. In this chapter, we would like to design a clustering algorithm based on local patterns. Thus, the above algorithms might not be suitable in this situation.

Ali et al. (1997) have proposed a partial classification technique using association rules. The clustering of frequent items using local association rules might not be a good idea. The number of frequent itemsets obtained from a set of association rules might be much less than the number of frequent itemsets extracted using apriori algorithm. Thus, efficiency of the clustering process would become low. A measure of similarity between two databases, $simi_2$, is proposed in Adhikari and Rao (2008c) and an algorithm is designed to cluster databases using measure $simi_2$.

Chen et al. (2012) have proposed a new algorithm to cluster multiple and parallel data streams using spectral component similarity analysis. The algorithm performs auto-regressive modeling to measure the lag correlation between the data streams and uses it as the distance metric for clustering. The algorithm uses a sliding window model to continuously report the most recent clustering results and to dynamically adjust the number of clusters. Malinen and Fränti (2012) have shown that clustering is also an optimization problem for an analytic function. The mean squared error or the squared error can be expressed as an analytic function. With an analytic function one can get benefit from the existence of standard optimization methods. Initially the gradient of this function is calculated and then the descent method is used to minimize the function.

Identifying clusters of arbitrary shapes remains a challenge in the field of data clustering. Lee and Ólafsson (2011) have proposed a new measure of cluster quality based on minimizing the penalty of disconnection between objects that would be ideally clustered together. This disconnectivity is based on analysis of nearest neighbors and the principle that an object should be in the same cluster as its nearest neighbors. An algorithm called MinDisconnect is proposed that heuristically minimizes disconnectivity and numerical results are presented that indicate that the new algorithm can effectively identify clusters of complex shapes and is robust in finding clusters of arbitrary shapes.

Mampaey and Vreeken (2013) have proposed an approach to build a summary by clustering attributes that strongly correlate, and uses the minimum description length principle to identify the best clustering, without requiring a distance measure between attributes.

11.3 Problem Statement

Based on the number of data sources, patterns in multiple databases could be classified into three categories. They are local patterns, global patterns, and patterns that are neither local nor global. Local patterns are used for local data analyses. On the other hand, global patterns are based on all the databases. They are used for global data analyses (Adhikari and Rao 2008d; Wu and Zhang 2003).

In the context of mining multiple databases using local pattern analysis, some itemsets are reported from many databases. They could be termed as HFISs. In this chapter, HFISs are synthesized based on local itemsets. We measure association among items in HFISs. Then we cluster frequent items based on associations obtained among items in HFISs. Highly associated items could be put in the same class. The motivation of proposed clustering technique is given as follows.

An approach for grouping databases, called existing technique, has been proposed in Adhikari and Rao (2008c), Wu et al. (2005). The grouping technique can be described as follows. It finds association between every pair of items (databases). A set of m arbitrary items forms a class, if the $^mC_2$ pair-wise associations corresponding to $^mC_2$ pairs of items are close. The level of association among the items in this class is assumed as the minimum of $^mC_2$ associations. If the number of items in a class is more than two, then we observe that this technique might fail to estimate correctly the association among the items in a group. Then accuracy of the entire clustering process becomes low. The proposed clustering technique follows a

Table 11.1 Symbols/abbreviations used in the study

Symbol/ abbreviation	Meaning/full form
α	User-defined minimum support level
$FI(D \mid \alpha)$	Set of frequent items in database D at a given α
$FIS(D \mid \alpha)$	Set of frequent itemsets in database D at a given α
$FI(1, n, \alpha)$	$\cup_{i=1}^{n} FI(D_i \mid \alpha)$
$FIS(1, n, \alpha)$	$\cup_{i=1}^{n} FIS(D_i \mid \alpha)$
γ	Minimum threshold of number of extractions of an itemset
n	Number of databases
$HFIS$	High frequency itemset
$SHFIS$	Synthesized $HFIS$
AIS	Two dimensional array such that $AIS(i)$ is the array of itemsets extracted from D_i
IS	Set of all itemsets in n databases
$S(X, D_1)$	Support of itemset in databases D_1
SA	Synthesized association
SD	Synthesized dispersion
δ	Level of association

different approach and it clusters local frequency items at a higher level of similarity as compared to existing technique.

Before stating our problem formally, we incorporate prime notations in the following table for easy reference (Table 11.1).

Each item in $FI(D_i \mid \alpha)$ is a local frequency item, $i = 1, 2, 3, \ldots$. We apply a measure of association and a multi-database mining technique for the purpose of clustering local frequency items in multiple databases. The proposed problem could be stated as follows.

There are n different databases D_i, i = 1, 2, ..., n. Find the best non-trivial partition (if it exists) of FI(*1, n, α) induced by FIS(1, n, α).*

A partition (Liu 1985) is a specific type of clustering. A formal definition of a non-trivial partition is given in Sect. 11.5.

11.4 Synthesizing Support of an Itemset

We have designed technique PFT for mining multiple large databases (Adhikari et al. 2010b). It improves the quality of global pattern as compared to existing techniques. We shall use this technique here for mining multiple large databases. Let the itemset X be extracted from k out of n databases using PFT, $0 < k \leq n$. Let γ be the minimum threshold of number of extractions of an itemset, $0 < \gamma \leq 1$. We would be interested in an itemset if it has been extracted from at least $n \times \gamma$ databases. Such itemsets are called *HFIS*s. If an itemset X has high frequency then another itemset $Y \subseteq X$ also has high frequency, for $Y \neq \phi$. We define a high frequency itemset as follows.

Definition 11.1 Let there be n databases. Let X be an itemset extracted from k ($\leq n$) databases. Then X has high frequency if $\frac{k}{n} \geq \gamma$, where γ is the minimum threshold of the number of extractions of an itemset, $0 < \gamma \leq 1$.

A high frequency itemset might not be frequent in all the databases. A high frequency itemset may not be extracted naturally when we apply PFT. After applying PFT, we synthesize supports of *HFIS*s. In (Adhikari and Rao 2008a), A_2 measure has been proposed to find association among items in a database. A detailed study on existing measures of association is completed for computing association among items in a database. It has been shown that the measure A_2 is effective in computing association among items in an itemset. In Example 11.1, we illustrate the procedure for synthesizing support of a *HFIS*.

Example 11.1 Consider a multi-branch company that has four branches. Let D_i be the database corresponding to ith branch, $i = 1, 2, 3, 4$. The branch databases are given as follows. $D_1 = \{\{a, b\}, \{a, b, c\}, \{a, b, c, d\}, \{c, d, e\}, \{c, d, f\}, \{c, d, i\}\}$; $D_2 = \{\{a, b\}, \{a, b, g\}, \{g\}\}$; $D_3 = \{\{a, b, d\}, \{a, c, d\}, \{c, d\}\}$, $D_4 = \{\{a\}, \{a, b, c\}, \{c, d\}, \{c, d, i\}\}$. Assume that $\alpha = 0.4$, and $\gamma = 0.6$. Let $X(\eta)$ denote the fact that the itemset X has support η in the database. We sort databases in a non-increasing order on database size (expressed in bytes). The sorted databases are given as follows: D_1, D_4, D_3, D_2. By applying PFT, we obtain the following itemsets in different local databases:

$LPB(D_1, \alpha) = \{\{a\}(0.5), \{b\}(0.5), \{c\}(0.833), \{d\}(0.667), \{a, b\}(0.5), \{c, d\}$ $(0.667)\}$, $LPB(D_4, \alpha) = \{\{a\}(0.667), \{b\}(0.25), \{c\}(0.75), \{d\}(0.25), \{a, b\}$ $(0.333), \{c, d\}(0.667)\}$, $LPB(D_3, \alpha) = \{\{a\}(0.667), \{b\}(0.333), \{c\}(0.667), \{d\}$ $(1.0), \{a, b\}(0.333), \{c, d\}(0.667)\}$, and $LPB(D_2, \alpha) = \{\{a\}(0.667), \{b\}(0.667), \{c\}$ $(0.0), \{d\}(0.0), \{a, b\}(0.667), \{c, d\}(0.0), \{g\}(0.667)\}$.

Let D be the union of D_1, D_2, D_3 and D_4. Synthesized *HFIS*s in D are given as follows: $SHFIS(D, 0.4, 0.6) = \{\{a\}(0.563), \{b\}(0.438), \{c\}(0.563), \{d\}(0.563), \{a, b\}(0.438), \{c, d\}(0.5)\}$.

Since we apply PFT for mining multiple databases, the *RuleSynthesizing* algorithm (Wu and Zhang 2003) is not applicable here. Moreover, PFT is a direct approach for estimating support of a high frequency itemset.

The collection of *SHFIS*s of size greater than one forms the basis of the proposed clustering technique. We present below an algorithm to form synthesized association among items in each *SHFIS* of size greater than one. Let N be the number of itemsets in the given n databases. Let *AIS* be a two dimensional array such that *AIS* (i) is the array of itemsets extracted from D_i, $i = 1, 2, …, n$. Also, let *IS* be the set of all itemsets in n databases. An itemset could be described by the following attributes: *itemset*, *supp*, and *did*. Here, *itemset*, *supp* and *did* denote the itemset, support and database identification of an itemset, respectively. All the synthesized itemsets are kept in *SIS*, an array of synthesized itemset. Each synthesized itemsets has the following attributes: *itemset*, *ss*, and *sa*. Here, *ss* and *sa* denote the synthesized support and synthesized association of the itemset, respectively. In the following algorithm we synthesize association among items of each *SHFIS* (Adhikari 2013).

Algorithm 11.1. Synthesize association among items of each *SHFIS* of size greater than one.
procedure *SynthesizingAssociation* (*n, AIS, size, γ*)
Input:
n: number of databases
AIS: two-dimensional array of itemsets extracted during mining multiple databases
size: array of number of transactions in input databases
γ: threshold for minimum number of extractions of an itemset
Output:
Synthesized association among items of each *SHFIS*
01:   store all the local itemsets into array *IS*;
02:   sort itemsets of *IS* based on *itemset* attribute;
03:   add sizes of all branch databases into variable *totalTransactions*;
04:   **let** *nSynItemSets* = 0; **let** *i* = 1;
05:   **if** (*i* ≤ |*IS*|) **then**
06:      **let** *j* = *i*; **let** *count* = 0;
07:      **while** (*j* ≤ *i* + *n*) **do**
08:         **if** (*IS*(*j*).*itemset* = *IS*(*i*).*itemset*) **then**
09:            process support of *IS*(*i*);
10:            increase *count* by 1; increase *j* by 1;
11:         **else** go to line 14;
12:         **end if**
13:      **end while** {07}
14:      *synSupp* = *supp*$_s$(*IS*(*i*).*itemset*, *D*);
15:      **if** (*count* / *n* ≥ *γ*) **then**
16:         *SIS*(*nSynItemSets*). *ss* = *synSupp*;
17:         *SIS*(*nSynItemSets*). *itemset* = *IS*(*i*).*itemset*;
18:      **end if**
19:      **let** *i* = *j*;
20:      increase *nSynItemSets* by 1;
21:      go to line 5;
22:   **end if** {05}
23:   **for** *j* = 1 to *nSynItemSets* **do**
24:      **if** (|*SIS*(*nSynItemSets*). *itemset*| ≥ 2) **then**
25:         *SIS*(*nSynItemSets*). *sa* = A_2(*SIS*(*nSynItemSets*). *itemset*, *D*);
26:      **end if**
27:   **end for** {23}
end procedure

We sort itemsets of *IS*, so that the same itemset extracted from different data sources comes consecutive. It helps processing itemsets easier. We find total number of transactions in different databases into variable *totalTransactions*. The variables *nSynItemSets* and *i* keep track of the number synthesized itemsets and the current itemset of *IS*, respectively. The algorithm segment in lines 5–22 is repeated *N* times. An itemset gets processed in an iteration. An itemset occurs maximum

n times, since there are n databases. Thus, the *while*-loop in lines 7–13 repeats maximum n times. The variable *count* keeps track of number of times an itemset gets extracted. Based on variable *count* we could determine whether an itemset has high frequency. If an itemset is highly frequent then we store the details into array *SIS*, and increase *nSynItemSets* by 1. We update variable i by j for processing the next itemset. We go back to line 5 for processing the next itemset. Using lines 23–27, we calculate synthesized association (Adhikari and Rao 2008a), for each synthesized itemset of size greater than one. In the next paragraph, we determine the time complexity of above algorithm.

Line 1 takes $O(N)$ time, since there are N itemsets in n databases. Line 2 takes $O(N \times \log(N))$ time to sort N itemsets. Line 3 takes $O(n)$ time, since there are n databases. The *while* loop at line 7 repeats maximum n times. The *if* statement at line 5 repeats N times. Thus, time complexity of program segment in lines 5–22 is $O(n \times N)$. Line 23 takes $O(N)$ time. Let the average size of an itemset be p. The time complexity for searching an itemset in *IS* is $O(N)$. The time-complexity for computing association of an itemset is $O(N \times p^2)$. Thus, the time complexity of program segment in lines 23–27 is $O(N^2 \times p^2)$ time. Thus, the time complexity of procedure *SynthesizingAssociation* is equal to maximum $\{O(N^2 \times p^2), O(N \times \log(N)), O(n \times N)\} = O(N^2 \times p^2)$, since $N > \log(N)$ and $N > n$.

11.5 Clustering Local Frequency Items

Local frequency items are the basic components of many important patterns in multiple databases. The main objective of this study is to cluster local frequency items. The existing techniques (Wu et al. 2005; Adhikari and Rao 2008c) of clustering multiple databases work as follows. A measure of similarity between two databases is proposed. Let there be m databases to be clustered. Then the similarities for $^{m}C_2$ pairs of databases are computed. An algorithm is designed based on the measure of similarity. Then based on a level of similarity, the databases are clustered into different classes.

For the purpose of clustering databases, Wu et al. (2005) have proposed the following measure of similarity between two databases.

$$sim_1(D_1, D_2) = \frac{|I(D_1) \cap I(D_2)|}{|I(D_1) \cup I(D_2)|} \qquad (11.6)$$

Here $I(D_i)$ is the set of items in the database D_i, $i = 1, 2$. Thus, sim_1 is the ratio of number of items common to databases and the total number of items in the two databases. Measure sim_1 could also be used to find similarity between two items in a database. Also, we observe that the measure sim_1 could be obtained from A_2 (Adhikari and Rao 2008a). In other words, the measure sim_1 is a special case of measure A_2. In the following example, we show that association among items of an itemset could not be determined correctly using associations of all possible subsets of size 2. In particular, association among items of $\{a, b, c\}$ could not be correctly estimated by the associations between the items in $\{a, b\}$, $\{a, c\}$, and $\{b, c\}$. We explain this issue by using Example 11.2.

Example 11.2 Let $D_5 = \{\{a, b, c, d\}, \{a, b, c, e\}, \{a, b, d\}, \{a, e, f\}, \{b, c, e\}, \{d, e, g\}, \{d, f, g\}, \{e, f, g\}, \{e, f, h\}, \{g, h, i\}\}$. Also, let α be 0.2. The supports of relevant frequent itemsets are given as follows. $S(\{a\}, D_5) = 0.4$, $S(\{b\}, D_5) = 0.4$, $S(\{c\}, D_5) = 0.3$, $S(\{a, b\}, D_5) = 0.3$, $S(\{a, c\}, D_5) = 0.2$, $S(\{b, c\}, D_5) = 0.3$, $S(\{a, b, c\}, D_5) = 0.2$. Now, $sim_1(\{a, b\}, D_5) = 0.6$, $sim_1(\{a, c\}, D_5) = 0.4$, $sim_1(\{b, c\}, D_5) = 0.75$. Using sim_1, the items a, b, and c could be put in the same class at the level of similarity 0.4, i.e., $minimum\{0.6, 0.4, 0.75\}$. Using A_2, we have $A_2(\{a, b, c\}, D_5) = 0.67$. Thus, the items a, b, and c could be put in the same class at the level 0.67. We observe that the subset of transactions $\{\{a, b, c, d\}, \{a, b, c, e\}, \{a, b, d\}, \{a, e, f\}, \{b, c, e\}\}$ of D_5 that results the amount of association among a, b, and c. Three out of five transactions contain at least two items of $\{a, b, c\}$. Two out of five transactions contain all the items of $\{a, b, c\}$. More the number of items of $\{a, b, c\}$ occur together, higher is the association among items of $\{a, b, c\}$. Thus, we observe that the level of association among the items of $\{a, b, c\}$ is close to 0.67 rather than 0.4. Thus, we fail to measure association correctly among the items of $\{a, b, c\}$ based on the similarities between items of $\{a, b\}$, $\{a, c\}$, and $\{b, c\}$.

Example 11.2 shows that the existing clustering technique (Wu et al. 2005; Adhikari and Rao 2008c) might cluster a set of items at a low accuracy. This is due to the fact that we are unable to estimate the association among the items in an itemset. If we cluster the items at a given similarity level using the existing technique, then items in a class become more similar than the given level of similarity. Thus, the degree of similarity among items in a class is not true representative of the similarity level of a clustering algorithm. The proposed clustering algorithm improves the accuracy of clustering, and it is more visible in dense databases. Based on the above discussion, the following observations are made.

Observation 11.1 Let $X = \{x_1, x_2, ..., x_m\}$ be an itemset in database D. The *BestClassification* algorithm (Wu et al. 2005) might put items of X in a class at the level $minimum \{sim_1(x_i, x_j, D) : 1 \leq i < j \leq m\}$.

Observation 11.2 The proposed clustering technique might put items of X in a class at the level of association $A_2(\{x_1, x_2, ..., x_m\}, D)$.

A clustering of items results in a set of classes of items. A class of frequent items over $FI(1, n, \alpha)$ could be defined as follows.

Definition 11.1 A class $class^\delta$ formed at a level of association δ under the measure of association A_2 over $FI(1, n, \alpha)$ in database D is defined as $X \subseteq FI(1, n, \alpha)$ such that $A_2(X, D) \geq \delta$, and one of the following conditions is satisfied: (i) $X \in SHFIS(1, n, \alpha, \gamma)$, for $|X| \geq 2$, (ii) $X \subseteq FI(1, n, \alpha)$, for $|X| = 1$.

Definition 11.2 enables us to obtain and define a clustering of frequent items over $FI(1, n, \alpha)$ as follows.

Definition 11.2 Let π^δ be a clustering of local frequency items over $FI(1, n, \alpha)$ at level of association δ under the measure of association A_2. Then, $\pi^\delta = \{X: X$ is a class of type $class^\delta$ over $FI(1, n, \alpha)\}$.

We symbolize the ith class of π^δ as $CL_i^{\delta,\alpha}$, $i = 1, 2, ..., |\pi^\delta|$. A clustering might not include all the local frequency items. We might be interested in clustering all local frequency items. A complete clustering of local frequency items over $FI(1, n, \alpha)$ is defined as follows.

Definition 11.3 A clustering $\pi^\delta = \left\{ CL_1^{\delta,\alpha}, CL_2^{\delta,\alpha}, ..., CL_m^{\delta,\alpha} \right\}$ is complete, if $\bigcup_{i=1}^{m} CL_i^{\delta,\alpha} = FI(1, n, \alpha)$, where $CL_i^{\delta,\alpha}$ is a class of type $class^\delta$ over $FI(1, n, \alpha)$, $i = 1, 2, ..., m$.

Two classes in a clustering might not be mutually exclusive. We might be interested in determining a mutually exclusive clustering. A mutually exclusive clustering over $FI(1, n, \alpha)$ could be defined as follows.

Definition 11.4 A clustering $\pi^\delta = \left\{ CL_1^{\delta,\alpha}, CL_2^{\delta,\alpha}, ..., CL_m^{\delta,\alpha} \right\}$ is mutually exclusive if $CL_i^{\delta,\alpha} \cap CL_j^{\delta,\alpha} = \varphi$, $CL_i^{\delta,\alpha}$ and $CL_j^{\delta,\alpha}$ are classes of type $class^\delta$ over $FI(1, n, \alpha)$, $i \neq j$, $i, j = 1, 2, ..., m$.

One might be interested in finding out such a mutually exclusive and complete clustering. The objective of this task is to find the best non-trivial partition, if it exists, of local frequency items. First, we define a partition of local frequency items as follows.

Definition 11.5 A complete and mutually exclusive clustering is called a partition.

A clustering is not necessarily to be a partition. In most of the cases, a trivial partition might not be of interest to us. We define a non-trivial partition of local frequency items as follows.

Definition 11.6 A partition π is non-trivial if $1 < |\pi| < n$, where n is the number of items in different databases.

A partition is based on *SHFIS*s and associations among items in these itemsets. For this purpose, we need to synthesize association among items of every *SHFIS* of size greater than one. We define a synthesized association among items of a *SHFIS* as follows.

Definition 11.7 Let there be n different databases. Let $X \in SHFIS(D)$ such that $|X| \geq 2$. Synthesized association among the items of X, computed using A_2, is denoted by $SA(X, D| \alpha, \gamma)$.

To determine goodness of a partition, we need to measure dispersion among items in a 2-item *SHFIS*. We define synthesized dispersion, *SD* of an itemset of size 2, as follows.

Definition 11.8 Let there be n different databases. Let $X \in SHFIS(D)$ such that $|X| = 2$. Synthesized dispersion *SD* between the items of X is given by

$$SD(X, D|\alpha, \gamma) = 1 - SA(X, D|\alpha, \gamma). \tag{11.7}$$

We calculate synthesized associations of all *SHFIS*s of size greater than one. In Example 11.3, we calculate synthesized associations of itemsets in *SHFIS(D)*.

Example 11.3 We continue here the discussion of Example 11.1. Synthesized associations among items of relevant *SHFIS*s are given as follows: $SA(\{a, b\}, D) = 0.78$, $SA(\{c, d\}, D) = 0.80$. We arrange *SHFIS*s of size greater than one in non-increasing order on synthesized association. The arranged *SHFIS*s are given as follows: $\{c, d\}, \{a, b\}$. Also, $FI((1, n, \alpha)) = \{a, b, c, d, g\}$. There exist two non-trivial partitions. They are given as follows: $\pi^{0.80} = \{\{a\}, \{b\}, \{g\}, \{c, d\}\}$, and $\pi^{0.78} = \{\{g\}, \{a, b\}, \{c, d\}\}$.

A local frequency item that is extracted only from a few databases forms a singleton class. In above partitions, g is an example of such an item.

11.5.1 Finding the Best Non-trivial Partition

In Example 11.3, we observe the existence of two non-trivial partitions at the levels of association 0.80 and 0.78. We would like to find the best partition among the available non-trivial partitions. The best partition is based on the principle of maximizing the intra-class association and maximizing inter-class dispersion. Intra-class association and inter-class dispersion are defined as follows.

Definition 11.9 The *intra-class association* of partition π at the level of association δ under the measure of synthesized association *SA* is defined as follows.

$$intra\text{-}class\,association(\pi^{\delta}) = \sum_{C \in \pi, |C| \geq 2} SA(C|\alpha, \gamma). \tag{11.8}$$

In partition π, there may exist classes of size greater than or equal to 2. Intra-class association of π is obtained by adding synthesized associations of those classes.

Definition 11.10 The *inter-class dispersion* of partition π at the level of association δ under the measure of synthesized dispersion *SD* is defined as follows.

$$inter\text{-}class\,dispersion(\pi^{\delta}) = \sum_{C_p, C_q \in \pi; p \neq q} \sum_{a \in C_p, b \in C_q; \{a,b\} \in SHFIS} SD(\{a, b\}|\alpha, \gamma)$$

$$\tag{11.9}$$

Between the two classes C_p and C_q in π, the synthesized dispersion is obtained by adding synthesized dispersion of $\{a, b\}$, where $a \in C_p$, and $b \in C_q$, $p \neq q$. Then inter-class dispersion of π is obtained by adding synthesized dispersion between items in C_p and C_q, where C_p, $C_q \in \pi$, $p \neq q$. We would like to define goodness measure of a partition for the purpose of finding the best partition among available non-trivial partitions. We define goodness measure of a partition as follows.

Definition 11.11

$$goodness(\pi^\delta) = intra\text{-}class\ association(\pi^\delta) + inter\text{-}class\ dispersion(\pi^\delta) - |\pi^\delta|.$$
$$(11.10)$$

We have subtracted $|\pi^\delta|$ from the sum of intra-class association and inter-class dispersion to remove the bias of goodness value of a partition. A better partition is obtained at a higher goodness value. In Example 11.4, we calculate the goodness value of each partition obtained in Example 11.3.

Example 11.4 For the first partition, we get *intra-class association* $(\pi^{0.80}) = 0.80$, *inter-class dispersion* $(\pi^{0.80}) = 0.22$, and *goodness* $(\pi^{0.80}) = 1.02$. For the second partition, we get *intra-class association* $(\pi^{0.78}) = 1.58$, *inter-class dispersion* $(\pi^{0.78}) = 0.0$, and *goodness* $(\pi^{0.78}) = 1.58$. The goodness value of the second partition is more than that of the first partition. Thus, the best non-trivial partition of $FI(1, n, \alpha)$ is $\{\{a, b\}, \{c, d\}, \{g\}\}$, and it exists at level 0.78.

Let us look back at the databases given in Example 11.1. Items a and b appear together in most of the transactions, whenever one of them is present in a transaction. Also, items c and d appear together in most of the transactions, whenever item c or d is present in a transaction. Thus, we find that partition $\pi^{0.78}$ matches the ground reality better than partition $\pi^{0.80}$, and the output of clustering is consistent with the input. It validates the clustering technique proposed in this chapter. In the following lemma, we provide a set of necessary and sufficient conditions for the existence of a non-trivial partition.

Lemma 11.1 *Let there be n different databases. There exits a non-trivial partition of FI(1, n, α) if and only if there exists an itemset X∈SHFIS(1, n, α, γ) such that (i) | X| ≥ 2, and (ii) SA(Y, D) ≠ SA(Z, D), for all Y, Z ∈ SHFIS(1, n, α, γ), and |Y|, |Z| ≥ 2.*

Proof We sort *SHFIS*s in non-increasing order on synthesized association. Let *SA* (M, D) = maximum $\{SA(X, D): X \in HFIS(1, n, \alpha, \gamma),\ and\ |X| \geq 2\}$. Before the itemset M, there exists no itemset in *SHFIS*, since it has the maximum synthesized association. Thus, it is trivially mutually exclusive with the previous *SHFIS*s. Due to condition (ii), there exists a partition at the level $SA(M, D)$. In addition, the partition is non-trivial due to condition (i). This non-trivial partition contains a single class M having size ≥ 2. The remaining classes of this partition are singleton.

At two different levels of association δ_1, and δ_2 ($\neq \delta_1$), one may obtain the same partition.

Definition 11.12 Let $C \subseteq FI(1, n, \alpha)$ be a class such that $C \neq \phi$. Two partitions π^{δ_1} and π^{δ_2} are the same if the following statement is true: $C \in \pi^{\delta_1}$ if and only if $C \in \pi^{\delta_2}$, for $\delta_1 \neq \delta_2$.

There are $2^{|S|}$ elements in the power set of S. Also, there are two trivial partitions for a non-null set viz., (i) The partition containing singleton elements, and (ii) The partition containing only one set of all elements. Thus, the number of distinct non-trivial partitions of a non-null set is always less than or equal to $2^{|S|} - 2$, for any non-null set S. In Lemma 11.2, we find the upper bound of the number of non-trivial partitions.

Lemma 11.2 *Let there be n different databases. Then the number of distinct non-trivial partitions is less than or equal to $|\{X: |X| \geq 2$ and $X \in SHFIS(D)\}|$. Equality holds if and only if the following conditions are true.*

(i) *There does not exist a $X \in SHFIS(D)$, for $|X| \geq 3$.*
(ii) *$Y \cap Z = \phi$, for all $Y, Z \in SHFIS(D)$, and $|Y|, |Z| \geq 1$.*
(iii) *$SA(Y, D) \neq SA(Z, D)$, for all $Y, Z \in SHFIS(D)$, and $|Y|, |Z| \geq 2$.*

Proof We arrange *SHFISs* in a non-increasing order based on the synthesized association, for all *SHFISs* of size greater than one. Let the arranged *SHFISs* be X_1, X_2, ...,X_m, $m \geq 2$. There exists a partition at $SA(X_1)$, if conditions (ii) and (iii) are satisfied when $Y = X_1$ (Lemma 11.1). In general, there exists another partition at $SA(X_k, D)$, if conditions (ii) and (iii) are satisfied for $X_1, X_2, ..., X_{k-1}$. If X is a *SHFIS* then $Y \subset X$ is also a *SHFIS*, for $Y \neq \phi$. Two partitions could not exist at $SA(X, D)$ and $SA(Y, D)$, since $Y \subset X$. Thus, condition (i) is necessary at the equality. The lemma follows. □

Corollary 11.1 *Let there be n different databases. The set of all non-trivial partitions of $FI(1,n, \alpha)$ is $\{\pi^{SA(X, D)}: X \in SHFIS(D), |X| \geq 2$, and $\pi^{SA(X, D)}$ exists$\}$.*

Based on Observation 11.1 and Corollary 11.1, one could find the level of difference in clustering levels stated as follows:

Observation 11.3 Let π be clustering of $FI(1, n, \alpha)$ using the proposed technique. Let $X = \{x_1, x_2, ..., x_m\} \in SHFIS(D)$ such that $\delta = SA(X, D)$, for $|X| \geq 2$. The amount of difference in clustering levels between the existing technique and the proposed technique is equal to

$$\delta - minimum\{sim_1(x_i, x_j, D) : 1 \leq i \leq j \leq m\}. \quad (11.11)$$

Using Algorithm 11.1, we obtain synthesized association among items of each *SHEIS* of size greater than one. We use this information for finding the best non-trivial partition of local frequency items in multiple databases (Adhikari 2013).

Algorithm 11.2. Finding the best non-trivial partition (if it exists) of local frequency items multiple databases.

procedure *BestPartition* (m, S)
Input:
m: number of *SHFIS*s of size greater than one
S: array of *SHFIS*s of size greater than one
Output:
The best non-trivial partition (if it exists) of local frequency items in multiple databases
01:   arrange the elements of S in non-increasing order on synthesized association;
02:   **let** $S(m+1) = \phi$; **let** $SA(S(m+1), D) = 0$;
03:   **if** $(m = 1)$ **then**
04:      form a class using items in $S(1)$;
05:      **for** each item in $FI(1, n, \alpha) - S(1)$ **do**
06:         form a singleton class;
07:      **end for**
08:      a partition is formed at level $SA(S(1), D)$;
09:      return the partition;
10:   **end if**
11:   **let** $temp = \phi$; **let** $mutualExcl$ = false;
12:   **for** $i = 1$ to m **do**
13:      **if** $(temp \cap S(i) \neq \phi)$ **then** $mutualExcl$ = true; **end if**
14:      **if** $((mutualExcl = \text{false})$ and $(SA(S(i), D) \neq SA(S(i+1), D)))$ **then**
15:         **for** $j = 1$ to i **do**
16:            construct a class using items in $S(j)$;
17:            $temp = temp \cup S(j)$;
18:         **end for**
19:         **for** each item in $(FI(1, n, \alpha) - temp)$ **do**
20:            construct a singleton class;
21:         **end for**
22:         store the classes formed and the level of partition as $SA(S(i), D)$;
23:      **else if** $(mutualExcl = \text{true})$ **then** go to line 26; **end if**
24:      **end if**
25:   **end for**
26:   return the partition having the maximum goodness;
end procedure

If m is equal to 1 then we have only one non-trivial partition. All the items of *SHFIS* form a class and each of the remaining frequent items forms a singleton class. The partition is formed at the level of synthesized association among items in the *SHFIS*. The variable *temp* accumulates all the items in previous *SHFIS*s. The variable *mutualExcl* is used to check the mutually exclusiveness among the current *SHFIS* and all the previous *SHFIS*s. Also, we need to check another condition expressing whether the synthesized association of current *SHFIS* different from the synthesized association among items of the next *SHFIS*. The conditions for existence of a partition are checked at line 14. If a partition exists at the current level then the items in each of the previous *SHFIS*s form a class. Each of the remaining items forms a singleton class. If the current *SHFIS* is not mutually exclusive with each of the previous *SHFIS*s of *S* then no more partitions exist. Some useful explanations are also given in Lemma 11.3.

Line 1 requires $O(m \times \log(m))$ time. The *for* loop at line 5 takes $O(|FI|)$ time. The *for* loop at line 12 repeats *m* times. Let the average size of a class be *p*. The *for* loop at line 15 takes $O(m \times p)$ time. The *for*-loop at line 19 takes $O(|FI|)$ time. Each of the statements at line 13 and 17 takes $O(p \times |FI|)$ time for a single execution of line 11. Thus, the time complexity of the program segment lines 15–18 is $O(m \times p \times |FI|)$, for each iteration of *for*-loop at line 12. Also, the *for* loop at 19 takes $O(|FI|)$ time. Thus, the time complexity of program segment 12–25 is $O(m^2 \times p \times |FI|)$. Therefore, the time complexity of algorithm *BestPartition* is $O(m^2 \times p \times |FI|)$.

Lemma 11.3 *Algorithm BestPartition finds the best partition, if such a partition exists.*

Proof We arrange *SHFIS*s of size greater than one in non-increasing order on synthesized association. Existence of only one *SHFIS* of size greater than one implies the existence of only one non-trivial partition (Lemma 11.1). By default, it will be the best non-trivial partition.

Let there be *m* ($m \geq 2$) *SHFIS*s of size greater than one. Let the arranged *SHFIS*s be $X_1, X_2, \ldots, X_m$. Then we need to check for the existence of partitions only at levels *SA* (X_i, D), $i = 1, 2, \ldots, m$ (Corollary 11.1). So, we have used a *for* loop at line 12 to check for partitions at *m* discrete levels $SA(X_i, D)$, $i = 1, 2, \ldots, m$. At the *j*th iteration of *for* loop at line 12, we check the mutually exclusiveness of the itemset X_j with previous itemset X_k, for $k = 1, 2, \ldots, j - 1$. If each of $X_1, X_2, \ldots,$ and X_{j-1} is mutually exclusive with X_j and $SA(X_j, D) \neq SA(X_{j+1}, D)$ then the current partition is recorded. On the other hand, if each of $X_1, X_2, \ldots,$ and X_{j-1} is mutually exclusive with X_j and $SA(X_j, D) = SA(X_{j+1}, D)$ then the current partition is not recorded. At this point, we are not sure whether X_{j+1} is mutually exclusive with X_k, for $k = 1, 2, \ldots, j$. At the next iteration $i = j + 1$, the partition is recorded (if it exists), and it contains the itemsets $X_1, X_2, \ldots, X_{j+1}$ as classes of size greater than one and each of the remaining frequent items forms a singleton class, provided $SA(X_{j+1}, D) > SA(X_{j+2}, D)$. At the *j*th iteration of *for* loop at line 12, if each of $X_1, X_2, \ldots,$ and X_{j-1} is not mutually exclusive with X_j then no more partition exists. Thus, the algorithm works correctly.   □

11.5.2 Error Analysis

To evaluate the proposed clustering technique we have measured the error occurred in an experiment. Let the number of *SHFIS*s be *m*. Supports of all *SHFIS*s have been synthesized during the clustering process. There are several ways one could define error of an experiment. We have adopted the following two types of errors.

1. *Average Error* (AE):

$$AE(D, \alpha, \gamma) = \frac{1}{m} \sum_{i=1}^{m} |SS(X_i, D) - S(X_i, D)| \qquad (11.12)$$

2. *Maximum Error* (ME):

$$ME(D, \alpha, \gamma) = \text{maximum}\{ |SS(X_i, D) - S(X_i, D)|, \quad i = 1, 2, \ldots, m\} \quad (11.13)$$

$SS(X, D)$ and $S(X, D)$ denote synthesized support and (apriori) support of X in D respectively. Error of the experiment is relative to the number of transactions, number of items, and the length of a transaction in local databases. Thus, the error of the experiment needs to be expressed along with the average number of transactions (*ANT*), average number of items (*ANI*), and the average length of a transaction (*ALT*) in all branch databases.

11.6 Experimental Results

Experimental results are based on three real and two synthetic databases. The database *retail* (Frequent itemset mining dataset repository)[2] is obtained from an anonymous Belgian retail supermarket store. The database *mushroom* is also available in (See footnote 2). The database *ecoli* is a subset of *ecoli database* (UCI ML repository content summary)[3] and has been processed for the purpose of conducting experiment. Also, we have omitted non-numeric fields of *ecoli database* for the purpose of conducting experiments. Databases *random500* and *random1000* are generated synthetically for the purpose of conducting experiments. Let *NT, ALT, AFI,* and *NI* denote the number of transactions, ALT, average frequency of an item and number of items in the database, respectively. We present the characteristics of these databases in Table 11.2.

Each of the above databases is divided into 10 databases for the purpose of conducting experiments. The databases obtained from $R, M, E, R5$ and $R1$ are named as $R_i, M_i, E_i, R5_i,$ and $R1_i$ respectively, $i = 0, 1, \ldots, 9$. The databases $R_i, M_i, E_i, R5_i,$ and $R1_i$ are called input databases, $i = 0, 1, \ldots, 9$. Some characteristics of these input databases are presented in Tables 11.3 and 11.4. We present below the results of the

---

[2] Frequent itemset mining dataset repository, http://fimi.cs.helsinki.fi/data/.

[3] UCI ML repository content summary, http://www.ics.uci.edu/~mlearn/MLSummary.html.

Table 11.2 Dataset characteristics

Database(D)	NT	ALT	AFI	NI
Retail (R)	88,162	11.306	99.674	10,000
Mushroom (M)	8,124	24.000	1,624.800	120
Ecoli (E)	336	7.000	25.565	92
random500 (R5)	10,000	6.470	109.400	500
random1000 (R1)	10,000	12.486	111.858	1,000

Table 11.3 Characteristics of input databases generated from R, M and E

D	NT	ALT	AFI	NI	D	NT	ALT	AFI	NI
R_0	9,000	11.244	12.070	8,384	R_5	9,000	10.856	16.710	5,847
R_1	9,000	11.209	12.265	8,225	R_6	9,000	11.200	17.416	5,788
R_2	9,000	11.337	14.597	6,990	R_7	9,000	11.155	17.346	5,788
R_3	9,000	11.490	16.663	6,206	R_8	9,000	11.997	18.690	5,777
R_4	9,000	10.957	16.040	6,148	R_9	7,162	11.692	15.348	5,456
M_0	812	24.000	295.273	66	M_5	812	24.000	221.455	88
M_1	812	24.000	286.588	68	M_6	812	24.000	216.533	90
M_2	812	24.000	249.846	78	M_7	812	24.000	191.059	102
M_3	812	24.000	282.435	69	M_8	812	24.000	229.271	85
M_4	812	24.000	259.840	75	M_9	816	24.000	227.721	86
E_0	33	7.000	4.620	50	E_5	33	7.000	3.915	59
E_1	33	7.000	5.133	45	E_6	33	7.000	3.500	66
E_2	33	7.000	5.500	42	E_7	33	7.000	3.915	59
E_3	33	7.000	4.813	48	E_8	33	7.000	3.397	68
E_4	33	7.000	3.397	68	E_9	39	7.000	4.550	60

Table 11.4 Characteristics of input databases generated from $R5$ and $R1$

D	NT	ALT	AFI	NI	D	NT	ALT	AFI	NI
$R5_0$	1,000	6.367	10.734	500	$R5_5$	1,000	6.338	10.676	500
$R5_1$	1,000	6.502	11.004	500	$R5_6$	1,000	6.624	11.248	500
$R5_2$	1,000	6.402	10.804	500	$R5_7$	1,000	6.415	10.830	500
$R5_3$	1,000	6.523	11.046	500	$R5_8$	1,000	6.579	11.158	500
$R5_4$	1,000	6.298	10.596	500	$R5_9$	1,000	6.652	11.304	500
$R1_0$	1,000	6.421	5.437	997	$R1_5$	1,000	6.444	5.454	998
$R1_1$	1,000	6.414	5.436	996	$R1_6$	1,000	6.477	5.493	997
$R1_2$	1,000	6.556	5.578	996	$R1_7$	1,000	6.477	5.487	998
$R1_3$	1,000	6.529	5.534	999	$R1_8$	1,000	6.538	5.554	997
$R1_4$	1,000	6.500	5.544	992	$R1_9$	1,000	6.500	5.561	989

experiments based on the above database. We observe that some local frequency items might not get reported for some combination of input databases, α, and γ.

11.6.1 Overall Output

Using Algorithm 11.2, we have shown below the best non-trivial partition of local frequency items for each of the given databases.

(a) *Experiment with retail*: The set of frequent items in different databases are given as follows: $FI(0, 9, 0.1) = \{\{0\}, \{1\}, \{2\}, \{3\}, \{4\}, \{5\}, \{6\}, \{7\}, \{8\}, \{9\}, \{32\}, \{38\}, \{39\}, \{41\}, \{48\}\}$. It gives high frequency items in local databases generated from *retail* support level 0.1. *SHFIS*s of size greater than one along with their synthesized associations are given as follows: $\{39, 48\}$ (0.444), $\{39, 41, 48\}$ (0.394), $\{39, 41\}$ (0.264), $\{41, 48\}$ (0.251), $\{38, 39\}$ (0.181). The best non-trivial partition is given as follows: $\pi^{0.444} = \{\{0\}, \{1\}, \{2\}, \{3\}, \{4\}, \{5\}, \{6\}, \{7\}, \{8\}, \{9\}, \{32\}, \{38\}, \{41\}, \{39, 48\}\}$. We observe that items 39 and 48 are highly associated, and their estimated association is 0.444. Also, there exists high association among items 39, 41 and 48.

(b) *Experiment with mushroom*: The set of frequent items in different databases are given as follows: $FI(0, 9, 0.5) = \{\{1\}, \{2\}, \{3\}, \{6\}, \{7\}, \{9\}, \{10\}, \{11\}, \{23\}, \{24\}, \{28\}, \{29\}, \{34\}, \{36\}, \{37\}, \{38\}, \{39\}, \{48\}, \{52\}, \{53\}, \{54\}, \{56\}, \{58\ \}, \{59\}, \{61\}, \{63\}, \{66\}, \{67\}, \{76\}, \{85\}, \{86\}, \{90\}, \{93\}, \{94\}, \{95\}, \{99\}, \{101\}, \{102\ \}, \{110\}, \{114\}, \{116\}, \{117\}, \{119\}\}$. We have taken α quite high, since the database is dense. Items are repeated in many transactions. As a result, associations among some items are very high. Top ten *SHFIS*s of size greater than one along with their synthesized associations are given as follows: $\{34, 90\}$ (0.9999), $\{34, 86\}$ (0.9995), $\{34, 85\}$ (0.9956), $\{34, 36, 85\}$ (0.9897), $\{34, 36, 90\}$ (0.9879), $\{34, 85, 90\}$ (0.9801), $\{34, 36, 86\}$ (0.9778), $\{34, 86, 90\}$ (0.9774), $\{34, 85, 86\}$ (0.9683), $\{85, 86\}$ (0.9627). The transactions in different databases are highly similar, since two transactions in a database have many common items. Therefore, the best partition has been reported at very high level of association. The best non-trivial partition is given as follows: $\pi^{0.9999} = \{\{1\}, \{2\}, \{3\}, \{6\}, \{7\}, \{9\}, \{10\}, \{11\}, \{23\}, \{24\}, \{28\}, \{29\}, \{36\}, \{37\}, \{38\}, \{39\}, \{48\}, \{52\}, \{53\}, \{54\}, \{56\}, \{58\ \}, \{59\}, \{61\}, \{63\}, \{66\}, \{67\}, \{76\}, \{85\}, \{86\}, \{93\}, \{94\}, \{95\}, \{99\}, \{101\}, \{102\ \}, \{110\}, \{114\}, \{116\}, \{117\}, \{119\}, \{34, 90\}\}$.

(c) *Experiment with ecoli*: Dataset *ecoli* has fewer items. Also, the number of transaction is less. Therefore, the supports of different itemsets are relatively higher. The frequent items in different databases are given as follows: $FI(0, 9, 0.1) = \{\{0\}, \{20\}, \{23\}, \{24\}, \{25\}, \{26\}, \{27\}, \{28\}, \{29\}, \{30\}, \{31\}, \{32\}, \{33\}, \{34\}, \{35\}, \{36\}, \{37\}, \{38\}, \{39\}, \{40\}, \{41\}, \{42\}, \{43\}, \{44\}, \{45\}, \{46\}, \{47\}, \{48\}, \{49\}, \{50\}, \{51\}, \{52\}, \{54\}, \{56\}, \{57\}, \{58\}, \{59\}, \{61\}, \{63\}, \{64\}, \{65\}, \{66\}, \{67\}, \{68\}, \{69\}, \{70\}, \{71\},$

{72}, {73}, {74}, {75}, {76}, {77}, {78}, {79}, {80}, {81}, {92}, {100}}. Some itemsets such as {37}, {48}, {50}, {52}, and {48, 50} has very high support. Top ten *SHFISs* of size greater than one along with their synthesized associations are given as follows: {48, 50}(0.8036), {37, 48, 50}(0.2321), {48, 50, 52}(0.2292), {40, 48, 50}(0.2262), {44, 48, 50}(0.2262), {46, 48, 50}(0.1935), {37, 48}(0.1905), {44, 50}(0.1905), {48, 50, 51}(0.1905), {40, 48}(0.1845). The best non-trivial partition is given as follows: $\pi^{0.8036}$ = {{0}, {20}, {23}, {24}, {25}, {26}, {27}, {28}, {29}, {30}, {31}, {32}, {33}, {34}, {35}, {36}, {37}, {38}, {39}, {40}, {41}, {42}, {43}, {44}, {45}, {46}, {47}, {49}, {51}, {52}, {54}, {56}, {57}, {58}, {59}, {61}, {63}, {64}, {65}, {66}, {67}, {68}, {69}, {70}, {71}, {72}, {73}, {74}, {75}, {76}, {77}, {78}, {79}, {80}, {81}, {92}, {100}, {48, 50}}. The best partition has occurred at very high level.

(d) *Experiment with random500*: Items in *random500* are sparsely distributed. Therefore, a very few items are reported even at support level 0.02. The set of frequent items in different databases are given as follows: *FI*(0, 9, 0.02) = {{3}, {4}, {8}, {25}, {32}, {101}, {145}, {178}, {221}, {234}, {256}, {289}, {320}, {442}}. *SHFISs* of size greater than one along with their synthesized associations are given as follows: {3, 25} (0.073), {3, 4} (0.068), {8, 25} (0.057), {25, 32} (0.048), {101, 178} (0.035). The best non-trivial partition is given as follows: $\pi^{0.073}$ = {{4}, {8}, {32}, {101}, {145}, {178}, {221}, {234}, {256}, {289}, {320}, {442}, {3, 25}}. The best partition occurs at a low level of 0.073.

(e) *Experiment with random1000*: Dataset *random1000* is also sparsely distributed. So, very few items are reported at support level 0.01. The set of frequent items in different databases are given as follows: *FI*(0, 9, 0.01) = {{1}, {6}, {9}, {20}, {25}, {46}, {79}, {95}, {122}, {136}, {180}, {234}, {268}, {291}, {325}, {378}, {406}, {432}}. *SHFISs* of size greater than one along with their synthesized associations are given as follows: {1, 20} (0.052), {1, 9} (0.034), {25, 46} (0.030), {46, 95} (0.027), {122, 180} (0.021), {180, 234} (0.016), {291, 378} (0.015). The best non-trivial partition is given as follows: $\pi^{0.052}$ = {{6}, {9}, {25}, {46}, {79}, {95}, {122}, {136}, {180}, {234}, {268}, {291}, {325}, {378}, {406}, {432}, {1, 20}}. The best partition occurs at a low level of 0.052.

11.6.2 Synthesis of High Frequency Itemsets

Partitions are based on HFISs in multiple databases. An itemset may not get reported from all the *n* databases. Thus, it is required to estimate its support in the union of *n* databases. In the Tables 11.5, 11.6, 11.7, 11.8 and 11.9, we have presented a few itemsets and their errors of synthesizing supports in the union on all databases.

Table 11.5 Error in synthesizing supports of *HFIS*s in *retail* at $\alpha = 0.1$ and $\gamma = 0.6$ (least 15)

HFIS X	$|SS(X, R) - S(X, R)|$	HFIS X	$|SS(X, R) - S(X, R)|$	HFIS X	$|SS(X, R) - S(X, R)|$
{32}	0.0000	{39, 48}	0.0000	{3}	0.0020
{39}	0.0000	{9}	0.0017	{4}	0.0020
{38}	0.0000	{5}	0.0019	{7}	0.0020
{48}	0.0000	{6}	0.0019	{8}	0.0020
{38, 39}	0.0000	{2}	0.0019	{1}	0.0021

Table 11.6 Error in synthesizing supports of *HFIS*s in *mushroom* at $\alpha = 0.5$ and $\gamma = 0.6$ (least 15)

HFIS X	$|SS(X, M) - S(X, M)|$	HFIS X	$|SS(X, M) - S(X, M)|$	HFIS X	$|SS(X, M) - S(X, M)|$
{67}	0.0010	{53, 85}	0.0011	{86, 90}	0.0013
{34, 67}	0.0010	{53, 90}	0.0011	{39}	0.0018
{53, 86}	0.0010	{53}	0.0012	{24, 90}	0.0022
{34, 39}	0.0011	{85, 86}	0.0013	{24, 85}	0.0033
{34, 90}	0.0011	{85, 90}	0.0013	{24, 86}	0.0033

Table 11.7 Error in synthesizing supports of *HFIS*s in *ecoli* at $\alpha = 0.1$ and $\gamma = 0.6$ (least 15)

HFIS X	$|SS(X, E) - S(X, E)|$	HFIS X	$|SS(X, E) - S(X, E)|$	HFIS X	$|SS(X, E) - S(X, E)|$
{40, 50}	0.0013	{40}	0.0018	{50, 51}	0.0024
{50, 52}	0.0013	{48, 50, 52}	0.0018	{37, 48, 50}	0.0030
{37}	0.0014	{44, 48, 50}	0.0021	{40, 48, 50}	0.0033
{52}	0.0014	{44}	0.0024	{46, 48, 50}	0.0035
{48, 51}	0.0015	{46, 48}	0.0024	{48, 50, 51}	0.0037

(a) *Experiment with retail*: In Table 11.5, we present errors in synthesizing *HFIS*s. The dataset retail has a unique characteristic that some itemsets such as {32}, {38}, {39}, {48}, {38, 39} and {39, 48} are extracted from every branch database. Thus, the error in synthesizing support of each of the above *HFIS*s is zero. Their supports are exact in the union of all the branch databases.

(b) *Experiment with mushroom*: The transactions in local databases are highly similar, in the sense that two transactions in a database have many common items. Thus, we get many frequent itemsets in local databases even at a high value of α. The errors in synthesizing some *HFIS*s are presented in Table 11.6. No itemset has been reported from all the local databases. But, there exist many itemsets that are extracted from eight/nine out of ten local databases.

(c) *Experiment with ecoli*: The average size of databases and the average length of transactions in different databases from *ecoli* are smaller. Thus, the error of synthesizing a *HFIS* is relatively higher. The errors in synthesizing some *HFIS*s are presented in Table 11.7. No itemset has been reported from all the local databases. But, there exist many itemsets that are extracted from eight/nine out of ten local databases.

Table 11.8 Error in synthesizing supports of HFISs in *random500* at $\alpha = 0.02$ and $\gamma = 0.6$ (least 15)

HFIS X	|SS(X, R) − S(X, R)|	HFIS X	|SS(X, R) − S(X, R)|	HFIS X	|SS(X, R) − S(X, R)|
{4}	0.0009	{178}	0.0023	{3, 4}	0.0026
{3}	0.0011	{221}	0.0025	{256}	0.0028
{8}	0.0014	{234}	0.0025	{3, 25}	0.0028
{25}	0.0020	{101}	0.0026	{25, 32}	0.0031
{3, 25}	0.0023	{8, 25}	0.0026	{101, 178}	0.0035

Table 11.9 Error in synthesizing supports of HFISs at *random1000* at $\alpha = 0.02$ and $\gamma = 0.6$ (least 15)

HFIS X	|SS(X, R) − S(X, R)|	HFIS X	|SS(X, R) − S(X, R)|	HFIS X	|SS(X, R) − S(X, R)|
{9}	0.0000	{291}	0.0018	{25, 46}	0.0028
{1}	0.0008	{268}	0.0021	{46, 95}	0.0028
{79}	0.0012	{136}	0.0023	{122, 180}	0.0030
{406}	0.0015	{46}	0.0026	{291, 378}	0.0033
{122}	0.0018	{1, 9}	0.0028	{180, 234}	0.0033

(d) *Experiment with random500*: No itemset has been reported from all the local databases. The databases generated from *random500* are uniform. Many synthesized itemsets extracted have similar supports in multiple databases. In Table 11.8, we present errors in synthesizing *HFISs* in *random500*.

(e) *Experiment with random1000*: The databases generated from *random1000* are uniform. Many synthesized itemsets extracted have similar supports in multiple databases. We observe that itemset {9} is extracted from every branch database. Thus, the error in synthesizing support of {9} is zero. In Table 11.9, we present errors in synthesizing *HFISs* in *random1000*.

11.6.3 Error Quantification

The error obtained in the experiment is based on the definitions given in (11.11) and (11.1). The errors reported in different experiments are presented in Table 11.10.

If the ANT in different databases increases then the average error of synthesizing *HEISs* is likely to decrease, provided the average length of transactions and the ANI remain constant. If the average length of transactions in different databases increases then the average error of synthesizing *HEISs* is likely to increase, provided the ANT and the ANI remain constant. Lastly, if the ANI in different databases increases then the error of synthesizing *HEISs* is likely to decrease, provided the ANT and the average length of transactions remain constant.

Table 11.10 Error of the experiments

D	α	γ	AE (ANT, ALT, ANI)	ME (ANT, ALT, ANI)
$\bigcup_{i=0}^{9} R_i$	0.1	0.7	0.0012	0.0029
			(8,816.2, 11.31, 5,882.1)	(8,816.2, 11.31, 5,882.1)
$\bigcup_{i=0}^{9} M_i$	0.5	0.7	0.0013	0.0038
			(812.4, 24.00, 80.7)	(812.4, 24.00, 80.7)
$\bigcup_{i=0}^{9} E_i$	0.5	0.7	0.0013	0.0041
			(33.6, 7.00, 56.5)	(33.6, 7.00, 56.5)
$\bigcup_{i=0}^{9} R5_i$	0.02	0.7	0.0010	0.0042
			(1,000, 6.47, 500)	(1,000, 6.47, 500)
$\bigcup_{i=0}^{9} R1_i$	0.01	0.7	0.0009	0.0038
			(1,000, 6.48, 995.9)	(1,000, 6.48, 995.9)

11.6.4 Average Error Versus γ

Here γ represents threshold of minimum number of extractions from different data sources. We have conducted experiments to study the behaviour of AE over different γs. The purpose of the experiment would be lost if we keep γ at a high value, since the number of *HEIS*s also decreases as γ increases. Thus, a decision based on *HEIS*s would have low validity at a high value of γ. In Figs. 11.1, 11.2, 11.3, 11.4 and 11.5, we present graphs of AE plotted versus γ obtained in different experiments.

Fig. 11.1 AE versus γ at $\alpha = 0.1$ (*retail*)

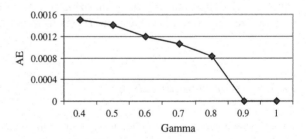

Fig. 11.2 AE versus γ at $\alpha = 0.5$ (*mushroom*)

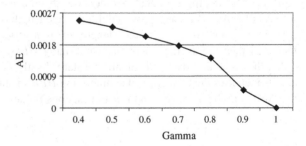

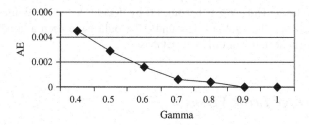

Fig. 11.3 AE versus γ at $\alpha = 0.1$ (*ecoli*)

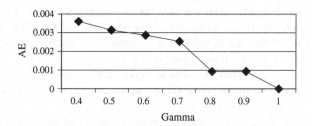

Fig. 11.4 AE versus γ at $\alpha = 0.02$ (*random500*)

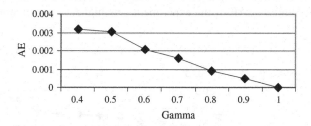

Fig. 11.5 AE versus γ at $\alpha = 0.01$ (*random1000*)

For databases generated from *retail*, the same itemsets are reported from all the local databases when γ is greater than or equal to 0.9. As a result, the AE becomes 0. In that case, there is no need to estimate the support of an itemset.

For datasets *retail* and *mushroom*, the values of AE decrease gradually when $\gamma \in$ [0.4, 0.7]. But, AE falls at a faster rate when $\gamma \in$ [0.8, 0.9].

Also, for databases generated from *ecoli* dataset, the same itemsets are reported from all the local databases when γ is greater than or equal to 0.9. AE falls gradually when γ varies from 0.4 to 0.9.

The behaviours of graph AE versus γ remain more or less the same for databases generated from *random500 random1000*. In general, we find that AE decreases as γ increases. In an extreme case, when γ is 1.0, no itemset has become frequent in all the local databases. Thus, AE is becomes 0 by default. From the figures presented above, we find that the value of γ around 0.7 would have been a good choice for clustering frequent items in different databases.

11.6.5 Average Error Versus α

Here α is user-defined level of minimum support. We have also conducted experiments to study the behaviour of AE over different αs. In Figs. 11.6, 11.7, 11.8, 11.9 and 11.10, we present graphs of AE versus α for different experiments.

In each of these experiments, we have taken around ten values of α. In dataset *retail*, itemsets are reported at a moderate value of α. Here we have started α at 0.05.

As noted earlier, the dataset *mushroom* is dense. Itemsets are reported even at a higher value of α. Here we have started α at 0.1.

Dataset *ecoli* is small. Thus, itemsets are reported at a higher value of α. Here we have started α at 0.1. In all the above graphs, we observe a slow and steady increase of the values of AE when α increases.

The behaviour of graph AE versus α remains more or less the same for databases generated from *random500* and *random1000*, since all these datasets are synthetic. As α increases, the larger number of databases would fail to extract an itemset. Thus, the error of synthesizing an itemset is likely to increase with the increase of α. In general, we find that AE of the experiment gets higher as α increases.

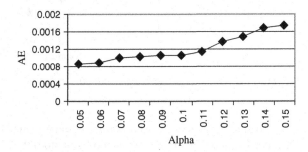

Fig. 11.6 AE versus α at $\gamma = 0.7$ (*retail*)

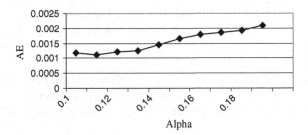

Fig. 11.7 AE versus α at $\gamma = 0.7$ (*mushroom*)

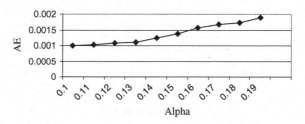

Fig. 11.8   AE versus α at $\gamma = 0.7$ (*ecoli*)

Fig. 11.9   AE versus α at $\gamma = 0.7$ (*random500*)

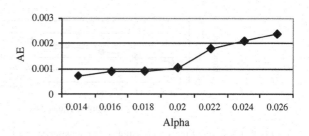

Fig. 11.10   AE versus α at $\gamma = 0.7$ (*random1000*)

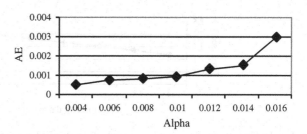

11.6.6   Clustering Time Versus Number of Databases

We have also studied the behaviour of clustering time required over the number of databases used in an experiment. In Figs. 11.11, 11.12, 11.13, 11.14 and 11.15, we present graphs of clustering time versus number of databases used in different experiments.

Although the number of transactions is less for *mushroom* as compared to *retail*, but the number of items in a transaction in *mushroom* is almost double than that of *retail*. As a result, graphs in Figs. 11.11 and 11.12 become comparable.

The sizes of local databases generated from *ecoli* are the smallest. Thus, the increment of clustering time is the least among all the five experiments. There is an increment of 0.17 s, on an average, for processing an extra database generated from *ecoli*.

The graph of clustering time using dataset *random1000* is steeper than that of dataset *random500*, since the local databases generated from *random1000* are

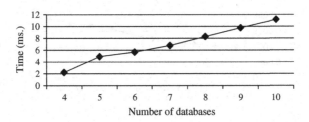

Fig. 11.11 Clustering time versus number of databases at $\alpha = 0.1$ and $\gamma = 0.7$ (*retail*)

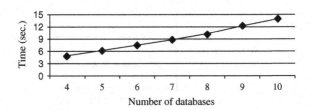

Fig. 11.12 Clustering time versus number of databases at $\alpha = 0.5$ and $\gamma = 0.7$ (*mushroom*)

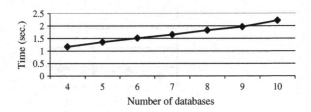

Fig. 11.13 Clustering time versus number of databases at $\alpha = 0.5$ and $\gamma = 0.7$ (*ecoli*)

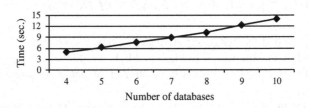

Fig. 11.14 Clustering time versus number of databases at $\alpha = 0.02$ and $\gamma = 0.7$ (*random500*)

larger. As the number of databases increases, the number of frequent itemsets also increases. In general, we find that clustering time increases as the number of databases increases.

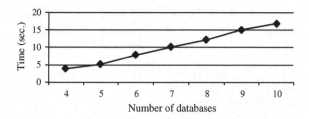

Fig. 11.15 Clustering time versus number of databases at $\alpha = 0.01$ and $\gamma = 0.7$ (*random1000*)

11.6.7 Comparison with Existing Technique

We have explained in Example 11.2 why the existing technique (Wu et al. 2005; Adhikari and Rao 2008c) could not cluster the local frequency items with high accuracy. Also, we derived an expression for difference occurring at clustering levels, see Observation 11.3. We observe that the proposed clustering technique is likely to cluster local frequency items at higher level, if a high frequency itemset of size greater than two has been synthesized. In each of Tables 11.11, 11.12, 11.13, 11.14 and 11.15, we present and compare partitions using the proposed technique and the existing technique, where δ is the level of clustering.

In dataset *retail*, partitions are reported when $\alpha = 0.1$ and $\gamma = 0.7$. Best partition is judged using the goodness measure as stated in (11.13). The dataset *retail* reports synthesized 3-itemsets even at a moderate value of α. In most of the cases, the synthesized support is higher than the maximum of synthesized supports of all the three 2-itemsets generated from it.

The dataset *mushroom* is dense. Many synthesized 3-itemsets are reported from multiple databases generated from *mushroom*. Thus, partitions exists at high value of $\alpha = 0.5$.

The dataset *ecoli* is small. Many synthesized 3-itemsets are reported from multiple databases generated from *ecoli*. Thus, partitions exists even at moderate value of $\alpha = 0.1$.

The datasets *random500* and *random1,000* are sparse. No k-itemset is reported ($k \geq 3$) when $\alpha = 0.1$ and $\gamma = 0.7$. We observe that the existing approach (Wu et al. 2005; Adhikari and Rao 2008c) and proposed approach produce the same result when a partition contains a class of maximum size 2. As a result, both the clustering techniques cluster items at the same level.

Table 11.11 A sample clustering of local frequency items in multiple databases (*retail*)

α	γ	Partition	δ (existing approach)	δ (proposed approach)	Difference in clustering levels
0.1	0.7	{{0}, {1}, {2}, {3}, {4}, {5}, {6}, {7}, {8}, {9}, {32}, {38, 39}, {39, 41, 48}}	0.251	0.394	0.143

Table 11.12 A sample clustering of frequent items in multiple databases (*mushroom*)

α	γ	Partition	δ (existing approach)	δ (proposed approach)	Difference in clustering levels
0.5	0.7	{{1}, {2}, {3}, {6}, {7}, {9}, {10}, {11}, {23}, {24}, {28}, {29}, {37}, {38}, {39}, {48}, {52}, {53}, {54}, {56}, {58}, {59}, {61}, {63}, {66}, {67}, {76}, {86}, {90}, {93}, {94}, {95}, {99}, {101}, {102}, · {110}, {114}, {116}, {117}, {119}, {34, 90}, {34, 86}, {34, 36, 85}}	0.843	0.990	0.147

Table 11.13 A sample clustering of frequent items in multiple databases (*ecoli*)

α	γ	Partition	δ (existing approach)	δ (proposed approach)	Difference in clustering levels
0.1	0.7	{{0}, {20}, {23}, {24}, {25}, {26}, {27}, {28}, {29}, {30}, {31}, {32}, {33}, {34}, {35}, {36}, {38}, {39}, {40}, {41}, {42}, {43}, {44}, {45}, {46}, {47}, {49}, {51}, {52}, {54}, {56}, {57}, {58}, {59}, {61}, {63}, {64}, {65}, {66}, {67}, {68}, {69}, {70}, {71}, {72}, {73}, {74}, {75}, {76}, {77}, {78}, {79}, {80}, {81}, {92}, {100}, {37, 48, 50}}	0.179	0.232	0.053

Table 11.14 A sample clustering of frequent items in multiple databases (*random500*)

α	γ	Partition	δ (existing approach)	δ (proposed approach)	Difference in clustering levels
0.1	0.7	{{4}, {8}, {32}, {101}, {145}, {178}, {221}, {234}, {256}, {289}, {320}, {442}, {3, 25}}	0.073	0.073	0.0

Table 11.15 A sample clustering of frequent items in multiple databases (*random1000*)

α	γ	Partition	δ (existing approach)	δ (proposed approach)	Difference in cluster-ing levels
0.1	0.7	{{6}, {9}, {25}, {46}, {79}, {95}, {122}, {136}, {180}, {234}, {268}, {291}, {325}, {378}, {406}, {432}, {1, 20}}	0.052	0.052	0.0

11.7 Conclusions

Clustering relevant objects is an important task in many decision support systems. In this chapter, we have proposed a new clustering technique for clustering local frequency items in multiple databases. For the purpose of applying the new clustering technique, we have proposed an algorithm for synthesizing HFISs. We have observed that an existing clustering technique might cluster local frequency items at a low level even when the associations among items in the itemsets are high. But the proposed clustering technique might cluster local frequency items at a higher level by capturing association among items using measure A_2. Especially, when the items in a database are highly associated, the proposed clustering technique clusters items at a higher level. The main problem with an existing clustering technique is that it might not be able to estimate association among items in a class with high accuracy. Thus, it might fail to cluster local frequency items at the higher level. We have shown that the proposed clustering method is sound. In case of dense databases, there is a considerable difference in clustering levels between the partitions made by the existing method and the proposed method. This is evident from the experimental results presented in Tables 11.10, 11.11 and 11.12. To apply the proposed technique, it does not matter whether the databases are required to be generated by real applications or synthetic database generators, since the algorithm is not dependent on how the databases were formed. The experimental results show that the proposed clustering technique is effective.

References

Adhikari A (2012) Synthesizing global exceptional patterns in different data sources. J Intell Syst 21(3):293–323

Adhikari A (2013) Clustering local frequency items in multiple databases. Inf Sci 237:221–241

Adhikari A, Rao PR (2008a) Capturing association among items in a database. Data Knowl Eng 67 (3):430–443

Adhikari A, Rao PR (2008b) Mining conditional patterns in a database. Pattern Recogn Lett 29 (10):1515–1523

Adhikari A, Rao PR (2008c) Efficient clustering of databases induced by local patterns. Decis Support Syst 44(4):925–943

Adhikari A, Rao PR (2008d) Synthesizing heavy association rules from different real data sources. Pattern Recogn Lett 29(1):59–71

Adhikari J, Rao PR, Adhikari A (2009) Clustering items in different data sources induced by stability. Int Arab J Inf Technol 6(4):394–402

Adhikari A, Ramachandrarao P, Pedrycz W (2010a) Developing multi-database mining applications. Springer, London

Adhikari A, Rao PR, Prasad B, Adhikari J (2010b) Mining multiple large data sources. Int Arab J Inf Technol 7(2):243–251

Adhikari A, Ramachandrarao P, Pedrycz W (2011a) Study of select items in different data sources by grouping. Knowl Inf Syst 27(1):23–43

Adhikari J, Rao PR, Pedrycz W (2011b) Mining icebergs in time-stamped databases. In: Proceedings of Indian international conferences on artificial intelligence, pp 639–658

Aggarwal C, Yu P (1998) A new framework for itemset generation. In: Proceedings of principles of database systems (PODS), pp 18–24

Agrawal R, Shafer J (1996) Parallel mining of association rules. IEEE Trans Knowl Data Eng 8 (6):962–969

Agrawal R, Srikant R (1994) Fast algorithms for mining association rules. In: Proceedings of the international conference on very large data bases, pp 487–499

Agrawal R, Imielinski T, Swami A (1993) Mining association rules between sets of items in large databases. In: Proceedings of SIGMOD conference on management of data, pp 207–216

Ali K, Manganaris S, Srikant R (1997) Partial classification using association rules. In: Proceedings of the 3rd international conference on knowledge discovery and data mining, pp 115–118

Brin S, Motwani R, Ullman JD, Tsur S (1997) Dynamic itemset counting and implication rules for market basket data. In: Proceedings of SIGMOD conference, pp 255–264

Chattratichat J, Darlington J, Ghanem M, Guo Y, Hüning H, Köhler M, Sutiwaraphun J, To HW, Yang D (1997) Large scale data mining: challenges and responses. In: Proceedings of the third international conference on knowledge discovery and data mining, pp 143–146

Chen L, Zou L, Tu L (2012) A clustering algorithm for multiple data streams based on spectral component similarity. Inf Sci 183(1):35–47

Cheung D, Ng V, Fu A, Fu Y (1996) Efficient mining of association rules in distributed databases. IEEE Trans Knowl Data Eng 8(6):911–922

Duan L, Street WN (2009) Finding maximal fully-correlated itemsets in large databases. In: Proceedings of ninth international conference on data mining (ICDM), pp 770–775

Estivill-Castro V, Yang J (2004) Fast and robust general purpose clustering algorithms. Data Min Knowl Disc 8(2):127–150

Han J, Kamber M (2011) Data mining: concepts and techniques, 3rd edn. Morgan Kauffmann Publishers, Los Altos

Han J, Pei J, Yiwen Y (2000) Mining frequent patterns without candidate generation. In: Proceedings of SIGMOD conference on management of data, pp 1–12

He D, Wu X, Zhu X (2010) Rule synthesizing from multiple related databases. In: Proceedings of advances in knowledge discovery and data mining (PAKDD)(2), pp 201–213

Hershberger SL, Fisher DG (2005) Measures of association, encyclopedia of statistics in behavioral science. Wiley, New Jersey

Jain AK, Murty MN, Flynn PJ (1999) Data clustering: a review. ACM Comput Surv 31 (3):264–323

Lee J-S, Ólafsson S (2011) Data clustering by minimizing disconnectivity. Inf Sci 181(4):732–746

Liu CL (1985) Elements of discrete mathematics. McGraw-Hill, New York

Liu H, Lu H, Yao J (2001) Toward multi-database mining: identifying relevant databases. IEEE Trans Knowl Data Eng 13(4):541–553

Malinen MI, Fränti P (2012) Clustering by analytic functions. Inf Sci 217:31–38

Mampaey M, Vreeken J (2013) Summarizing categorical data by clustering attributes. Data Min Knowl Disc 26(1):130–173

Piatetsky-Shapiro G (1991) Discovery, analysis, and presentation of strong rules. In: Proceedings of knowledge discovery in databases, pp 229–248

Savasere A, Omiecinski E, Navathe S (1995) An efficient algorithm for mining association rules in large databases. In: Proceedings of the 21st international conference on very large data bases, pp 432–443

Tan P-N, Kumar V, Srivastava J (2003) Selecting the right interestingness measure for association patterns. In: Proceedings of international conference on knowledge discovery and data mining (SIGKDD), pp 32–41

Wu X, Zhang S (2003) Synthesizing high-frequency rules from different data sources. IEEE Trans Knowl Data Eng 14(2):353–367

Wu X, Zhang C, Zhang S (2005) Database classification for multi-database mining. Inf Syst 30 (1):71–88

Yakut I, Polat H (2012) Privacy-preserving hybrid collaborative filtering on cross distributed data. Knowl Inf Syst 30(2):405–433

Zhang T, Ramakrishnan R, Livny M (1997) BIRCH: a new data clustering algorithm and its applications. Data Min Knowl Disc 1(2):141–182

Zhang S, Wu X, Zhang C (2003) Multi-database mining. IEEE Comput Intell Bull 2(1):5–13

Zhang C, Liu M, Nie W, Zhang S (2004a) Identifying global exceptional patterns in multi-database mining. IEEE Comput Intell Bull 3(1):19–24

Zhang S, Zhang C, Wu X (2004b) Knowledge discovery in multiple databases. Springer, London

Zhou W, Xiong H (2009) Efficient discovery of confounders in large data sets. In: Proceedings of ninth international conference on data mining (ICDM), pp 647–656

Chapter 12
Mining Patterns of Select Items in Different Data Sources

A number of important decisions are based on a set of specific items in a database called *select items*. Thus the analysis of select items in multiple databases becomes of primordial relevance. In this chapter, we focus on the following issues. First, a model of mining global patterns of select items from multiple databases is presented. Second, a measure of quantifying an overall association between two items in a database is discussed. Third, we present an algorithm that is based on the proposed overall association between two items in a database for the purpose of grouping the frequent items in multiple databases. Each group contains a select item called the *nucleus item* and the group grows while being centered around the nucleus item. Experimental results are concerned with some synthetic and real-world databases.

12.1 Introduction

In Chap. 6, we have presented a generalized multi-database mining technique, MDMT: PFT + SPS, by combining pipelined feedback technique (PFT) and simple pattern synthesizing (SPS) algorithm. We have noted that one could develop a multi-database mining application using MDMT: PFM + SPS which performs reasonably well. The following question arises as to whether MDMT: PFM + SPS is the most suitable technique for mining multiple large databases in all situations. In many applications, one may need to extract true non-local patterns of a set of specific items present in multiple large databases. In such applications, MDMT: PFM + SPS could not be fully endorsed as it may return approximate non-local patterns. In this chapter, we present a technique that extracts genuine global patterns of a set of specific items from multiple large databases.

Many decisions are based on a set of specific items called *select items*. Let us highlight several decision support applications where the decisions are based on the performance of select items.

© Springer International Publishing Switzerland 2015
A. Adhikari and J. Adhikari, *Advances in Knowledge Discovery in Databases*,
Intelligent Systems Reference Library 79, DOI 10.1007/978-3-319-13212-9_12

- Consider a set of items (products) that are profit making. We could consider them as the select items in this context. Naturally, the company would like to promote them. There are various ways one could promote an item. An indirect way of promoting a select item is to promote items that are positively associated with it. The implication of positive association between a select item P and another item Q is that if Q is purchased by a customer then P is likely to be purchased by the same customer at the same time. In this way, item P becomes indirectly promoted. It is important to identify the items that are positively associated with a select item.
- Each of the select items could be of high standard. Thus, they bring goodwill for the company. They help promoting other items. Therefore it is essential to know how the sales of select items affect other items. Before proceeding with such analyses, one may need to identify the items that are positively associated with the select items.
- Again, each of the select items could be a low-profit making product. From this perspective, it is important to know how they promote the sales of other items. Otherwise, the company could stop dealing with those products.

In general, the performance of select items could affect many decision making problems. Thus a better, more comprehensive analysis of select items might lead to better decisions. We study the select items based on the frequent itemsets extracted from multiple databases. The first question is whether a "traditional" data mining technique could provide a good solution when dealing with multiple large databases. The "traditional" way of mining multiple databases might not provide a sound solution due to several reasons:

- The company might have to employ parallel hardware and software to deal with a sheer volume of data.
- A single computer might take unreasonable amount of time to mine a large volume of data. In some extreme cases, it might not be feasible to carry data mining.
- A traditional data mining algorithm might extract a large number of patterns comprising many irrelevant items. Thus the processing of patterns could be complex and time consuming.

Therefore, the traditional way of mining multiple databases could not provide an efficient solution to the problem. In this situation, one could apply local pattern analysis (Zhang et al. 2003). Given this model of mining multiple databases, each branch of a company requires to mine its local database by utilizing some traditional data mining technique. Afterwards, each branch forwards the pattern base to the central office. The central office processes such pattern bases collected from different branches and synthesizes the global patterns and eventually makes decisions. Due to the reasons stated above, the local pattern analysis would not be a judicious choice to solve the proposed problem.

Each local pattern base might contain a large number of patterns consisting of many irrelevant items. Under these circumstances, the data analysis becomes complicated and time consuming. A pattern of a select item might be absent in

some local pattern bases. One may be required to estimate or ignore some patterns
in certain databases. Therefore we may fail to report the true global patterns of
select items in the union of all local databases. All in all, we conclude that the local
pattern analysis alone might not provide a good solution to the problem.

Due to difficulties identified above, we aim at developing a technique that mines
true global patterns of select items in multiple databases. There are two apparent
advantages of using such technique. First, the synthesized global patterns are exact.
In other words, there is no necessity to estimate patterns in some databases. Second,
one could avoid dealing with huge volumes of data.

12.2 Mining Global Patterns of Select Items

In Fig. 12.1, we visualize an essence of the technique of mining global patterns of
select items in multiple databases (Adhikari et al. 2011). It consists of the following
steps:

1. Each branch constructs the database and sends it to the central office.
2. Also, each branch extracts patterns from its local database.
3. The central office amalgamates these forwarded databases into a single database
   FD.
4. A traditional data mining technique is applied to extract patterns from *FD*.
5. The global patterns of select items could be extracted effectively from local
   patterns and the patterns extracted from *FD*.

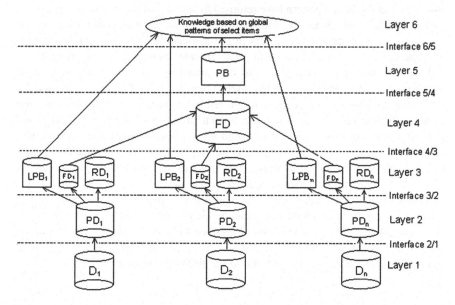

Fig. 12.1 A multilevel process of mining global patterns of select items in multiple databases

In Sect. 12.4, we will explain steps 1–5 with the help of a specific illustrative example. The local databases are located at the bottom level of the figure. We need to process these databases as they may not be at the appropriate state to realize the mining task. Various data preparation techniques (Pyle 1999) like data cleaning, data transformation, data integration, data reduction, etc. are applied to the data present in local databases. We produce local processed database PD_i for the ith branch, for $i = 1, 2, ..., n$. The proposed model comes with a set of interfaces combined with a set of layers. Each interface delivers a set of operations that produces dataset(s) (or knowledge) based on the dataset(s) available at the lower level. There are five interfaces in the proposed model. The functions of the interfaces are described below.

Interface 2/1 is used to clean/transform/integrate/reduce data present at the lowest level. By applying these procedures we construct database resulting from the original database. These operations are carried out at the respective branch. We apply an algorithm (located at interface 3/2) to partition a local database into two parts: forwarded database and remaining database. It is easy to find the forwarded database corresponding to a given database. In the following paragraph, we discuss how to construct FD_i, from D_i, $i = 1, 2, ..., n$.

Initially, FD_i is kept empty. Let T_{ij} be the jth transaction of D_i, $j = 1, 2, ..., |D_i|$. For D_i, a for-loop on j would run $|D_i|$ times. At the jth iteration, the transaction T_{ij} is tested. If T_{ij} contains at least one of the select items then FD_i is updated, resulting in the union $FD_i \cup \{T_{ij}\}$. At the end of the for-loop completed for j, FD_i is constructed.

A transaction related to select items might contain items other than those being selected. A traditional data mining algorithm could be applied to extract patterns from FD. Let PB be the pattern base returned by a traditional data mining algorithm (at the interface 5/4). Since the database FD is not large, one could reduce further the values of user-defined parameters of the association rules, like minimum support and minimum confidence, so that PB contains more patterns of select items. A better analysis of select items could be realized by using more patterns. If we wish to study the association between a select item and other frequent items then the exact support values of other items might not be available in PB. In this case, the central office sends a request to each branch office to forward the details (like support values) of some items that would be required to study the select items. Hence each branch applies a "traditional" mining algorithm (at interface 3/2) which is completed on its local database and forwards the details of local patterns requested by the central office. Let LPB_i be the details of ith local pattern base requested by the central office, $i = 1, 2, ..., n$. A global pattern mining application of select items might be required to access the local patterns and the patterns in PB. A global pattern mining application (interface 6/5) is developed based on the patterns present in PB and LPB_i, $i = 1, 2, ..., n$. The technique of mining global patterns of select items is efficient due to the following reasons:

- One could extract more patterns of select items by lowering further the parameters of association rule such as the minimum support and minimum confidence, based on the level of data analysis of select items, since *FD* is reasonably small.
- We get true global patterns of select items as there is no need to estimate them.

In light of these observations, we can anticipate that the quality of global patterns is high, since there is no need to estimate them.

To evaluate the effectiveness of the above technique, we present a problem of multi-database mining. We show how the data mining technique presented above could be used in finding the solution to the problem. We start with the notion of overall association between two items in a database (Adhikari et al. 2011).

12.3 Overall Association Between Two Items in a Database

Let $I(DB)$ be the set of items in database DB. A common measure of similarity between two objects could be used as a measure of positive association between two items in a database (Wu et al. 2005; Xin et al. 2005). We define positive association between two items in a database as follows:

$$PA(x, y, DB)$$
$$= \frac{\#\text{transaction containing both } x \text{ and } y, DB}{\#\text{transaction containing at least one of } x \text{ and } y, DB}, \quad \text{for } x, y \in I(DB)$$

(12.1)

where, $\# P, DB$ is the number of transactions in DB that satisfy predicate P. PA measures only positive association present between two items in a database. It does not measure negative association between two items in a database. In the following example, we show that PA fails to compute an overall association between two items.

Example 12.1 Let us consider four branches of a multi-branch company. Let D_i be the database corresponding to the ith branch of the company, $i = 1, 2, 3, 4$. The company is interested in analyzing globally a set of select items (*SI*). Let $SI = \{a, b\}$. The content of different databases is given as follows: $D_1 = \{\{a, e\}, \{b, c, g\}, \{b, e, f\}, \{g, i\}\}$; $D_2 = \{\{b, c\}, \{f, h\}\}$; $D_3 = \{\{a, b, c\}, \{a, e\}, \{c, d\}, \{g\}\}$; $D_4 = \{\{a, e\}, \{b, c, g\}\}$. Initially, we wish to measure the association between two items in a single database, say D_1. Now, $PA(a, b, D_1) = 0$, since there is no transaction in D_1 containing both the items a and b. In these transactions, if one of the items of $\{a, b\}$ is present then the other item of $\{a, b\}$ is not present. Thus, the transactions $\{a, e\}$, $\{b, c, g\}$ and $\{b, e, f\}$ in D_1 imply that the items a and b are negatively associated. We need to define a measure of negative association between two items in a database.

Similarly to the measure of positive association, one defines a measure of negative association between two items in a database as follows:

$$NA(x, y, DB)$$
$$= \frac{\# \text{ transaction containing exactly one of } x \text{ and } y, \ DB}{\# \text{ transaction containing at least one of } x \text{ and } y, \ DB}, \quad \text{for } x, y \in I(DB).$$

$$(12.2)$$

Now, $NA(a, b, D_1) = 1$. We note that $PA(a, b, D_1) < NA(a, b, D_1)$. Overall, we state that the items a and b are negatively associated, and the level of overall association between the items a and b in D_1 is $PA(a, b, D_1) - NA(a, b, D_1) = -1.0$. The accuracy of association analysis might be low if we consider only the positive association between two items.

The analysis of relationships among variables is a fundamental task being at the heart of many data mining problems. For example, metrics such as support, confidence, lift, correlation, and collective strength have been used extensively to evaluate the interestingness of association patterns (Klemettinen et al. 1994; Silberschatz and Tuzhilin 1996; Aggarwal and Yu 1998; Silverstein et al. 1998; Liu et al. 1999). These metrics are defined in terms of the frequency counts tabulated in a 2 × 2 contingency table as shown in Table 12.1. Tan et al. (2002) presented an overview of twenty one interestingness measures proposed in statistics, machine learning and data mining literature. We continue our discussion with the examples cited in Tan et al. (2002) and show that none of the proposed measures is effective in finding the overall association by considering both positive and negative associations between two items in a database.

From the examples shown in Table 12.2, we notice that the overall association level between two items could be negative as well as positive. In fact, a measure of overall association between two items in a database produces results in $[-1, 1]$. We consider the following five out of twenty one interestingness measures, since the association between two items calculated using one of these five measures lies in $[-1, 1]$. Thus, we study their usefulness for the specific requirement of the proposed problem. These five measures are included in Table 12.3.

In Table 12.4, we rank the contingency tables by using each of the above measures.

Also, we rank the contingency tables based on the concept of overall association explained in Example 12.1. In Table 12.5, we present the ranking of contingency tables using overall association.

Table 12.1 A 2 × 2 contingency table for variables x and y

	y	$\neg y$	Total
x	f_{11}	f_{10}	$f_{1.}$
$\neg x$	f_{01}	f_{00}	$f_{0.}$
Total	$f_{.1}$	$f_{.0}$	$f_{..}$

Table 12.2 Examples of contingency tables

Example	f_{11}	f_{10}	f_{01}	f_{00}
E1	8,123	83	424	1,370
E2	8,330	2	622	1,046
E3	9,481	94	127	298
E4	3,954	3,080	5	2,961
E5	2,886	1,363	1,320	4,431
E6	1,500	2,000	500	6,000
E7	4,000	2,000	1,000	3,000
E8	4,000	2,000	2,000	2,000
E9	1,720	7,121	5	1,154
E10	61	2,483	4	7,452

Table 12.3 Selected interestingness measures for association patterns

Symbol	Measure	Formula
ϕ	ϕ-coefficient	$\dfrac{P(\{x\} \cup \{y\}) - P(\{x\}) \times P(\{y\})}{\sqrt{P(\{x\}) \times P(\{y\}) \times (1 - P(\{x\})) \times (1 - P(\{y\}))}}$
Q	Yule's Q	$\dfrac{P(\{x\} \cup \{y\}) \times P(\neg(\{x\} \cap \{y\})) - P(\{x\} \cup \neg\{y\}) \times P(\neg\{x\} \cup \{y\})}{P(\{x\} \cup \{y\}) \times P(\neg(\{x\} \cap \{y\})) - P(\{x\} \cup \neg\{y\}) \times P(\neg\{x\} \cup \{y\})}$
Y	Yule's Y	$\dfrac{\sqrt{P(\{x\} \cup \{y\}) \times P(\neg(\{x\} \cap \{y\}))} - \sqrt{P(\{x\} \cup \neg\{y\}) \times P(\neg\{x\} \cup \{y\})}}{\sqrt{P(\{x\} \cup \{y\}) \times P(\neg(\{x\} \cap \{y\}))} - \sqrt{P(\{x\} \cup \neg\{y\}) \times P(\neg\{x\} \cup \{y\})}}$
κ	Cohen's κ	$\dfrac{P(\{x\} \cup \{y\}) + P(\neg\{x\} \cup \neg\{y\}) - P(\{x\}) \times P(\{y\}) - P(\neg\{x\}) \times P(\neg\{y\})}{1 - P(\{x\}) \times P(\{y\}) - P(\neg\{x\}) \times P(\neg\{y\})}$
F	Certainty factor	$\max\left(\dfrac{P(\{y\}|\{x\}) - P(\{y\})}{1 - P(\{y\})}, \dfrac{P(\{x\}|\{y\}) - P(\{x\})}{1 - P(\{x\})}\right)$

Table 12.4 Ranking of contingency tables using above interestingness measures

Example	E1	E2	E3	E4	E5	E6	E7	E8	E9	E10
ϕ	1	2	3	4	5	6	7	8	9	10
Q	3	1	4	2	8	7	9	10	5	6
Y	3	1	4	2	8	7	9	10	5	6
κ	1	2	3	5	4	7	6	8	9	10
F	4	1	6	2	9	7	8	10	3	5

The ranks given in Table 12.5 and the ranks given for each of the five measures in Table 12.4 are not similar. In other words, none of the above five measures satisfies the requirement formulated in the proposed problem. Based on the above discussion, we propose the following measure *OA* as an overall association between two items in a database.

Table 12.5 Ranking of
contingency tables using
overall association

Example	Overall association	Rank
E1	0.76	3
E2	0.77	2
E3	0.93	1
E4	0.09	5
E5	0.02	6
E6	−0.10	8
E7	0.10	4
E8	0	7
E9	−0.54	10
E10	−0.24	9

Definition 12.1

$$OA(x, y, DB) = PA(x, y, DB) - NA(x, y, DB), \quad \text{for } x, y \in I(DB).$$

If $OA(x, y, DB) > 0$ then the items x and y are positively associated in DB. If $OA(x, y, DB) < 0$ then the items x and y are negatively associated in DB. The problem is concerned with the association between a nucleus item and another item in a database. Thus, we are not concerned about the association between two items in a group, where none of them is a nucleus item. In other words, it could be considered as a problem of grouping rather than a problem of classification or clustering.

12.4 An Application: Study of Select Items in Multiple Databases Through Grouping

As before, let us consider a multi-branch company having n branches. Each branch maintains a separate database for the transactions made in that particular branch. Let D_i be the database corresponding to the ith branch of the multi-branch company, $i = 1, 2, …, n$. Also, let D be the union of all branch databases. A large section of a local database might be irrelevant to the current problem. Thus, we divide database D_i into FD_i and RD_i, where FD_i and RD_i are called the *forwarded database* and *remaining database* corresponding to the ith branch, respectively, $i = 1, 2, …, n$. We are interested in the forwarded databases, since every transaction in a forwarded database contains at least one select item. The database FD_i is forwarded to the central office for mining global patterns of select items, $i = 1, 2, …, n$. All the local forwarded databases are amassed into a single database (FD) for the purpose of mining task. We note that the database FD is not overly large as it contains transactions related to select items. Before proceeding with the detailed discussion, we first offer some definitions.

A set of items is referred to as an *itemset*. An itemset containing k items is called a *k-itemset*. The *support* (*supp*) (Agrawal et al. 1993) of an itemset refers to the fraction of transactions containing this itemset. If an itemset satisfies the user-specified minimum support (α) criterion, then it is called a *frequent itemset* (*FIS*). Similarly, if an item satisfies the user-specified minimum support criterion, then it is called a *frequent item* (*FI*). If a k-itemset is frequent then every item in the k-itemset is also frequent. In this chapter, we study the items in *SI*. Let $SI = \{s_1, s_2, ..., s_m\}$. We wish to construct m groups of frequent items in such a way that the ith group grows by being centered around the nucleus item s_i, $i = 1, 2, ..., m$. Let FD be the union of FD_i, $i = 1, 2, ..., n$. Furthermore let $FIS_k(DB \mid \alpha)$ be the set of frequent k-itemsets in database DB at the minimum support level α, $k = 1, 2$. We state our problem as follows:

Let G_i be the ith group of frequent items containing the nucleus item $s_i \in SI$, $i = 1, 2, ..., m$. Construct G_i using $FIS_2(FD \mid \alpha)$ and local patterns in D_i such that $x \in G_i$ implies $OA(s_i, x, D) > 0$, for $i = 1, 2, ..., m$.

Two groups may not be mutually exclusive, as our study involves identifying pairs of items such that the following conditions are true: (i) the items in each pair are positively associated between each other in D, and (ii) one of the items in a pair is a select item. Our study is not concerned with the association between a pair of items in a group such that none of them is a select item. The above problem actually results in $m + 1$ groups where $(m + 1)$-th group G_{m+1} contains the items that are not positively associated with any one of the select items. The proposed study is not concerned with the items in G_{m+1}.

The crux of the proposed problem is to determine the supports of the relevant frequent itemsets in multiple large databases. A technique of estimating support of a frequent itemset in multiple real databases has been proposed by Adhikari and Rao (2008a). To estimate the support of an itemset in a database, this technique makes use of the trend of supports of the same itemset in other databases. The trend approach for estimating support of an itemset in a database could be described as follows:

Let the itemset X gets reported from databases $D_1, D_2, ..., D_m$. Also let $supp(X, \cup_{i=1}^{m} D_i)$ be the support of X in the union of $D_1, D_2, ..., D_m$. Let D_k be a database that does not report X, for $k = m + 1, m + 2, ..., n$. Then the support of X in D_k could be estimated by $\alpha \times supp(X, \cup_{i=1}^{m} D_i)$. Given an itemset X, some local supports of X are estimated and the remaining local supports of X are obtained using a traditional data mining technique. The global support of X is obtained by combining these local supports with the numbers of transactions (i.e., sizes) of the respective databases. The proposed technique synthesizes true supports of relevant frequent itemsets in multiple databases.

In the previous chapters, we have discussed the limitations of the traditional way of mining multiple large databases. We have observed that local pattern analysis alone could not provide an effective solution to this problem. The mining technique visualized in Fig. 12.1 offers a viable solution. A pattern based on all the databases is called a *global pattern*. A global pattern containing at least one select item is called a *global pattern of select item*.

12.4.1 Properties of Different Measures

If the itemset $\{x, y\}$ is frequent in DB then $OA(x, y, DB)$ is not necessarily be positive, since the number of transactions containing only one of the items of $\{x, y\}$ could be more than the number of transactions containing both the items x and y. $OA(x, y, DB)$ could attain maximum value for an infrequent itemset $\{x, y\}$ also. Let $\{x, y\}$ be infrequent. The distributions of x and y in DB are such that no transaction in DB contains only one item of $\{x, y\}$. Thus, $OA(x, y, DB) = 1.0$. In what follows, we discuss a few properties of different measures.

Lemma 12.1 (i) $0 \leq PA(x, y, DB) \leq 1$; (ii) $0 \leq NA(x, y, DB) \leq 1$; (iii) $-1 \leq OA(x, y, DB) \leq 1$; (iv) $PA(x, y, DB) + NA(x, y, DB) = 1$; for $x, y \in I(DB)$.

$PA(x, y, DB)$ could be considered as a similarity between x and y in DB. Thus, $1 - PA(x, y, DB)$ i.e., $NA(x, y, DB)$ could be considered as a distance between x and y in DB. A characteristic of a good distance measure is that it satisfies metric properties (Barte 1976) over the concerned domain.

Lemma 12.2 $NA(x, y, DB) = 1 - PA(x, y, DB)$ is a metric over $[0, 1]$, for $x, y \in I(DB)$.

Proof We prove only the property of triangular inequality, since the remaining two properties of the metric are obvious. Let $I(DB) = \{a_1, a_2, \ldots, a_N\}$. Let ST_i be the set of transactions containing item $a_i \in I(DB)$, $i = 1, 2, \ldots, N$.

$$1 - PA(a_p, a_q, DB) = 1 - \frac{|ST_p \cap ST_q|}{|ST_p \cup ST_q|} \geq \frac{|ST_p - ST_q| + |ST_q - ST_p|}{|ST_p \cup ST_q \cup ST_r|} \quad (12.3)$$

Thus,

$$1 - PA(a_p, a_q, DB) + 1 - PA(a_q, a_r, DB)$$
$$\geq \frac{|ST_p - ST_q| + |ST_q - ST_p| + |ST_q - ST_r| + |ST_r - ST_q|}{|ST_p \cup ST_q \cup ST_r|} \quad (12.4)$$

$$= \frac{|ST_p \cup ST_q \cup ST_r| - |ST_p \cap ST_q \cap ST_r| + |ST_p \cap ST_r| + |ST_q| - |ST_p \cap ST_q| - |ST_q \cap ST_r|}{|ST_p \cup ST_q \cup ST_r|}$$
$$\quad (12.5)$$

$$= 1 - \frac{|ST_p \cap ST_q \cap ST_s| - |ST_p \cap ST_s| - |ST_q| + |ST_p \cap ST_q| + |ST_q \cap ST_s|}{|ST_p \cup ST_q \cup ST_s|} \quad (12.6)$$

$$= 1 - \frac{\{|ST_p \cap ST_q \cap ST_s| + |ST_p \cap ST_q| + |ST_q \cap ST_s|\} - \{|ST_p \cap ST_s| + |ST_q|\}}{|ST_p \cup ST_q \cup ST_s|} \quad (12.7)$$

Let the number of elements in the shaded region of Fig. 12.2c, d be N_1 and N_2, respectively. Then the expression (12.7) becomes

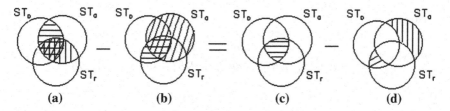

Fig. 12.2 Simplification using Venn diagram

$$1 - \frac{N_1 - N_2}{\left|ST_p \cup ST_q \cup ST_r\right|} \geq \begin{cases} 1 - \frac{N_1 - N_2}{\left|ST_p \cup ST_q \cup ST_r\right|}, & \text{if } N_1 \geq N_2 \quad (\text{case 1}) \\[2mm] 1 - \frac{\left|ST_p \cap ST_r\right|}{\left|ST_p \cup ST_q \cup ST_r\right|}, & \text{if } N_1 < N_2 \quad (\text{case 2}) \end{cases} \qquad (12.8)$$

In case 1, the expression remains the same. In case 2, a positive quantity $ST_p \cap ST_r$ has been put in place of a negative quantity $N_1 - N_2$. Thus the expression (12.8) reads as

$$\geq \begin{cases} 1 - \frac{N_1 - N_2}{\left|ST_p \cup ST_r\right|}, & \text{if } N_1 \geq N_2 \\[2mm] 1 - \frac{\left|ST_p \cap ST_r\right|}{\left|ST_p \cup ST_r\right|}, & \text{if } N_1 < N_2 \end{cases} \geq \begin{cases} 1 - \frac{N_1}{\left|ST_p \cup ST_r\right|}, & \text{if } N_1 \geq N_2 \\[2mm] 1 - \frac{\left|ST_p \cap ST_r\right|}{\left|ST_p \cup ST_r\right|}, & \text{if } N_1 < N_2 \end{cases}$$

$$\geq \begin{cases} 1 - \frac{\left|ST_p \cap ST_r\right|}{\left|ST_p \cup ST_r\right|}, & \text{if } N_1 \geq N_2 \\[2mm] 1 - \frac{\left|ST_p \cap ST_r\right|}{\left|ST_p \cup ST_r\right|}, & \text{if } N_1 < N_2 \end{cases} \qquad (12.9)$$

where $N_1 = \left|ST_p \cap ST_q \cap ST_r\right| \leq \left|ST_p \cap ST_r\right|$. Therefore, irrespectively of the relationship between N_1 and N_2, $1 - PA(a_p, a_q, DB) + 1 - PA(a_q, a_r, DB) \geq 1 - PA(a_p, a_r, DB)$. Thus, $1 - PA(x, y, DB)$ satisfies the requirement of triangular inequality. $\qquad \square$

To compute an overall association between two items, we need to express OA in terms of supports of frequent itemsets.

Lemma 12.3 *For any two items* $x, y \in I(DB)$, $OA(x, y, DB)$ *can be expressed as follows:*

$$OA(x, y, DB) = \frac{3 \times supp(\{x, y\}, DB) - supp(\{x\}, DB) - supp(\{y\}, DB)}{supp(\{x\}, DB) + supp(\{y\}, DB) - supp(\{x, y\}, DB)} \qquad (12.10)$$

Proof

$$OA(x, y, DB) = PA(x, y, DB) - NA(x, y, DB)$$

Now,

$$PA(x, y, DB) = \frac{supp(\{x, y\}, DB)}{supp(\{x\}, DB) + supp(\{y\}, DB) - supp(\{x, y\}, DB)} \quad (12.11)$$

Also,

$$NA(x, y, DB) = \frac{supp(\{x\}, DB) + supp(\{y\}, DB) - 2 \times supp(\{x, y\}, DB)}{supp(\{x\}, DB) + supp(\{y\}, DB) - supp(\{x, y\}, DB)}$$
$$(12.12)$$

Thus, the lemma follows.   □

12.4.2 Grouping Frequent Items

For the purpose of explaining the grouping process, we continue our discussion of Example 12.1.

Example 12.2 Based on *SI*, the forwarded databases are given as follows:

$$FD_1 = \{\{a, e\}, \{b, c, g\}, \{b, e, f\}\}$$
$$FD_2 = \{\{b, c\}\}$$
$$FD_3 = \{\{a, b, c\}, \{a, e\}\}$$
$$FD_4 = \{\{a, e\}, \{b, c, g\}\}$$

Let $size(DB)$ be the number of transactions in DB. Then $size(D_1) = 4$, $size(D_2) = 2$, $size(D_3) = 4$, and $size(D_4) = 2$. The union of all forwarded databases is given as

$$FD = \{\{a, e\}, \{b, c, g\}, \{b, e, f\}, \{b, c\}, \{a, b, c\}, \{a, e\}, \{a, e\}, \{b, c, g\}\}.$$

The transaction $\{a, e\}$ has been shown three times, since it has originated from three data sources. We mine the database FD and get the following set of frequent itemsets:

$$FIS_1(FD|1/14) = \{\{a\}\,(4/12), \{b\}\,(5/12)\}$$
$$FIS_2(FD|1/14) = \{\{a, b\}\,(1/12), \{a, c\}(1/12), \{a, e\}(3/12), \{b, c\}(4/12),$$
$$\{b, e\}(1/12), \{b, f\}(1/12), \{b, g\}(2/12)\}$$

where $X(\eta)$ denotes the fact that the frequent itemset X has support η. All the transactions containing item x not belonging to SI might not be available in FD. Thus other frequent itemsets of size one could not be mined correctly from FD. They are not shown in $FIS_1(FD)$. Each frequent itemset extracted from FD contains an item from SI. The collection of patterns in $FIS_1(FD \mid 1/14)$ and $FIS_2(FD \mid 1/14)$ could be considered as PB with reference to Fig. 6.1. Using the frequent itemsets in $FIS_1(FD \mid \alpha)$ and $FIS_2(FD \mid \alpha)$ we might not be able to compute the value of OA between two items. The central office of the company requests each branch for the supports of the relevant items (RIs) to calculate the overall association between two items. Such information would help the central office to compute exactly the value of the overall association in the union of all databases. Relevant items are the items in $FIS_1(FD \mid \alpha)$ that do not belong to SI. In this example, RIs are c, e, f and g. The supports of relevant items in different databases are given below:

$$RI(D_1) = \{\{c\}\,(1/4),\,\{e\}\,(2/4),\,\{f\}\,(1/4),\,\{g\}\,(2/4)\}$$
$$RI(D_2) = \{\{c\}\,(1/2),\,\{e\}\,(0),\,\{f\}\,(1/2),\,\{g\}\,(0)\}$$
$$RI(D_3) = \{\{c\}(2/4),\,\{e\}(1/4),\,\{f\}(0),\,\{g\}(1/4)\}$$
$$RI(D_4) = \{\{c\}\,(1/2),\,\{e\}\,(1/2),\,\{f\}(0),\,\{g\}\,(1/2)\}.$$

$RI(D_i)$ could be considered as LPB_i with reference to Fig. 12.1, $i = 1, 2, \ldots, n$. We follow here a grouping technique based on the proposed measure of overall association OA. If $OA(x, y, D) > 0$ then y could be placed in the group of x, for $x \in SI = \{a, b\}, y \in I(D)$. We explain the procedure of grouping frequent items with the help of following example.

Example 12.3 Here we continue the discussion of Example 12.2. Based on the available supports of local 1-itemsets, we synthesize 1-itemsets in D as mentioned in Table 12.6.

We note that the supports of $\{a\}$ and $\{b\}$ are not required to be synthesized, since they could be determined exactly from mining FD. The values of OA corresponding to itemsets of FIS_2 are presented in Table 12.7.

Table 12.6 Supports of relevant 1-itemsets in D

Itemset ($\{x\}$)	$\{a\}$	$\{b\}$	$\{c\}$	$\{e\}$	$\{f\}$	$\{g\}$
$supp(\{x\}, D)$	4/12	5/12	5/12	4/12	2/12	4/12

Table 12.7 Overall association between two items in a frequent 2-itemset in FD

Itemset ($\{x, y\}$)	$\{a, b\}$	$\{a, c\}$	$\{a, e\}$	$\{b, c\}$	$\{b, e\}$	$\{b, f\}$	$\{b, g\}$
$OA(x, y, D)$	−3/4	−3/4	1/5	1/3	−3/4	−2/3	−3/7

In Table 12.7, we find that the items a and e are positively associated. Thus, item e could be placed in the group containing nucleus item a. Items b and c are positively associated as well. Item c could be put in the group containing nucleus item b. Thus, the output grouping π using the proposed technique comes in the form:

$$\pi(FIS_1(D) \mid \{a, b\}, 1/12) = \{\text{Group 1, Group 2}\},$$

where

$$\text{Group 1} = \{(a, 1.0), (e, 0.2)\}$$
$$\text{Group 2} = \{(b, 0.1), (c, 0.33)\}.$$

Each item in a group is associated with a real number, which represents the strength of an overall association between the item and the nucleus item of the group. Using the proposed grouping technique we also construct the third group of items, i.e., $\{f, g\}$. The proposed study is not concerned with the items in $\{f, g\}$.

Each group grows being centered around a select item. The ith group (G_i) grows centering around the ith select item s_i, $i = 1, 2, \ldots, m$. With respect to group G_i, the item s_i is called the nucleus item of G_i, $i = 1, 2, \ldots, m$. We define a group as follows.

Definition 12.2 The ith group is a collection of frequent items a_j and the nucleus item $s_i \in SI$ such that $OA(s_i, a_j, D) > 0$, $j = 1, 2, \ldots, |G_i|$, and $i = 1, 2, \ldots, m$.

Let us describe the data structures used in the algorithm for finding groups. The set of frequent k-itemsets is maintained in an array $FISk$, $k = 1, 2$. After finding OA value between two items in a 2-itemset, it is kept in array $IS2$. Thus, the number of itemsets in $IS2$ is equal to the number of frequent 2-itemsets extracted from FD. A two-dimensional array $Groups$ is maintained to store m groups. The ith row of $Groups$ stores the ith group, for $i = 1, 2, \ldots, m$. The first element of ith row contains the ith select item, for $i = 1, 2, \ldots, m$. In general, the jth element of the ith row contains a pair ($item$, $value$), where $item$ refers to the jth item of the ith group and $value$ refers to the amount of OA value between the ith select item and $item$, for $j = 1, 2, \ldots, |G_i|$. The grouping algorithm can be outlined as follows.

Algorithm 12.1. Construct m groups of frequent items in D such that i-th group grows being centered around the i-th select item, for $i = 1, 2, ..., m$.

procedure m-grouping $(m, SI, N_1, FIS1, N_2, FIS2, GSize, Groups)$
Input: $m, SI, N_1, FIS1, N_2, FIS2$
m: the number of select items
SI: set of select items
N_k: number of frequent k-itemsets
$FISk$: set of frequent k-itemsets
Output: *GSize, Groups*
GSize: array of number of elements in each group
Groups: array of m groups
01: **for** $i = 1$ to N_2 **do**
02:      $IS2(i).value = OA(FIS2(i).item1, FIS2(i).item2, D)$;
03:      $IS2(i).item1 = FIS2(i).item1$; $IS2(i).item2 = FIS2(i).item2$;
04: **end for**
05: **for** $i = 1$ to m **do**
06:      $Groups(i)(1).item = SI(i)$; $Groups(i)(1).value = 1.0$; $GSize(i) = 1$;
07: **end for**
08: **for** $i = 1$ to N_2 **do**
09:    **for** $j = 1$ to m **do**
10:      **if** $((IS2(i).item1 = SI(j))$ **and** $(IS2(i).value > 0))$ **then**
11:        $GSize(j) = GSize(j) + 1$; $Groups(j)(GSize(j)).item = IS2(i).item2$;
12:        $Groups(j)(GSize(j)).value = IS2(i).value$;
13:      **end if**
14:      **if** $((IS2(i).item2 = SI(j))$ **and** $(IS2(i).value > 0))$ **then**
15:        $GSize(j) = GSize(j) + 1$; $Groups(j)(GSize(j)).item = IS2(i).item1$;
16:        $Groups(j)(GSize(j)).value = IS2(i).value$;
17:      **end if**
18:    **end for**
19: **end for**
20: **for** $i = 1$ to m **do**
21:    sort items of group i in non-increasing order on *OA* value;
22: **end for**
end procedure

The algorithm works as follows. Using (12.10), we compute the value of *OA* for each itemset in *FIS2*. After computing *OA* value for a pair of items, we store the items and *OA* value in *IS2*. The algorithm performs these tasks using the *for*-loop shown in lines 01–04. We initialize each group with the corresponding nucleus item as shown in lines 05–07. A relevant item or an item in *SI* could belong to one or more groups. Thus, we check for the possibility of including each of the relevant items and items in *SI* to each group using the *for*-loop (lines 09–18). All the relevant items and items in *SI* are covered using *for*-loop present in lines 08–19. For the purpose of better presentation, we finally sort items of ith group in non-increasing order on *OA* value, $i = 1, 2, ..., m$.

Assume that the frequent itemsets in *FIS1* and *FIS2* are sorted on items in the itemset. Thus, the time complexities for searching an itemset in *FIS1* and *FIS2* are $O(\log(N_1))$ and $O(\log(N_2))$, respectively. The time complexity of computing present at line 02 is $O(\log(N_1))$, since $N_1 > N_2$. The time complexity of calculations carried out in lines 01–04 is $O(N_2 \times \log(N_1))$. Lines 05–07 are used to complete all necessary initialization. The time complexity of this program segment is $O(m)$. Lines 08–19 process frequent 2-itemsets and construct m groups. If one of the two items in a frequent 2-itemset is a select item, then other item could be placed in the group of the select item, provided the overall association between them is positive. The time complexity of this program segment is $O(m \times N_2)$. Lines 20–22 present groups in sorted order. Each group is sorted in non-increasing order with respect to the OA value. The association of nucleus item with itself is 1.0. Thus the nucleus item is kept at the beginning of the group (line 06). Let the average size of a group be k. Then the time complexity of this program segment is $O(m \times k \times \log(k))$. The time complexity of the procedure *m-grouping* is *maximum* $\{O(N_2 \times \log(N_1)), O(m),$ $O(m \times N_2), O(m \times k \times \log(k))\}$, i.e., *maximum* $\{O(N_2 \times \log(N_1)), O(m \times N_2),$ $O(m \times k \times \log(k))\}$.

12.4.3 Experiments

A suite of experiments has been carried out to quantify the effectiveness of the above approach. We present the experimental results using four databases, viz., *retail* (Frequent Itemset Mining Dataset Repository 2004), *mushroom* (Frequent Itemset Mining Dataset Repository 2004), *T10I4D100K* (Frequent Itemset Mining Dataset Repository 2004), and *check*. The database *retail* obtained from an anonymous Belgian retail supermarket store and concerns a real-world problem. The database *mushroom* comes from the UCI databases. The database *T10I4D100K* is synthetic and was obtained using a generator from IBM Almaden Quest research group. The database *check* is artificial whose grouping is already known. We have experimented with database *check* to verify that our grouping technique works correctly. We present some characteristics of these databases in Table 12.8. Let *NT*, *AFI*, *ALT*, and *NI* denote the number of transactions, average frequency of an item, average length of a transaction, and number of items in the database, respectively.

We divide each of these databases into ten databases called here *input databases*. The input databases obtained from R, M, T and C are names as R_i, M_i, T_i, and C_i, respectively, $i = 1, 2, \ldots, 10$. We present some characteristics of the input databases in Table 12.9.

Table 12.8 Characteristics of databases used in the experiment

Database	NT	ALT	AFI	NI
retail (R)	88,162	11.31	99.67	10,000
mushroom (M)	8,124	24.00	1624.80	120
T10I4D100K (T)	1,00,000	11.10	1276.12	870
check (C)	40	3.03	3.10	39

Table 12.9 Characteristics of input databases obtained from, (a) *retail* and *mushroom*, (b) *T10I4D100K*

a

DB	NT	ALT	AFI	NI	DB	NT	ALT	AFI	NI
R_1	9,000	11.24	12.07	8,384	M_1	812	24.00	295.27	66
R_2	9,000	11.21	12.27	8,225	M_2	812	24.00	286.59	68
R_3	9,000	11.34	14.60	6,990	M_3	812	24.00	249.85	78
R_4	9,000	11.49	16.66	6,206	M_4	812	24.00	282.43	69
R_5	9,000	10.96	16.04	6,148	M_5	812	24.00	259.84	75
R_6	9,000	10.86	16.71	5,847	M_6	812	24.00	221.45	88
R_7	9,000	11.20	17.42	5,788	M_7	812	24.00	216.53	90
R_8	9,000	11.16	17.35	5,788	M_8	812	24.00	191.06	102
R_9	9,000	12.00	18.69	5,777	M_9	812	24.00	229.27	85
R_{10}	9,000	11.69	15.35	5,456	M_{10}	816	24.00	227.72	86

b

DB	ALT	AFI	NI	DB	ALT	AFI	NI
T_1	11.06	127.66	866	T_6	11.14	128.63	866
T_2	11.13	128.41	867	T_7	11.11	128.56	864
T_3	11.07	127.65	867	T_8	11.10	128.45	864
T_4	11.12	128.44	866	T_9	11.08	128.56	862
T_5	11.13	128.75	865	T_{10}	11.08	128.11	865

The input databases obtained from database *check* are given as follows:

$$C_1 = \{\{1,4,9,31\}, \{2,3,44,50\}, \{6,15,19\}, \{30,32,42\}\}$$
$$C_2 = \{\{1,4,7,10,50\}, \{3,44\}, \{11,21,49\}, \{41,45,59\}\}$$
$$C_3 = \{\{1,4,10,20,24\}, \{5,7,21\}, \{21,24,39\}, \{26,41,46\}\}$$
$$C_4 = \{\{1,4,10,23\}, \{5,8\}, \{5,11,21\}, \{42,47\}\}$$
$$C_5 = \{\{1,4,10,34\}, \{5,49\}, \{25,39,49\}, \{49\}\}$$
$$C_6 = \{\{1,3,44\}, \{6,41\}, \{22,26,38\}, \{45,49\}\}$$
$$C_7 = \{\{1,2,3,10,20,44\}, \{11,12,13\}, \{24,35\}, \{47,48,49\}\}$$
$$C_8 = \{\{2,3,20,39\}, \{2,3,20,44,50\}, \{32,49\}, \{42,45\}\}$$
$$C_9 = \{\{2,3,20,44\}, \{3,19,50\}, \{5,41,45\}, \{21\}\}$$
$$C_{10} = \{\{2,20,45\}, \{5,7,21\}, \{11,19\}, \{22,30,31\}\}$$

In Table 12.10, we present some relevant details regarding different experiments. We have chosen the first 10 frequent items as the select items, except for the last experiment. One could choose select items as the items whose data analyses are needed to be performed.

The first experiment is based on database *retail*. The grouping of frequent items in *retail* is given below:

Table 12.10 Some relevant information regarding experiments

Database	α	SI
R	0.03	{0, 1, 2, 3, 4, 5, 6, 7, 8, 9}
M	0.05	{1, 3, 9, 13, 23, 34, 36, 38, 40, 52}
T	0.01	{2, 25, 52, 240, 274, 368, 448, 538, 561, 630}
C	0.07	{1, 2, 3}

$$\pi(FI(retail)|SI, \alpha) = \{\underline{0}(1.00); \underline{1}(1.00); \underline{2}(1.00); \underline{3}(1.00); \underline{4}(1.00); \underline{5}(1.00);$$
$$\underline{6}(1.00); \underline{7}(1.00); \underline{8}(1.00); \underline{9}(1.00)\}$$

Two resulting groups are separated by semicolon (;). The nucleus item in each group is underlined. Each item in a group is associated with a real number shown in bracket. This value represents the strength of the overall association between the item and the nucleus item. The groups are shaded for the purpose of clarity of visualization. We observe that no item in database *retail* is positively associated with the select items using the measure *OA*. This does not necessarily mean that the level of AE or ME for the experiment is zero. There may exist frequent itemsets of size two such that overall association between two items in each of the itemsets is non-positive and at least one of the two items belongs to the set of select items.

The second experiment is based on database *mushroom*. The grouping of frequent items in *mushroom* is given below:

$$\pi(FI(mushroom)|SI, \alpha) = \{\underline{1}(1.00), 24(0.23, 110(0.12), 29(0.10), \ 36(0.10),$$
$$61(0.10), 38(0.06), 66(0.06), 90(0.01); \underline{3}(1.00);$$
$$\underline{9}(1.000000); \underline{13}(1.00); \underline{23}(1.00), 93(0.53), 59(0.22),$$
$$2(0.14), 39(0.01), \ 63(0.15); \underline{34}(1.00), 86(0.99),$$
$$85(0.95), \ 90(0.80), 36(0.63), 39(0.33), 59(0.23),$$
$$63(0.17), 53(0.16), 67(0.13), 24(0.12), 76(0.11);$$
$$\underline{36}(1.00), 85(0.68), 90(0.65), 86(0.63), \underline{34}(0.63),$$
$$59(0.17), 39(0.16), \ 63(0.11), \ 110(0.10), 1(0.10);$$
$$\underline{38}(1.00), 48(0.38), 102 \ (0.19), 58(0.14), 1(0.06),$$
$$94(0.05), 110(0.01); \underline{40}(1.00); \underline{52}(1.00)\}$$

We observe that some frequent items are not included in any of these groups, since their overall associations with each of the select items are non-positive.

The third experiment is based on database *T10I4D100K*. The grouping of frequent items in *T10I4D100K* is given below:

$$\pi(FI(T10I4D100K)|SI, \alpha) = \{2(1.00); \underline{25}(1.00); \underline{52}(1.00); \underline{240}(1.00); \underline{274}(1.00);$$
$$\underline{368}(1.00); \underline{448}(1.00); \underline{538}(1.00); \underline{561}(1.00); \underline{630}(1.00)\}$$

We observe that databases *retail* and *T10I4D100K* are sparse. Thus, the grouping contains groups of singleton item for these two databases. The overall association between a nucleus item and itself is 1.0. Otherwise, the overall association between a frequent item and a nucleus item is non-positive for these two databases.

The fourth experiment is based on database *check*. This database is constructed artificially to verify the following existing grouping.

$$\pi(FI(check)|SI, \alpha) = \{(\underline{1}, 1.00), (4, 0.43), (10, 0.43); (\underline{2}, 1.00), (20, 0.43),$$
$$(3, 0.11); (\underline{3}, 1.00), (44, 0.50), (2, 0.11)\}.$$

We have calculated average errors using both trend and the proposed approaches. Figures 12.3, 12.4 and 12.5 show the graphs of AE versus the number databases for the first three databases. The proposed model enables us to find actual supports of all the relevant itemsets in a database. Thus, the AE of an experiment for the proposed approach remains 0. As the number of databases increases, the relative presence of a frequent itemset normally decreases, the error of synthesizing an itemset also increases. Overall, the AE of the experiment using trend approach is likely to increase as the number of databases increases. We observe this phenomenon in Figs. 12.3, 12.4 and 12.5.

12.5 Related Work

Recently, multi-database mining has been recognized as an important and timely research area in the KDD community. The work reported so far could be classified broadly into two categories: mining/synthesizing patterns in multiple databases and

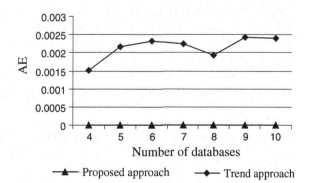

Fig. 12.3 AE versus the number of the databases from *retail*

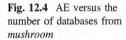

Fig. 12.4 AE versus the number of databases from *mushroom*

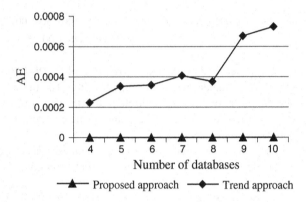

Fig. 12.5 AE versus the number of databases from *T10I4D100K*

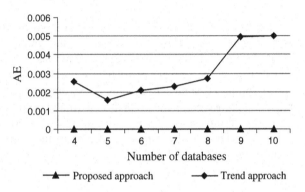

post processing of local patterns. We mention some work related to the first category. Wu and Zhang (2003) have proposed a weighting method for synthesizing high-frequency rules in multiple databases. Zhang et al. (2004a) have developed an algorithm to identify global exceptional patterns in multiple databases. When it comes to the second category, Wu et al. (2005) have proposed a technique for classifying multiple databases for multi-database mining. Using local patterns, we have proposed an efficient technique for clustering multiple databases (Adhikari and Rao 2008b). Lin et al. (2013) have introduced the notion of stable items based on minimum support and variance. Authors have provided a measure of similarity between stable items based on gray relational analysis, and presented a hierarchical gray clustering method for mining stable patterns.

In the context of estimating support of itemsets in a database, Jaroszewicz and Simovici (2002) have proposed a method using Bonferroni-type inequalities (Galambos and Simonelli 1996). The maximum-entropy approach to support estimation of a general Boolean expression is proposed by Pavlov et al. (2000). But these support estimation techniques are suitable for a single database only.

Zhang et al. (2004b), Zhang (2002) have studied various strategies for mining multiple databases. Proefschrift (2004) has studied data mining on multiple relational databases.

Existing parallel mining techniques (Agrawal and Shafer 1999; Chattratichat et al. 1997; Cheung et al. 1996) could also be used to deal with multi-databases. These techniques, however, might provide expensive solutions for studying select items in multiple databases.

12.6 Conclusions

The proposed measure of overall association *OA* is effective as it considers both positive and negative association between two items. Association analysis of select items in multiple market basket databases is an important as well as highly promising issue, since many data analyses of a multi-branch company are based on select items. One could also apply one of the multi-database mining techniques discussed in Chap. 4. Each technique, except partition algorithm, returns approximate global patterns. On the other hand, the partition algorithm scans each database twice. Therefore, the proposed model of mining global patterns of select items from multiple databases is efficient, since one does not need to estimate the patterns in multiple databases. Moreover, it does not fully scan each database two times.

References

Adhikari A, Ramachandrarao P, Pedrycz W (2011) Study of select items in different data sources by grouping. Knowl Inf Syst 27(1):23–43

Adhikari A, Rao PR (2008a) Synthesizing heavy association rules from different real data sources. Pattern Recogn Lett 29(1):59–71

Adhikari A, Rao PR (2008b) Efficient clustering of databases induced by local patterns. Decis Support Syst 44(4):925–943

Aggarwal C, Yu P (1998) A new framework for itemset generation. In: Proceedings of the 17th symposium on principles of database systems, pp 18–24

Agrawal R, Imielinski T, Swami A (1993) Mining association rules between sets of items in large databases. In: Proceedings of ACM SIGMOD conference, pp 207–216

Agrawal R, Shafer J (1999) Parallel mining of association rules. IEEE Trans Knowl Data Eng 8 (6):962–969

Barte RG (1976) The elements of real analysis, 2nd edn. Wiley, Hoboken

Chattratichat J, Darlington J, Ghanem M, Guo Y, Hüning H, Köhler M, Sutiwaraphun J, To HW, Yang D (1997) Large scale data mining: challenges, and responses. In: Proceedings of the third international conference on knowledge discovery and data mining, pp 143–146

Cheung D, Ng V, Fu A, Fu Y (1996) Efficient mining of association rules in distributed databases. IEEE Trans Knowl Data Eng 8(6):911–922

Frequent Itemset Mining Dataset Repository (2004) http://fimi.cs.helsinki.fi/data

Galambos J, Simonelli I (1996) Bonferroni-type inequalities with applications. Springer, New York

Jaroszewicz S, Simovici DA (2002) Support approximations using bonferroni-type inequalities. In: Proceedings of sixth European conference on principles of data mining and knowledge discovery, pp 212–223

Klemettinen M, Mannila H, Ronkainen P, Toivonen T, Verkamo A (1994) Finding interesting rules from large sets of discovered association rules. In: Proceedings of the 3rd international conference on information and knowledge management, pp 401–407

Lin Y, Hu X, Li X, Wu X (2013) Mining stable patterns in multiple correlated databases. Decis Support Syst 56:202–210

Liu B, Hsu W, Ma Y (1999) Pruning and summarizing the discovered associations. In: Proceedings of the 5th international conference on knowledge discovery and data mining, pp 125–134

Pavlov D, Mannila H, Smyth P (2000) Probabilistics models for query approximation with large sparse binary data sets. In: Proceedings of sixteenth conference on uncertainty in artificial intelligence, pp 465–472

Proefschrift (2004) Multi-relational data mining, Ph.D thesis, Dutch Graduate School for Information and Knowledge Systems, Aan de Universiteit Utrecht

Pyle D (1999) Data preparation for data mining. Morgan Kufmann, San Francisco

Silberschatz A, Tuzhilin A (1996) What makes patterns interesting in knowledge discovery systems. IEEE Trans Knowl Data Eng 8(6):970–974

Silverstein C, Brin S, Motwani R (1998) Beyond market baskets: generalizing association rules to dependence rules. Data Min Knowl Disc 2(1):39–68

Tan P-N, Kumar V, Srivastava J (2002) Selecting the right interestingness measure for association patterns. In: Proceedings of SIGKDD conference, pp 32–41

Wu X, Zhang S (2003) Synthesizing high-frequency rules from different data sources. IEEE Trans Knowl Data Eng 14(2):353–367

Wu X, Zhang C, Zhang S (2005) Database classification for multi-database mining. Inf Syst 30 (1):71–88

Xin D, Han J, Yan X, Cheng H (2005) Mining compressed frequent-pattern sets. In: Proceedings of the 31st VLDB conference, pp 709–720

Zhang S (2002) Knowledge discovery in multi-databases by analyzing local instances, Ph D thesis, Deakin University

Zhang C, Liu M, Nie W, Zhang S (2004a) Identifying global exceptional patterns in multi-database mining. IEEE Comput Intell Bull 3(1):19–24

Zhang S, Wu X, Zhang C (2003) Multi-database mining. IEEE Comput Intell Bull 2(1):5–13

Zhang S, Zhang C, Wu X (2004b) Knowledge discovery in multiple databases. Springer, Berlin

Chapter 13
Mining Calendar-Based Periodic Patterns in Time-Stamped Data

A large class of problems is concerned with temporal data. Identifying temporal patterns in these datasets is a fully justifiable as well as an important task. Recently, researchers have reported an algorithm for finding calendar-based periodic pattern in a time-stamped data and introduced the concept of certainty factor in association with an overlapped interval. In this chapter, we have extended the concept of certainty factor by incorporating support information for effective analysis of overlapping intervals. We have proposed a number of improvements of the algorithm for identifying calendar-based periodic patterns. In this direction we have proposed a hash based data structure for storing and managing patterns. Based on this modified algorithm, we identify full as well as partial periodic calendar-based patterns. We provide a detailed data analysis incorporating various parameters of the algorithm and make a comparative analysis with the existing algorithm, and show the effectiveness of our algorithm. Experimental results are provided on both real and synthetic databases.

13.1 Introduction

A large amount of data being collected every day exhibits a temporal connotation. For example, databases which originate from transactions in a supermarket, logs in a network, transactions in a bank, and events related to manufacturing industry are all inherently related to time. Data mining techniques could also be applied to these databases to discover various temporal patterns to understand the behavior of customers, markets, or monitored processes in different points of time. Temporal data mining is concerned with the analyses of data to find out patterns and regularities from a set of temporal data. In this context, sequential association rule (Agrawal and Srikant 1995), periodical association rule (Li and Deogun 2005), calendar association rule (Li et al. 2003), calendar-based periodic pattern (Mahanta et al. 2008), and up-to-date pattern (Hong et al. 2009) are some interesting temporal patterns reported in the recent time.

© Springer International Publishing Switzerland 2015
A. Adhikari and J. Adhikari, *Advances in Knowledge Discovery in Databases*,
Intelligent Systems Reference Library 79, DOI 10.1007/978-3-319-13212-9_13

For effective management of business activities, we often wish to discover knowledge from time-stamped data. There are several important aspects of mining time-stamped data including trend analysis, similarity search, forecasting and mining of sequential and periodic patterns. In a database from a retail store, the sales of ice cream in summer and the sales of blanket in winter could be higher than those of the other seasons. Such seasonal behaviour of specific items can only be discovered when a proper window size is chosen for the data mining process (Roddick and Spiliopoulou 2002). A supermarket manager may discover that turkey and pumpkin pie are frequently sold together in November in every year. Discovering such patterns may reveal interesting information that can be used for understanding the behaviour of customers, markets or monitored processes in different time periods. However, these types of seasonal patterns cannot be discovered by traditional non-temporal data mining approaches that treat all the data as one large segment with no attention paid to utilizing the time information of the transactions. If one looks into the entire dataset rather than the transactions that occur in November, it is likely that one will not be able to discover the pattern of turkey and pumpkin pie since the overall support for them will be evidently low. In general, a time-stamped database might exhibit some periodic behaviours. Length of a period might vary from one context to another context. For example, in case of sales of ice cream, the basic time interval could be of three months, since in many regions March, April and May together is considered as summer. Also, in case of sales of blanket, the basic time interval could be considered from November to February in every year. In addition, in many business applications, one might be interested in quarterly patterns over the years, where length of the period is equal to three months. A large amount of data is collected every day in the form of event time sequences. These sequences are valuable sources to analyze not only the frequencies of certain events, but also the patterns with which these events occur. For example, from data consisting of web clicks one may discover that a large number of web browsers who visit www.washingtonpost.com in morning hours also visit www.cnn.com. Using such information one can group users as daily morning users, daily evening users, weekly users, etc. This information might be useful for communicating to the users. Temporal patterns in a stock market, such as whether certain months, days of the week, time periods or holidays provide better returns than other time periods have received particularly a large amount of attention. Due to the presence of various types of applications in many fields, periodic pattern mining is an interesting area of study.

Mahanta et al. (2008) used set superimposition (Baruah 1999) to find the membership value of each fuzzy interval. The concept of set *superimposition* is defined as follows. If set A is superimposed over set B or set B is superimposed over set A then set superimposition operation can be expressed as $A(S)B = (A-B)$ $(+)\,(A \cap B)^{(2)}(+)\,(B-A)$, where (S) denotes the set superimposition operation. Here, the elements of $(A \cap B)^{(2)}$ are the elements of $(A \cap B)$ represented twice and $(+)$ represents union of disjoint sets. Authors have also designed an algorithm for mining calendar-based periodic patterns. While applying this concept authors have assumed intervals with equal membership grade, and accordingly the concept of certainty factor has been proposed for each sub-interval. Certainty factor of an interval over different time periods expresses the likelihood of reporting the pattern in that particular interval. If two intervals overlap then the certainty factor is more for the overlapped region than the non-overlapped region. When two intervals are superimposed, authors have assumed 1/2 membership grade for each interval. After superimposition, the membership grade fuzzy membership value for the overlapped region becomes 1. The membership grade for non-overlapped region remains 1/2. But these two intervals may have different supports for the pattern. The *support* (Agrawal et al. 1993) of a pattern represents a fraction of transactions containing the pattern. A pattern is *frequent* if its support is greater than equal to a user-defined threshold, *minsupp*. The certainty factor and support of a pattern in an interval are two different concepts. For an effective analysis of overlapped regions, these two concepts need to be introduced along with an overlapped region. Thus, in this chapter we propose an extended analysis of superimposed intervals. The main weak point of the aforementioned paper is that the concept of set superimposition is not necessary in the proposed algorithm. Therefore, we have proposed a modified algorithm for identifying full as well as partial calendar-based periodic patterns. We have also improved the proposed algorithm by introducing a hash based data structure for storing relevant information associated with intervals. In addition, we have suggested some other improvements in the proposed algorithm. Before concluding this section, we give an example of a time-stamped database that will be used for providing illustrative examples on various concepts.

Example 13.1 Consider the following database D of transactions. Each record contains items purchased as well as the date of the transaction (Table 13.1).

We have omitted the time of a transaction, since our data analysis is not associated with the time component of a transaction. We will refer to this database from time to time for the purpose of illustrating various concepts.

This chapter is organized as follows. We discuss related work in Sect. 13.2. In Sect. 13.3, we have discussed calendar-based periodic patterns and proposed an

Table 13.1 A sample time-stamped database

Time-stamp	Items	Time-stamp	Items	Time-stamp	Items
29/03/1990	a, b, c	07/04/1992	a, c, e, g, h	17/04/1993	a, c, f
06/04/1990	a, c, e	12/04/1992	c, e	06/04/1994	a, b, c,d
21/04/1990	a, d	14/04/1992	c, e, f	10/04/1994	g, h
25/04/1990	a, c, d	19/04/1992	f, g	13/04/1994	a, g
06/03/1991	a, c	04/03/1993	a, c	18/04/1994	g, h, i
12/03/1991	a, c, e	09/03/1993	a, c, g	20/04/1994	a, c, e, f
19/04/1991	f, g	01/04/1993	c, h, i		
03/03/1992	a, c, d	07/04/1993	c, d		

extended certainty factor of an interval. We have designed an algorithm for identifying calendar-based periodic patterns in Sect. 13.4. Experimental results are provided in Sect. 13.5. We conclude the paper in Sect. 13.6.

13.2 Related Work

A calendar time expression is composed of calendar units in a specific calendar and represents different time features, such as an absolute time interval and a periodic time over a specific time period. A calendar-based periodic pattern is associated with time hierarchy for calendar years. In this study, we have dealt with calendar dates over the years.

Verma et al. (2005) have proposed an algorithm H-Mine, where a header table H is created separately for each interval. Each frequent item entry has three fields viz., item-id, support count and a hyper-link. In order to deal with the patterns in time-stamped databases we have proposed a hash-based data structure where at the first index level we store distinct years that appear in the transactions. Then we keep an array of pointers corresponding to every year in the index table. The kth pointer of this array points to tables containing interesting itemsets of size k.

Lee et al. (2009) have proposed two data mining systems for discovering fuzzy temporal association rules and fuzzy periodic association rules. The mined patterns are expressed in fuzzy temporal and periodic association rules that satisfy the temporal requirements specified by the user. In the proposed algorithm the mined patterns are dependent on user inputs such as maximum gap between two intervals and minimum length of an interval.

Li et al. (2003) proposed two classes of temporal association rules, temporal association rules with respect to precise match and temporal association rules with respect to fuzzy match, to represent regular association rules along with their temporal patterns. Our work differs from it, since we identify frequent itemsets along with the associated intervals. Then we use match ratio to determine whether a pattern is full periodic or partial periodic. Subsequently, Zimbrao et al. (2002) reported a similar work. Authors incorporate multiple granularities of time intervals from which both cyclic and user-defined calendar patterns can be achieved. Ale and Rossi (2000) proposed an algorithm to discover temporal association rules. In this algorithm, support of an item is calculated only during its lifespan. In the proposed work we compute and store supports of itemsets when they satisfy the requirements of the user.

Lee et al. (2006) have proposed a technique for mining partial multiple periodic patterns without redundant rules. Without mining every period, authors checked the necessary period and used this information to do further mining. Instead of considering the whole database, the information needed for mining partial periodic patterns is transformed into a bit vector that can be stored in a main memory. This approach needs to scan the database at the most two times. Our approach extracts both partial and full periodic patterns together by scanning the database repeatedly to find the higher-level patterns as done using apriori algorithm (Agrawal and Srikant 1994).

In the context of support definition, Kempe et al. (2008) have proposed a new support definition that counts the number of pattern instances, handles multiple instances of a pattern within one interval sequence and allows time constraints on a pattern instance.

Lee et al. (2002) have proposed a new temporal data mining technique that can extract temporal interval relation rules from temporal interval data by using Allen's theory (Allen 1983). Authors designed a preprocessing algorithm for generalization of temporal interval data. Also, authors have proposed an algorithm for discovering a temporal interval relation. Although there are thirteen different types of relations between two intervals, in our work we have focused on only overlapped intervals to find locally frequent itemsets of larger size and detect periodicity of patterns.

Ozden et al. (1998) proposed a method of finding patterns having periodic nature where the period has to be specified by the user. Han et al. (1999) proposed several algorithms for mining partial periodic patterns by exploring some interesting properties such as the apriori property and the max-subpattern hit set property by shared mining of multiple periods.

13.3 Calendar-Based Periodic Patterns

In Sects. 13.1 and 13.2, some important applications of calendar-based periodic patterns are presented. A calendar-based periodic pattern is dependent on the schema of a calendar. There are various ways one could define the schema of a calendar. It is assumed that the schema of calendar-based pattern is based on day, month and year. This schema is also useful to determine weekly-based pattern, since first seven days of any month correspond to the first week, days 8–14 of any month correspond to the second week, and so on. Thus, one can have several types of calendar-based periodic patterns viz., daily, weekly, monthly and yearly. Based on a schema, some examples of calendar patterns are given as follows: every day of January, 1999; every 16th day of January in each year; second week of every month. Again, each of these periodic patterns could be of two types viz., partially periodic pattern and full periodic pattern. A problem related to periodicity could be of finding patterns occurring at regular time intervals. Thus it emphasizes on two aspects viz., pattern and interval.

A calendar pattern refers to a market cycle that repeats periodically on a consistent basis. Seasonality could be a major force in a marketplace. While calendar patterns are based on a framework of multiple time granularities viz., day, month and year, but the periodic patterns are defined in terms of a single granularity. Here patterns are dependent on the lifespan of an item in a database. Lifespan of an item (x) is a pair $(x, [t_1, t_2])$, where t_1 and t_2 denote the time that the item x appears in the database for the first time and last time, respectively. The problem of periodic pattern mining can be categorized into two types. One is full periodic pattern mining, where every point in time granularity (Bettini et al. 2000) contributes to a cyclic behavior of the pattern. The other and more general one is called partial periodic pattern mining, which specifies the behavior of the pattern at some but not all points of time granularity in the database. Partial periodicity is a looser form of periodicity than full periodicity, and it also occurs more commonly in a real world database. A pattern is associated with a real number m ($0 < m < 1$), called match ratio (Li et al. 2003) that reveals that a pattern holds with respect to fuzzy match satisfying at least $100m$ % of the time intervals. Match ratio is an important measure which determines whether a calendar-based pattern could be full periodic or partial periodic. When the match ratio is equal to 1 then it is a full periodic pattern. In case of partial periodic pattern the match ratio lies between 0 and 1. While finding yearly periodic patterns, Mahanta et al. (2008) have proposed match ratio in somewhat a different way. Authors have proposed match ratio as the number of intervals is divided by the number of years in the lifespan of the pattern for the purpose of mining yearly pattern. It might be difficult to work with this definition, since a mining algorithm returns itemsets and their intervals. A mining algorithm might not be concerned with reporting the first and last appearances of an itemset. Therefore, we will follow the definition proposed by Li et al. (2003).

We have discussed the concept of certainty factor in Sect. 13.1. Also we have noticed that the analysis of overlapped region using certainty factor might not be sufficient. Therefore, we propose its extension.

13.3.1 Extending Certainty Factor

The concept of certainty factor is based on the concept of set superimposition. If we are interested in yearly patterns, during the analysis of superimposed intervals the year component is ignored. We explain here the concept of set superimposition using the following example.

Example 13.2 Consider the database of Example 13.1. Itemset $\{a, c\}$ is present in 3 out of 4 transactions in the intervals [29/03/1990–25/04/1990]. Also, $\{a, c\}$ is present in 2 out of 3 transactions in the intervals and [06/03/1991–19/04/1991]. Therefore, $\{a, c\}$ is frequent in these intervals at minimum support level of 0.66. These two intervals are being superimposed where each of these intervals has fuzzy set membership value 1/2. The overlapped area of these two intervals is [29/03–19/04]. Based on the concept of set superimposition, an itemset reported in a non-overlapped region has the fuzzy set membership value 1/2. But, an itemset reported in the overlapped interval [29/03–19/04] has the set membership value equal to $1/2 + 1/2 = 1$.

For the purpose of mining periodic patterns, Mahanta et al. (2008) have proposed the use of the certainty factor. It is based on a set of overlapped intervals corresponding to a pattern occurring on a periodic basis. For example, one might be interested in identifying yearly periodic patterns in a database. Authors have considered all the intervals having equal membership grade. For example, if n intervals are superimposed then every interval has $1/n$ equal fuzzy membership grade and in an overlapped area the membership value will be added. The certainty of the pattern in the overlapped interval is more than the certainty in the other intervals. Let $[t_1, t'_1]$ and $[t_2, t'_2]$ be two overlapped intervals where a pattern X gets reported with certainty value $1/2$. When the two intervals are superimposed the certainty factors of X associated with the various subintervals are given as follows:

$$[t_1, t'_1]^{1/2}(S)[t_2, t'_2]^{1/2} \ = \ [t_1, t_2)^{1/2}[t_2, t'_1]^1 (t'_1, t'_2]^{1/2} \tag{13.1}$$

The notion of certainty factor seems to be an important contribution made by the authors. It represents the certainty of reporting a pattern in an interval by considering a sequence of periods. For example, we might be interested in knowing the certainty of pattern $\{a, c\}$ in the month of April with respect to the database in Example 13.1. It is an important statistical evidence of a pattern being in an interval over a sequence of years (periods). For example, one could say that the evidence of the pattern $\{a, c\}$ is certain in the month of April when the years viz., 1990, 1991, 1992 and 1993 are considered. But the concept of certainty factor does not convey the information regarding the frequency of a pattern in an overlapped region. In addition, it gives equal importance to all the intervals by considering them as equal fuzzy grade intervals. From the perspective of the evidence of a pattern, such assumption might be realistic. But from the perspective of the depth of evidence, such concept might not be sufficient. Thus, we propose an extension to the concept of certainty factor. In the proposed extension, we incorporate the information regarding support of a pattern in an interval. There are many ways one could keep the information regarding support. In Example 13.1, there are four overlapping intervals corresponding to the pattern $\{a, c\}$. There exists a region where all the intervals are overlapped, while some regions may not be overlapped at all. Apart from the certainty factor of a region, one could also keep the support information of the pattern in that interval. In general, a region could be overlapped by all intervals. Let there be n supports of a pattern corresponding to n intervals. Then the question comes to our mind, how to keep the support information of the pattern for n intervals. The answer to this question might not be agreeable to all. One might be interested in keeping the average support of the pattern along with the certainty factor for that interval. Some of us might be interested in keeping information regarding the minimum and maximum of n supports. In an extreme case, one might be interested in keeping all the n supports of the pattern corresponding to n intervals. Let us consider that we are interested in yearly pattern. Let the lifespan of a pattern be forty years. Then one has to keep a maximum of forty supports corresponding to an overlapped region. It might not be realistic to maintain all the forty supports. Let $s\text{-}info(X, [t_1, t_2])$ be the support information of the pattern X for the interval $[t_1, t_2]$.

Let a pattern X be frequent in time intervals $[t_i, t'_i]$, $i = 1, 2, \ldots, n$. Each of these intervals is taken from a different period of time such that $\bigcap_{i=1}^{n} [t_i, t'_i] \neq \varphi$.

In Example 13.1, patterns $\{a\}$, $\{c\}$ and $\{a, c\}$ get reported in the month of April in every year. By generalizing (13.1), the certainty factor of X in overlapped regions could be obtained as follows:

$$[t_1, t_1']^{1/n}(S)[t_2, t_2']^{1/n}(S)\ldots(S)[t_n, t_n']^{1/n} = [t^{(1)}, t^{(2)})^{1/n}[t^{(2)}, t^{(3)})^{2/n}[t^{(3)}, t^{(4)})^{3/n}\ldots[t^{(r)}, t^{(r+1)})^{r/n}$$
$$\ldots \times [t^{(n)}, t'^{(1)}]^1(t'^{(1)}, t'^{(2)}]^{n-1/n}\ldots(t'^{(n-2)}, t'^{(n-1)}]^{2/n}(t'^{(n-1)}, t'^{(n)}]^{1/n}$$

$$(13.2)$$

where $\{t^{(i)}\}_{i=1}^n$ is the sequence obtained from $\{t_i\}_{i=1}^n$ by sorting in ascending order and $\{t'^{(i)}\}_{i=1}^n$ is obtained from $\{t_i'\}_{i=1}^n$ by sorting in ascending order. We propose an extended certainty factor of X in the above overlapped intervals as follows.

When X is reported in $[t^{(n)}, t'^{(1)}]$ then the certainty value is 1 with support information $s\text{-}info(X, [t^{(n)}, t'^{(1)}])$. But, the certainty value of X for the outside of $[t^{(1)}, t'^{(n)}]$ is 0 with support information 0. When X is reported in $[t^{(r-1)}, t^{(r)})$, then the certainty value is $(r - 1)/n$ with support information $s\text{-}info(X, [t^{(r-1)}, t^{(r)}))$, for $r = 2$, 3, ..., n. Otherwise, the certainty value of X for $(t'^{(r-1)}, t'^{(r)}]$ is $(n - r + 1)/n$ with support information $s\text{-}info(X, (t'^{(r-1)}, t'^{(r)}])$, for $r = 3$, 4, ..., n.

Suppose we are interested in identifying yearly periodic patterns. So each time interval is taken from a year. From the perspective of n years, the pattern X gets reported in every year in the interval $[t^{(n)}, t'^{(1)}]$. So the certainty of X is 1 (the highest) in this interval. But, X is not frequent pattern outside of $[t^{(1)}, t'^{(n)}]$. Therefore, from the perspective of all the years the certainty of X is 0 (lowest) outside of the interval. The certainty factor also provides the information regarding how many intervals are overlapped on a sub-interval. For example, if the certainty factor of a sub-interval is 2/5, for given five intervals, then two intervals are overlapped on the sub-interval. On the other hand, $s\text{-}info$ provides the information regarding degree of frequency of X in an interval. To illustrate the above concept we consider the following example.

Example 13.3 The purpose of this example is to explain the proposed concept of extended certainty factor stated above. Let the years 1980, 1981, 1982 and 1983 be of our interest. We would like to check whether the pattern X is yearly periodic. Assume that the mining algorithm has reported X as frequent in the time intervals $[t_1, t'_1]$, $[t_2, t'_2]$, $[t_3, t'_3]$ and $[t_4, t'_4]$ for the years 1980, 1981, 1982 and 1983, respectively. Also, let the supports of X in $[t_1, t'_1]$, $[t_2, t'_2]$, $[t_3, t'_3]$ and $[t_4, t'_4]$ be 0.2, 0.15, 0.16 and 0.12, respectively. Based on the proposed extended concept, we wish to analyze the time interval $[t_1, t'_4]$ by overlapping these intervals corresponding to the four years. The overlapped intervals are depicted in Fig. 13.1.

While computing support information we use here the range measure for a set of values. One could use another support information depending on the requirement. An analysis of the overlapped intervals corresponding to X is presented in Table 13.2. Certainty of a sub-interval is based on the number of intervals overlapped with it. For example, $[t_1, t_2)$ comes with certainty of 1/4, since there is only one interval out of four intervals.

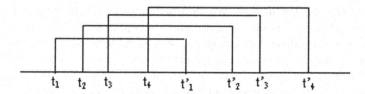

Fig. 13.1 Overlapped intervals for finding yearly pattern X

Table 13.2 An analysis of the overlapped intervals for finding yearly pattern X

Interval	Certainty factor	s-info	Interval	Certainty factor	s-info
$[t_1, t_2)$	1/4	0.2–0.2	$(t'_1, t'_2]$	3/4	0.12–0.15
$[t_2, t_3)$	1/2	0.15–0.2	$(t'_2, t'_3]$	1/2	0.12–0.16
$[t_3, t_4)$	3/4	0.15–0.2	$(t'_3, t'_4]$	1/4	0.12–0.12
$[t_4, t'_1]$	1	0.12–0.2			

Here *s-info* corresponding to interval $[t_3, t_4)$ represents the fact that the maximum and minimum supports of overlapped intervals are 0.2 and 0.15 respectively.

Certainty factor and support information are not the same. They represent two different aspects of a pattern in an interval. Certainty factor is normally associated with multiple time intervals. It expresses the likelihood of reporting a pattern in a sub-interval of the multiple overlapped intervals. But the concept of support is associated with a single time-interval. It is defined as the fraction of the transactions containing the pattern in a time-interval. Thus, for an effective analysis of a superimposed interval both the certainty factor and support information are needed in association with an interval.

13.3.2 Extending Certainty Factor with Respect to Other Intervals

In Fig. 13.1, we have shown four intervals overlapped corresponding to four different years. But in reality the scenario could be different. For four intervals, there may exist different combinations of overlapped intervals. But, whatever may be the case, the certainty factor of a sub-interval depends on the number of intervals overlapped in that sub-interval and *s-info* depends on the supports of the pattern in the intervals that are being overlapped on a sub-interval. Let us consider a sub-interval $[t, t']$, where m out of n intervals are overlapped on $[t, t']$. Based on certainty factor (Mahanta et al. 2008), we propose an extended certainty factor as follows:

When X is reported in $[t, t']$, then the certainty value is m/n with support information *s-info*$(X, [t, t'])$, where *s-info*$(X, [t, t'])$ is based on supports of X in the

m intervals overlapped on $[t, t']$. We illustrate this issue with the help of Example 13.4. Before that, we present a few definitions related to overlapped intervals. Let *maxgap* be the user-defined maximum gap (time units) between current time-stamp of a pattern and the time-stamp of the pattern when it was last seen. If the gap between current time-stamp of a pattern and the time-stamp of the pattern when it was last seen is greater than *maxgap* then a new interval is formed for the pattern with the current time-stamp as the start of the interval. Also, the previous interval of the pattern was ended when it was seen last time. Let *mininterval* be the minimum period length of a time interval. Each interval should be of sufficient length, otherwise a pattern appearing once in a transaction also becomes frequent in an interval. If two intervals are overlapped and the length of the overlapped region exceeds *mininterval* then the overlapped region could be interesting.

Example 13.4 We refer to the database of Example 13.1. Let the value of *maxgap* be 40 days. Then pattern $\{a, c\}$ gets reported in the following intervals: [29/03/ 1990–25/04/1990], [06/03/1991–12/03/1991], [03/03/1992–07/04/1992], [04/03/ 1993–17/04/1993], and [06/04/1994–20/04/1994]. Let the value of *mininterval* be 10 days. The interval [06/03/1991–12/03/1991] does not satisfy the criterion of *mininterval*. Also let the value of *minsupp* be 0.5. Then $\{a, c\}$ is not locally frequent in the interval [06/04/1994–20/04/1994]. We shall analyse the pattern $\{a, c\}$ in the following intervals: [29/03/1990–25/04/1990], [03/03/1992–07/04/ 1992], and [04/03/1993–17/04/1993]. After superimposition, we require to analyse the interval [03/03–25/04]. We present superimposed intervals in Fig. 13.2.

We present an analysis of the time interval [03/03–25/04] based on the concept of the extended certainty factor. Extended certainty factor of a pattern in an interval provides information of both the certainty factor and *s-info* for a pattern. In Table 13.3, we include an analysis of intervals for determining the yearly pattern $\{a, c\}$.

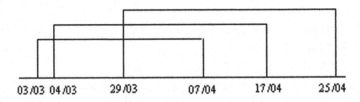

Fig. 13.2 Overlapped intervals for finding yearly pattern $\{a, c\}$

Table 13.3 An analysis of the time interval [03/03–25/04] for finding yearly pattern $\{a, c\}$

Interval	Certainty	s-info	Interval	Certainty	s-info
[03/03–04/03)	1/5	1.0–1.0	(07/04–17/04]	2/5	0.6–0.75
[04/03–29/03)	2/5	0.6–0.75	(17/04–25/04]	1/5	0.75–0.75
[29/03–07/04]	3/5	0.6–1.0			

In the above, we have presented an analysis of the time interval [03/03–25/04]. The subintervals [03/03–04/03] and [17/04–25/04] are also shown, but they do not satisfy the *mininterval* criterion. In the experimental results we have not presented such subintervals.

13.4 Mining Calendar-Based Periodic Patterns

Itemsets in transactions could be considered as a basic type of pattern in a database. Many interesting patterns such as association rules (Agrawal et al. 1993), negative association rules (Wu et al. 2004), Boolean expressions induced by itemset (Adhikari and Rao 2007) and conditional patterns (Adhikari and Rao 2008) are based on itemset patterns. Some itemsets are frequent in certain time intervals but may not be frequent throughout the lifespan of the itemsets. In other words, some itemsets may appear in the transactions for a certain time period and then disappear for a long period and then reappear. In view of making a data analysis involving various itemsets, it might be required to extract the itemsets together with the associated time-slots.

13.4.1 Improving Mining Calendar-Based Periodic Patterns

The goal of this paper is to study the existing algorithm, and to propose an effective algorithm by improving the limitations of the existing algorithm for mining calendar-based periodic patterns. As noted earlier that the concept of certainty factor of an interval does not provide good analysis of overlapped intervals. Therefore, the concept of extended certainty factor has been proposed. In view of designing an effective algorithm, we also need to understand the existing algorithm. Mahanta et al. (2005) have proposed an algorithm for finding all the locally frequent itemsets of size one. While studying the algorithm we have found that some variables contradict their definitions. Authors defined two variables *ptcount* and *ctcount* as follows. The variable *ptcount* is used to count the number of transactions in an interval in which the current item belongs. On the other hand, the variable *ctcount* is used to count the number of transactions in that interval. Therefore, the assignment *ptcount*[k] = *ctcount*[k] in the algorithm, seems to be not appropriate. Also, the variable *icount* is defined as the number of items present in the whole dataset. Therefore, the initialization, *icount* = 1, placed just before starting a new interval seems to be inappropriate. Moreover, the validity of the experimental results is low, since it is based on only one dataset. In view of improving the algorithm further we propose a number of modifications mentioned as follows: (i) The proposed algorithm makes corrections on the existing algorithm using the points noted above. (ii) It makes effective data analysis by incorporating extended certainty factor. (iii) We propose a hash-based data structure to improve the space efficiency of our algorithm. (iv) Also, we have improved the average time complexity of the algorithm. (v) We have completed a comparative analysis with the existing algorithm. (vi) In addition, we have improved the validity of the experimental results by conducting experiments on more datasets.

13.4.2 Data Structure

We discuss here the data structure used in the proposed algorithms for mining itemsets along with the time intervals in which they are frequent. We describe the data structure using Example 13.5 given below.

Example 13.5 Consider the database *D* of Example 13.1. Transactions consisting of items *a*, *b*, *c*, *d*, *e*, *f*, *g*, *h*, and *i* occurred in the years from 1990 to 1994. We propose Algorithm 13.1 to mine locally frequent itemsets of size one along with their intervals. The algorithm produces output as shown at level 1 of Fig. 13.3.

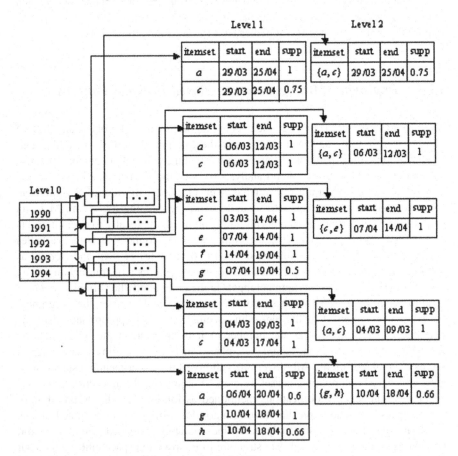

Fig. 13.3 Data structure used in the proposed algorithms

We assume here that *maxgap*, *mininterval* and *minsupp* assume values of 40 days, 5 days and 0.5, respectively. We are interested in identifying yearly periodic patterns. At the level 0 we have shown all the years that appeared in the transaction. The pointer corresponding to the year 1990 keeps all the locally frequent itemsets of size one, their supports and intervals. All the five years are stored in an index table at level 0. After level 0, we keep an array of pointers for every year. The first pointer corresponding to year 1990 points to a table containing interesting itemsets of size one, their intervals and local supports. The second pointer corresponding to year 1990, points to a table containing interesting itemsets of size two, their intervals and local supports, and so on. Here itemsets of size three corresponding to a year do not get reported. Different itemsets, their intervals, and supports are shown in Fig. 13.3.

13.4.3 A Modified Algorithm

As mentioned in Sect. 13.1, we have proposed a number of improvements to the algorithm proposed by Mahanta et al. (2005) for finding locally frequent itemsets of size one. We calculate the support of each item in an interval and store it whenever the item is frequent in that interval. Intervals that satisfy the user-defined constraints *mininterval* and *minsupp* are retained. The modification made seems to be significant from the overall viewpoint of apriori algorithm. We have used a hash-based data structure to improve efficiency of storing and accessing locally frequent itemsets of size one. We explain all the variables and their functions in the following paragraph.

Let *item* be an array of items in D. Also let the total number of items be n. We use index *level_0* to keep track of different years. It is a two-dimensional array containing 2 columns. First column of *level_0* contains the different years in increasing order. A two-dimensional array *itemset_addr* is used to store the addresses of tables containing itemsets. *itemset_addr*[*row*][*j*] contains the address of the table containing locally frequent itemsets of size j for the year *row*. The second column of *level_0* stores addresses of arrays pointing to these tables. Tables at *level_p* store the frequent itemsets of size p, $p = 1, 2, 3, ...$ Variables *row* and *row_p* are used to index arrays *itemset_addr* and *level_p* respectively, $p = 0, 1, 2,$ .... We consider a transaction as a record containing transaction date (*date*) and items purchased. Function *year*() is used to extract year from a given date. *firstseen* [*k*] and *lastseen*[*k*] specify the date when the kth item is seen for the first time and last time in an interval, respectively. Each item in the database is associated with the arrays *itemIntervalFreq* and *nTransInterval*. Cells *itemIntervalFreq*[*k*] and *nTransInterval*[*k*] are used to keep the number of transactions containing item k and total number of transactions in a time interval, respectively. Variable *nItemsTrans* is used to keep track of the number of items in the current transaction. The goal of the Algorithm 13.1 is to find all the locally frequent itemsets of size one, their intervals and supports. The algorithm is presented as follows (Adhikari and Rao 2013).

Algorithm 13.1. Mine locally frequent items and their intervals
procedure *MiningFrequentItems* (*D, maxgap, mininterval, minsupp*)
Inputs: *D, maxgap, mininterval, minsupp*
D: database to be mined
minsupp: as defined in Section 13.1
maxgap, mininterval: as defined in Section 13.3.2
Outputs:
Locally frequent items, their intervals and supports as mentioned in Figure 13.3
01: **let** *nItemsTrans* = 0; *row* = 1; *row_0* =1; *row_1* = 1;
02: **for** $k = 1$ to *n* **do**
03:    *lastseen*[*k*] = 0; *itemIntervalFreq*[*k*] = 0; *nTransInterval*[*k*] = 0;
04: **end for**
05: read a transaction $t \in D$;
06: *level_0*[*row_0*][1] = *year*(*t.date*);
07: *level_0*[*row_0*][2] = *itemset_addr*[*row*][1];
08: **while** not end of transaction in *D* **do**
09:    *transLength* = |*t*|;
10:    **if** (*level_0*[*row_0*][1] ≠ *year*(*t.date*)) **then**
11:       **for** $k = 1$ to *n* **do**
12:          **if** (|*lastseen*[*k*] − *firstseen*[*k*]| ≥ *mininterval*) **and**
             (*itemIntervalFreq*[*k*] / *nTransInterval*[*k*] ≥ *minsupp*) **then**
13:             store the *k*-th *item*, its *firstseen*, *lastseen* and local support at *level_1*[*row_1*];
14:             increase *row_1* by 1;
15:          **end if** {12}
16:       **end for** {11}
17:       *row_1* = 1;
18:       increase *row_0* by 1; increase *row* by 1;
19:       *level_0*[*row_0*][1] = *year*(*t.date*); *level_0*[*row_0*][2] = *itemset_addr*[*row*][1];
20:       **for** $k = 1$ to *n* **do**
21:          *lastseen*[*k*] = 0; *itemIntervalFreq*[*k*] = 0; *nTransInterval*[*k*] = 0;
22:       **end for**
23:    **end if** {10}
24:    **for** $k = 1$ to *n* **do**
25:       **if** (*item*[*k*] ∈ *t*) **then**
26:          increase *nItemsTrans* by 1;
27:          **if** (*lastseen*[*k*] = 0) **then**
28:             initialize both *lastseen*[*k*] and *firstseen*[*k*] by *t.date*;
                initialize both *itemIntervalFreq*[*k*] and *nTransInterval*[*k*] by 1;
29:          **else if** (| *t.date* − *lastseen*[*k*] | ≤ *maxgap*) **then**
30:                *lastseen*[*k*] = *t.date*;
31:                increase *itemIntervalFreq*[*k*] by 1; increase *nTransInterval*[*k*] by 1;
32:             **end if**
33:          **else if** (| *lastseen*[*k*] − *firstseen*[*k*] | ≥ *mininterval*) **and**
             (*itemIntervalFreq*[*k*] / *nTransInterval*[*k*] ≥ *minsupp*) **then**
34:             store the *k*-th *item*, its *firstseen*, *lastseen* and local support at *level_1*[*row_1*];
35:             increase *row_1* by 1;
36:             initialize both *lastseen*[*k*] and *firstseen*[*k*] by *t.date*;
37:             initialize both *itemIntervalFreq*[*k*] and *nTransInterval*[*k*] by 0;
38:             **end if** {33}
39:          **end if** {27}
40:       **else** increase *nTransInterval*[*k*] by 1;
41:       **end if** {25}
42:       **if** (*nItemsTrans* = *transLength*) **then** exit from *for-loop*; **end if**
43:    **end for** {24}
44:    read a transaction $t \in D$;
45: **end while** {08}
46: **for** $k = 1$ to *n* **do**
47:    **if** (|*lastseen*[*k*] − *firstseen*[*k*]| ≥ *mininterval*) **and**
       (*itemIntervalFreq*[*k*] / *nTransInterval*[*k*] ≥ *minsupp*) **then**
48:       store the *k*-th *item*, its *firstseen*, *lastseen* and local support at *level_1*[*row_1*];
49:       increase *row_1* by 1;
50:    **end if** {47}
51: **end for** {46}
52: sort arrays *level_1* on non-increasing order on primary key item and
    secondary key start date;
end procedure

At line 5 we read the first transaction of database. Afterwards the first row of the index *level_0* is initialized with the first year obtained from the transaction. The pointer field of the first row of *level_0* is initialized by the address of the first row of the table *itemset_addr*. Lines 8–45 are repeated until all the transactions are read. At line 10 we check whether the current transaction belongs to a different year. If it happens so then we close the last interval of different items using lines 11–16. We retain those intervals that satisfy criteria of *mininterval* and *minsupp*. Lines 17–22 assign the necessary initializations for a different year. Lines 25–41 are repeated for each item in the current transaction. Line 27 checks whether the item is first time seen in the transaction and the necessary assignment is done in line 28. Lines 29–32 determine whether the current transaction-date is coming under the current interval by comparing the difference between *t.date* and *lastseen* with *maxgap*. In Lines 33–38 we construct an interval and compute the local support. Line 42 avoids the unnecessary repetition by comparing the transaction length. Line numbers 46–51 close all the last intervals for last year. Line 52 sorts arrays *level_1* on non-increasing order on primary key item and secondary key start date.

The time complexity of the algorithm has been reduced significantly by computing the length of the current transaction (at line number 9) and putting a check at line number 25. Consider a database containing 10,000 items. Let the current transaction be of length 20 and these items are within the first 100 items. Then the *for-loop* at line number 24 need not have to continue for the remaining 9,900 items, but the worst-case complexity of the algorithm remains the same as before.

We now present below an algorithm that makes use of locally frequent itemsets obtained by Algorithm 13.1 and apriori property (Agrawal and Srikant 1994). We use array *level_1* to generate the candidate sets at the second level. Then array *level_2* is used to generate candidate sets at the third level, and so on. We apply pruning using conditions at line number 6 to eliminate some itemsets at the next level. This pruning step ensures that the size of the itemsets at the current level is one more than the size of an itemset at the previous level. Also we apply pruning using user-defined thresholds such as *maxgap*, *mininterval* and *minsupp*. The goal of the Algorithm 13.2 is to find all the locally frequent itemsets of size greater than one, their intervals and supports. The algorithm is presented as follows (Adhikari and Rao 2013).

Algorithm 13.2. Mine locally frequent itemsets at higher level and the associated intervals

procedure *MiningHigherLevelItemsets* (*D, S*)

Inputs: *D, S*

D: database to be mined

S: partially constructed data structure containing locally frequent itemsets of size one

Outputs: locally frequent itemsets at higher levels, the associated intervals and supports as mentioned in Figure 13.3

```
01: let L₁ = set of elements at level_1 of S; let k = 2;
02: while Lₖ₋₁ ≠ ∅ do
03:     Cₖ = ∅;
04:     for each itemset l₁∈ Lₖ₋₁ do
05:         for each itemset l₂∈ Lₖ₋₁ do
06:             if ((l₁[1] = l₂[1]) ∧ ... ∧ (l₁[k-2] = l₂[k-2]) ∧ (l₁[k-1] < l₂[k-1])) then
07:                 c = l₁⋈ l₂; Cₖ = Cₖ ∪ c;
08:             end if {06}
09:         end for {05}
10:     end for {04}
11:     for each element c ∈ Cₖ do
12:         construct intervals for c as mentioned in Algorithm 13.1;
13:         if the intervals corresponding to c satisfy maxgap, mininterval and minsupp then
14:             add c and the intervals to level_k of S;
15:         end if {13}
16:     end for {11}
17:     increase k by 1;
18:     let Lₖ = set of elements at level_k of S;
19: end while {02}
```

end procedure

Using Algorithms 13.1 and 13.2, one could construct the data structure S presented in Fig. 13.3. One can use S to determine whether an itemset pattern is fully/partially periodic.

13.5 Experimental Studies

Experiments are done using datasets *retail* (Frequent Itemset Mining Dataset Repository),[1] *BMS-WebView-1* (Frequent Itemset Mining Dataset Repository), and *T10I4D100K* (Frequent Itemset Mining Dataset Repository). Since the records in these databases contain only items purchased in transactions, we have attached time-stamps randomly as calendar date for the transactions. The characteristics of the databases are given in Table 13.4.

For the purpose of conducting experiments, each of the databases *retail*, *BMS-WebView-1* and *T10I4D100K* has been divided into 30 sub-databases, called yearly databases. The characteristics of these databases are given in Table 13.5. Let D, *NT*, *ALT*, *AFI*, and *NI* be the given database, the number of transactions, average length of a transaction, average frequency of an item, and the number of items, respectively. In Table 13.5 we have shown how the transactions have been time-stamped. The yearly databases obtained from *retail*, *BMS-WebView-1* and *T10I4D100K* are named as R_i, B_i and T_i respectively, $i = 1, ..., 30$. For simplicity, we have kept the number of transactions in each of the yearly databases fixed,

---

[1] Frequent itemset mining dataset repository, http://fimi.cs.helsinki.fi/data.

Table 13.4 Database characteristics

D	NT	ALT	AFI	NI	Size (megabytes)
retail	88,162	11.31	60.54	16,470	3.97
BMS-WebView-1	1,49,639	2.00	44.57	6,714	1.97
T10I4D100K	1,00,000	11.10	1276.12	870	3.83

Table 13.5 Characteristics of yearly databases

D	NT	Starting date, ending date	Average number of transactions per day
R_1	2,920	01/01/1961, 31/12/1961	8
...	...	...	...
R_{29}	2,920	01/01/1989, 31/12/1989	8
R_{30}	3,482	01/01/1990, 31/12/1990	9.54
B_1	5,110	01/01/1961, 31/12/1961	14
...	...	...	...
B_{29}	5,110	01/01/1989, 31/12/1989	14
B_{30}	1,449	01/01/1990, 31/12/1990	3.97
T_1	3,285	01/01/1961, 31/12/1961	9
...	...	...	...
T_{29}	3,285	01/01/1989, 31/12/1989	9
T_{30}	4,735	01/01/1990, 31/12/1990	12.97

except for the last database. We assume that the first and the last transactions occur on 01/01/1961 and 31/12/1990, respectively, and also assume that each year contains 365 days. In our experimental studies we report yearly periodic patterns and the associated periodicities. We also determine the certainty factor and the match ratio of a pattern with respect to overlapped intervals.

In addition to partial periodic patterns, we mine full periodic patterns in the above databases. Itemset patterns of size one and two of *retail* is shown in Tables 13.6 and 13.7 respectively. In *retail* the itemsets {39} and {48} occur in all the thirty years and they are periodic throughout the year. Therefore, these itemsets are full periodic in the interval [1/1–31/12]. Itemset {41} is partially periodic, since the match ratio is less than 1. Initially it becomes frequent for thirteen years and then it does not get reported, and again it becomes frequent for the last six years. The subintervals that do not satisfy the *mininterval* criterion are not shown. We have noticed some peculiarity in the mined patterns. For example, many patterns such as {0} and {1} are frequent throughout a year. Although, it is peculiar but it remains also an artificial phenomenon, since the time-stamps are enforced by us. There are many itemsets such as {16,217} are frequent in many years with non-overlapping intervals. In Table 13.6, we present itemsets of size one that are also part of interesting itemsets of size two as shown in Table 13.7. While computing the certainty factor of an itemset we have used lifespan of the itemset.

Table 13.6 Selected yearly periodic itemsets of size one (for *retail*)

retail (*minsupp* = 0.25, *mininterval* = 8, *maxgap* = 10)				
Itemset	Intervals	Certainty	*s-info*	Match ratio
{0}	[1/1–31/12]	2/2	0.35–0.66	1.0
{1}	[3/1–31/12]	2/3	0.57–0.66	0.67
{39}	[1/1–31/12]	30/30	0.52–0.63	1.0
{41}	[1/1–22/12]	13/30	0.26–0.32	0.43
{41}	[2/12–30/12]	6/30	0.27–0.32	0.20
{48}	[1/1–31/12]	30/30	0.43–0.53	1.0
{16,217}	[1/1–30/5]	1/1	0.87–0.87	1.0
{16,217}	[7/9–31/12]	1/1	0.97–0.97	1.0

Table 13.7 Yearly periodic itemsets of size two (for *retail*)

retail (*minsupp* = 0.25, *mininterval* = 8, *maxgap* = 10)				
Itemset	Intervals	Certainty	*s-info*	Match ratio
{0, 1}	[15/10–31/12]	1/1	0.46	1.0
{39, 41}	[1/1–30/12]	1/1	0.25	1.0
{39, 48}	[1/1–30/12]	30/30	0.28–0.38	1.0
{39, 16,217}	[1/1–30/5]	1/1	0.34	1.0
{48, 16,217}	[7/9–31/12]	1/1	0.27	1.0

For example, itemset {0} gets reported from two years and it becomes frequent in both the years. Therefore its certainty factor is 2/2 = 1.

Interesting itemset patterns of size one and two in *BMS-WebView-1* are shown in Tables 13.8 and 13.9, respectively. Here full periodic patterns are not reported since all the itemsets in *BMS-WebView-1* have the value of the match ratio lower than 1. Therefore, these patterns are partially periodic. Itemset {12,355} becomes frequent in 3 years but it has lifespan for 7 years. In this database the items are sparse. Therefore, one requires choosing a smaller *minsupp*. From Table 13.9 one could observe that itemset {33,449, 33,469} shows periodicity by appearing two times in 6 years and the remaining interesting itemsets are reported for a year only.

In Table 13.10 we present yearly periodic itemsets of size one for *T10I4D100K* database. In this database patterns with full periodicity are not available, since the

Table 13.8 Yearly periodic itemsets of size one (for *BMS-WebView-1*)

BMS-WebView-1(*minsupp* = 0.06, *mininterval* = 7, *maxgap* = 10)				
Itemset	Intervals	Certainty	*s-info*	Match ratio
{10,311}	[29/1–6/10]	2/6	0.063–0.86	0.33
{12,355}	[21/12–28/12]	3/7	0.060–0.061	0.43
{12,559}	[22/4–11/5]	1/2	0.064–0.066	0.5
{33,449}	[3/1–26/12]	5/7	0.063–0.08	0.71
{33,469}	[3/1–31/3]	5/7	0.067–0.08	0.71

Table 13.9 Yearly periodic itemsets of size two (for *BMS-WebView-1*)

BMS-WebView-1(minsupp = 0.06, mininterval = 7, maxgap = 10)				
Itemset	Intervals	Certainty	*s-info*	Match ratio
{10,311, 12,559}	[30/4–9/4]	1/1	0.06–0.06	1.0
{10,311, 33,449}	[3/3–11/4]	1/1	0.065	1.0
{33,449, 33,469}	[15/2–25/3]	2/6	0.061–0.064	0.33

Table 13.10 Selected yearly periodic itemsets of size one (for *T10I4D100K*)

T10I4D100K (minsupp = 0.13, mininterval = 7, maxgap = 10)					
Itemset	Interval	*s-info*	Itemset	Interval	*s-info*
{966}	[28/1/1961–19/2/1961]	0.16	{998}	[13/11/1964–22/11/1964]	0.17
{966}	[16/3/1961–23/3/1961]	0.17	{998}	[15/12/1964–27/12/1964]	0.16
{966}	[14/12/1961–25/12/1961]	0.16	{998}	[2/9/1965–13/9/1965]	0.14
{966}	[22/3/1964–13/4/1964]	0.15	{998}	[27/11/1966–8/12/1966]	0.14
{966}	[1/11/1975–8/11/1975]	0.16	{998}	[27/11/1973–11/12/1973]	0.13
{966}	[12/4/1981–19/4/1981]	0.17	{998}	[14/12/1983–23/12/1983]	0.17
{966}	[2/12/1988–12/12/1988]	0.15	{998}	[15/12/1984–23/12/1984]	0.16

intervals corresponding to an item are not overlapped. We have presented examples of such items in the following table. From interval column, one could observe that the itemsets are frequent for the short intervals, but do not appear at the same time for all the years. For example, itemset {966} appears in three intervals in 1961, but it does not show any periodicity since the intervals are not overlapped. It is interesting to note that the itemset {966} appears at the beginning, both in first and second months, of the year, then at the middle of the year i.e., for the third and fourth months, and finally at the end of the year (eleventh and twelfth month). This is also true for itemset {998}. Interesting itemset patterns of size two are not reported in this database.

An itemset that satisfies *minsupp*, *mininterval* criteria are reported. Also, a locally frequent itemset in two intervals for a particular year is also reported in the intervals, provided the intervals satisfy *maxgap* criterion. The number of interesting intervals could increase by lowering the thresholds. In the following paragraphs we have presented a study of these aspects.

13.5.1 Selection of Mininterval and Maxgap

The selection of *mininterval* and *maxgap* might be crucial since the process of data mining would depend on factors like seasonality, type of application and the data source. Some items are used for a particular season; while others are purchased throughout the year. When the items are purchased throughout the year, the choices of *mininterval* and *maxgap* do not have much significance in mining yearly patterns. This observation seems to be valid for the items in *retail* and

BMS-WebView-1. But the items in *T10I4D100K* are frequent in smaller intervals and therefore, *mininterval* and *maxgap* might have an impact on data mining. On the other hand, the requirement of an organization might determine an important parameter for mining calendar-based patterns. The distribution of items in databases also matters in selecting the right values of *mininterval* and *maxgap*. For a sparse database *maxgap* could be longer, and it could be even longer than *mininterval* provided *minsupp* remains low.

13.5.1.1 Mininterval

In the following experiments we would like to analyse the effect of *mininterval* for given *maxgap* and *minsupp*. We observe in Figs. 13.4, 13.5 and 13.6, the number of intervals decreases as *mininterval* increases. An itemset might be frequent in many intervals. The number of itemsets frequent in an interval decreases as the length of *mininterval* increases. Although the above observation is true in general, but the type of the graphs might differ from one data source to another. In *retail* many itemsets are locally frequent for longer period of time. In Fig. 13.4 we observe that there exists nearly 110 intervals for *mininterval* of 29 days. Whereas in *BMS-WebView*-1 and *T10I4D100K*, the itemsets are frequent for shorter duration. As a result, the number of intervals reduces significantly when *mininterval* remains small. Thus the choice of *mininterval* is an important issue.

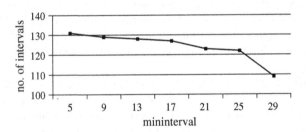

Fig. 13.4 *retail* (*minsupp* = 0.25, *maxgap* = 7)

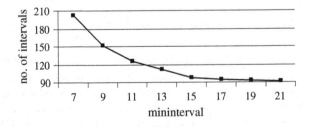

Fig. 13.5 *BMS-WebView*-1 (*minsupp* = 0.06, *maxgap* = 7)

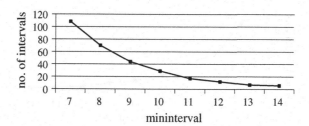

Fig. 13.6 *T10I4D100K* (*minsupp* = 0.13, *maxgap* = 7)

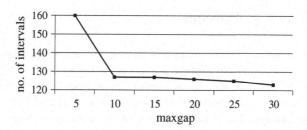

Fig. 13.7 *retail* (*minsupp* = 0.25, *mininterval* = 10)

13.5.1.2 Maxgap

In view of analyzing *maxgap* parameter, we present graphs of the number of intervals versus *maxgap* at given *minsupp* and *mininterval* in Figs. 13.7, 13.8, and 13.9. The graphs show that the number of intervals decreases as *maxgap* increases. In *retail* the number of intervals decreases rapidly when *maxgap* varies from 5 to 10. Afterwards the change is not so significant. In *BMS-WebView*-1 the decrement takes place almost at a uniform rate. Unlike *retail* and *BMS-WebView*-1, the number of intervals decreases faster at the smaller values of *maxgap* in *T10I4D100K* dataset.

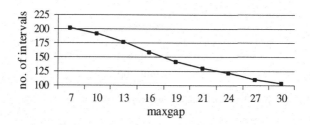

Fig. 13.8 *BMS-WebView*-1(*minsupp* = 0.06, *mininterval* = 7)

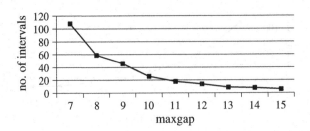

Fig. 13.9 *T10I4D100K* (*minsupp* = 0.13, *mininterval* = 7)

13.5.2 *Selection of* Minsupp

The number of intervals and minimum support are inversely related to given *maxgap* and *mininterval*. We observe this phenomenon in Figs. 13.10, 13.11 and 13.12. When the value of *maxgap* is smaller, the number of intervals reported is quite large. Initially the number of intervals decreased significantly with small decrement of *minsupp*. Later the decrement of the number of intervals is not so significant.

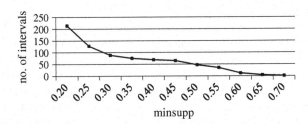

Fig. 13.10 *Retail* (*mininterval* = 10, *maxga*p = 12)

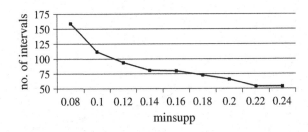

Fig. 13.11 *BMS-WebView-*1(*mininterval* = 7, *maxgap* = 10)

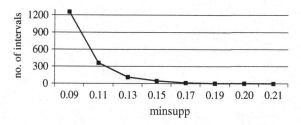

Fig. 13.12 *T10I4D100K (mininterval* = 7, *maxga*p = 9)

13.5.3 Performance Analysis

In this section, we present the performance of our algorithm and compare it with the performance of the existing algorithm for mining calendar-based periodic patterns. To quantify the performance, two experiments have been conducted. In the first experiment, we have measured the scalability of the two algorithms with respect to different database sizes. In the second experiment, we have measured the scalability of the two algorithms with respect to different support thresholds. In Figs. 13.13, 13.14 and 13.15, we have shown the relationship between the database size and execution time for mining periodic patterns. We observed that the number of patterns increases as the number of transactions increases. Thus, the execution time increases with the increase of the database size. Initially both the algorithms take almost equal amount of time. In Figs. 13.13 and 13.14 we observe that execution time for mining 88,162 transactions of *retail* and 149,639 transactions of *BMS-WebView*-1 take nearly equal amount of time. The reason is that the average length of transactions in *retail* is more than that of *BMS-WebView*-1. Therefore, the execution time is not only dependent on the size of the database, but also depends on the factors such as *ALT* and *NI*. The experimental results in Figs. 13.13, 13.14 and 13.15 show that our algorithm performs better than the existing algorithm.

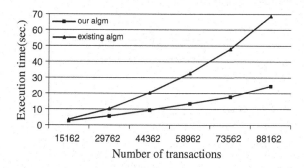

Fig. 13.13 Execution time versus size of database at *minsupp* = 0.25, *mininterval* = 8, *maxga*p = 10 (*retail*)

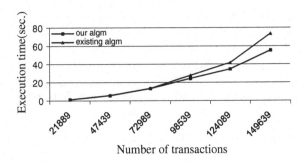

Fig. 13.14 Execution time versus size of database at *minsupp* = 0.1, *mininterval* = 7, *maxga*p = 10 (*BMS-WebView*-1)

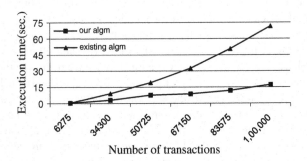

Fig. 13.15 Execution time versus size of database at *minsupp* = 0.13, *mininterval* = 7, *maxga*p = 10 (*T10I4D100K*)

In Figs. 13.16, 13.17, and 13.18, we have presented another comparison by considering *minsupp* threshold. When the minimum support increases, the number of frequent itemsets decreases and subsequently the execution time also decreases. The experimental results have shown that the execution time of both the algorithms decreases slowly when the support threshold increases, and our algorithm takes less time than the existing algorithm.

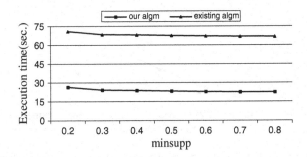

Fig. 13.16 Execution time versus *minsupp* (*mininterval* = 8, *maxga*p = 10) for *retail*

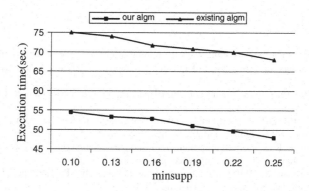

Fig. 13.17 Execution time versus *minsupp* (*mininterval* = 7, *maxga*p = 10) for *BMS-WebView*-1

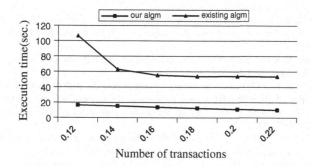

Fig. 13.18 Execution time versus *minsupp* (*mininterval* = 7, *maxga*p = 10) for *T10I4D100K*

13.6 Conclusions

In this chapter, we have proposed modifications to the existing algorithm for mining locally frequent itemsets along with the set of intervals and associated supports. We have also extended the concept of the certainty factor for a detailed analysis of overlapped intervals. We have proposed a number of improvements of the existing algorithm for finding calendar-based periodic patterns. For managing locally frequent itemsets effectively we have introduced a hash-based data structure. An extensive data analysis by involving constraints such as *mininterval, minsupp* and *maxgap* is presented. In addition we have compared our algorithm with the existing algorithm. Experimental results show that the proposed algorithm runs faster than the existing algorithm. Experimental results also report that whether a periodic pattern is full or partial. The proposed algorithm can also be used to extract yearly, monthly, weekly and daily calendar-based patterns.

References

Adhikari A, Rao PR (2007) A framework for mining arbitrary Boolean expressions induced by frequent itemsets. In: Proceedings of Indian international conference on artificial intelligence, pp 5–23

Adhikari A, Rao PR (2008) Mining conditional patterns in a database. Pattern Recogn Lett 29 (10):1515–1523

Adhikari J, Rao PR (2013) Identifying calendar-based periodic patterns. In: Jain LC, Howlett RJ, Ramanna S (eds) Emerging paradigms in machine learning. Springer, Berlin, pp 329–357

Agrawal R, Srikant R (1994) Fast algorithms for mining association rules. In: Proceedings of 20th very large databases (VLDB) conference, pp 487–499

Agrawal R, Srikant R (1995) Mining sequential patterns. In: Proceedings of international conference on data engineering (ICDE), pp 3–14

Agrawal R, Imielinski T, Swami A (1993) Mining association rules between sets of items in large databases. In: Proceedings of the ACM SIGMOD conference management of data, pp 207–216

Ale JM, Rossi GH (2000) An approach to discovering temporal association rules. In: Proceedings of ACM symposium on applied computing, pp 294–300

Allen JF (1983) Maintaining knowledge about temporal intervals. Commun ACM 26(11):832–843

Baruah HK (1999) Set superimposition and its application to the theory of fuzzy sets. J Assam Sci Soc 10(1–2):25–31

Bettini C, Jajodia S, Wang SX (2000) Time granularities in databases, data mining and temporal reasoning. Springer, Berlin

Han J, Dong G, Yin Y (1999) Efficient mining on partial periodic patterns in time series database. In: Proceedings of fifteenth international conference on data engineering, pp 106–115

Hong TP, Wu YY, Wang SL (2009) An effective mining approach for up-to-date patterns. Expert Syst Appl 36:9747–9752

Kempe S, Hipp J, Lanquillon C, Kruse R (2008) Mining frequent temporal patterns in interval sequences. Int J Uncertainty Fuzziness Knowl Based Syst 16(5):645–661

Lee JW, Lee YJ, Kim HK, Hwang BH, Ryu KH (2002) Discovering temporal relation rules mining from interval data. In: Proceedings of the 1st EuroAsian conference on advance in information and communication technology, pp 57–66

Lee G, Yang W, Lee JM (2006) A parallel algorithm for mining multiple partial periodic patterns. Inf Sci 176(24):3591–3609

Lee YJ, Lee JW, Chai D, Hwang B, Ryu KH (2009) Mining temporal interval relational rules from temporal data. J Syst Softw 82(1):155–167

Li D, Deogun JS (2005) Discovering partial periodic sequential association rules with time lag in multiple sequences for prediction. In: Hacid M-S, Murray NV, Raś ZW, Tsumoto S (eds) Foundations of intelligent systems, vol 3488, LNCS. Springer, Berlin, pp 1–24

Li Y, Ning P, Wang XS, Jajodia S (2003) Discovering calendar-based temporal association rules. Data Knowl Eng 44(2):193–218

Mahanta AK, Mazarbhuiya FA, Baruah HK (2005) Finding locally and periodically frequent sets and periodic association rules. In: Pal SK, Bandyopadhyay S, Biswas S (eds) Pattern recognition and machine intelligence, vol 3776, LNCS. Springer, Berlin, pp 576–582

Mahanta AK, Mazarbhuiya FA, Baruah HK (2008) Finding calendar-based periodic patterns. Pattern Recogn Lett 29(9):1274–1284

Ozden B, Ramaswamy S, Silberschatz A (1998) Cyclic association rules. In: Proceedings of 14th international conference on data engineering, pp 412–421

Roddick JF, Spiliopoulou M (2002) A survey of temporal knowledge discovery paradigms and methods. IEEE Trans Knowl Data Eng 14(4):750–767

Verma K, Vyas OP, Vyas R (2005) Temporal approach to association rule mining using T-tree and P-tree. In: Perner P, Imiya A (eds) Machine learning and data mining in pattern recognition, vol 3587, LNCS. Springer, Berlin, pp 651–659

Wu X, Zhang C, Zhang S (2004) Efficient mining of both positive and negative association rules. ACM Trans Inf Syst 22(3):381–405

Zimbrao G, de Souza JM, de Almeida VT, de Silva WA (2002) An algorithm to discover calendar-based temporal association rules with item's lifespan restriction. In: Proceedings of the eighth ACM SIGKDD international conference on knowledge discovery and data mining

Chapter 14
Measuring Influence of an Item in Time-Stamped Databases

Influence of items on some other items might not be the same as the association between these sets of items. Many tasks of data analysis are based on expressing influence of items on other items. In this chapter, we introduce the notion of an overall influence of a set of items on another set of items. We also discuss an extension to the notion of overall association between two items in a database. Using this notion, we have designed two algorithms of influence analysis involving specific items in a database. As the number of databases increases on a yearly basis, we have adopted incremental approach to these algorithms. Experimental results are reported on both synthetic and real-world databases.

14.1 Introduction

Every time a customer interacts with business, we have an opportunity to gain strategic knowledge. Transactional data contain a wealth of information about customers and their purchase patterns. In fact, these data could be one of the most valuable assets, when used wisely. This has been recognized a long time ago by many large organizations such as supermarkets, insurance companies, healthcare organizations, telecommunications, and banks. These organizations have spent significant resources for collecting and analyzing transactional data. Many applications are based on inherent knowledge present in a database. Such applications could be dealt with mining databases (Han et al. 2000; Agrawal and Srikant 1994; Savasere et al. 1995). As a database changes over time, the inherent knowledge also changes. Therefore in the competitive market, knowledge-based decisions are more appropriate. Data mining algorithms are effective tools to support making such decisions. Data mining algorithms often extract different patterns from a database. Some examples of patterns in a database are frequent item sets (Agrawal et al. 1993), association rules (Agrawal et al. 1993), negative association rules (Wu et al. 2004), Boolean expressions induced by itemset (Adhikari and Rao 2007) and conditional patterns (Adhikari and Rao 2008a). Nevertheless, there are some applications for which association-based analysis might be inappropriate. For

© Springer International Publishing Switzerland 2015
A. Adhikari and J. Adhikari, *Advances in Knowledge Discovery in Databases*,
Intelligent Systems Reference Library 79, DOI 10.1007/978-3-319-13212-9_14

example, an organization might deal with a large number of items with its customers. The company might be interested in knowing how the purchase of a particular item affects the purchase of some other item. In this chapter, we study such influences based on transactional time-stamped database.

Many companies transact a large number of products (items) with their customers. It might be required to perform data analyses involving different items. Such analyses might originate from different applications. One such analysis is identifying stable items in databases over time (Adhikari et al. 2009). It could be useful in devising strategies for a company. Little work has been reported on data analyses realized over time. In this chapter, we present another application involving different items in a database formed over time.

Consider a company that collects a huge amount of transactional data on a yearly basis. Let DT_i be the database corresponding to the ith year, $i = 1, 2, ..., k$. Each of these databases corresponds to a specific period of time. Thus, one could call these time databases. Each time database is mined using a traditional data mining technique (Adhikari et al. 2011). In this application, we will deal with itemsets in a database. An itemset is a set of items in the database. Let I be the set of all items in the time databases. Each itemset X in a database D is associated with a statistical measure, called *support* (Agrawal et al. 1993), denoted by $supp(X, D)$. The support of an itemset is defined as the fraction of transactions containing the itemset.

Solutions to many problems are based on the study of relationships among variables. We will see later that the study of influence of a set of variables on another set of variables might not be the same as studying the association between these two sets of variables. Association analysis among variables has been studied well (Agrawal et al. 1993; Adhikari et al. 2011; Brin et al. 1997; Shapiro 1991; Adhikari and Rao 2008 a, b). In the context of studying association among variables using association rules one could conclude that the confidence of the association rule gives positive influence of antecedent on the consequent of the association rule. Such positive influences might not be sufficient for many data analyses.

Consider an established company possessing data being collected over fifty consecutive years. Generally, the sales of a product vary from one season to another. Also, a season re-appears on a yearly basis. Thus, we divide the entire database into a sequence of yearly databases. In this context, a yearly database could be considered as a time database. In this study, we estimate the influence of item x on y, for $x, y \in I$ where I is the set of all items in database D. In Sect. 14.3, we define the concept of influence of an itemset on another itemset.

An itemset could be viewed as a basic type of pattern in a database. Different types of pattern in a database could be derived from itemset patterns. For example, frequent itemset, association rule, negative association rule, Boolean expression induced by itemset and conditional pattern are examples of derived patterns in a database. Few applications have been reported on analysis of patterns over time. In this chapter, we wish to study the influence of an item on a specific item/a set of specific items in a database.

Most of the association analyses are based on a positive association between variables. Such positive association gives rise to positive influence of variables on other variables. Most of the real databases are large and sparse. In such cases an association analysis using positive influence might not be appropriate, if the overall influence of former variable on latter variable becomes negative. Thus, the concept of overall influence needs to be introduced.

The chapter is organized as follows. In Sect. 14.2, we extend the notion of overall association between two items in a database. In Sect. 14.3, we introduce the notion of overall influence of an itemset on another itemset in a database. We study various properties of proposed measures. Also, we introduce the notion of overall influence of an item on a set of specific items in a database. In addition, we discuss the motivation of the proposed problem in this section. We state our problem in Sect. 14.4. We discuss work related to the proposed problem in Sect. 14.5. In Sect. 14.6, we design an algorithm to measure the overall influence of an item on another item (incrementally). In addition, we design another algorithm of overall influence of an item on a set of specific items (incrementally). Experimental results are provided in Sect. 14.7. We conclude the chapter in Sect. 14.8.

14.2 Association Between Two Itemsets

Adhikari et al. (2011) have proposed a measure denoted by OA, for computing an overall association between two items in a market basket data. Using positive association PA between two items (Adhikari et al. 2011), one could extend positive association between two itemsets in a database as follows:

$$PA(X, Y, D) = \frac{\#\ \textit{transaction}\ \text{containing both}\ X\ \text{and}\ Y, D}{\#\ \text{transaction containing at least one of}\ X\ \text{and}\ Y, D},$$

where X and Y are itemsets in database D and "$\#P, D$" is the number of transactions in D that satisfy the predicate P.

Similarly, negative association NA between two items (Adhikari et al. 2011) could be extended as follows:

$$NA(X, Y, D) = \frac{\#\ \text{transaction containing exactly one of}\ X\ \text{and}\ Y, D}{\#\ \text{transaction containing at least one of}\ X\ \text{and}\ Y, D},$$

where X and Y are itemsets in database D.

Using PA and NA, OA between two itemsets X and Y in database D could be defined as follows:

$$OA(X, Y, D) = PA(X, Y, D) - NA(X, Y, D) \tag{14.1}$$

$OA(X, Y, D)$ can be positive, negative or zero, and accordingly all the items in X together with all the items in Y together are positively, negatively or

Table 14.1 Supports of itemsets in D_1

Itemset($\{X\}$)	$\{a, b\}$	$\{c, d\}$	$\{a, c\}$	$\{b, d\}$	$\{d, e\}$	$\{e, g\}$
supp($\{X\}$, D_1)	4/9	4/9	3/9	3/9	4/9	2/9

Table 14.2 Overall association between two itemsets in D_1

Itemset($\{X, Y\}$)	$\{\{a, b\}, \{c, d\}\}$	$\{\{a, c\}, \{b, d\}\}$	$\{\{c\}, \{d, e\}\}$
OA(X, Y, D_1)	1/5	1	−3/7

independently associated in D. We illustrate different types of association using the following example.

Example 14.1 Let database D_1 contain the following transactions: $\{a, d, e\}$, $\{a, b, c, d, g\}$, $\{a, b, e, g\}$, $\{b, c, g\}$, $\{d, e, g\}$, $\{b, e, f\}$, $\{c, d, e, f\}$, $\{a, b, c, d, f, g\}$, and $\{a, b, c, d, e\}$. We find here overall association between itemsets X, and Y, for some X, Y in D_1. In Table 14.1, supports of some itemsets are given.

Here $PA(\{a, b\}, \{c, d\}, D_1) = 3/5$ and $NA(\{a, b\}, \{c, d\}, D_1) = 2/5$. Therefore, $OA(\{a, b\}, \{c, d\}, D_1) = 1/5$. In Table 14.2, we show overall associations.

Considering these results, we observe that the OA values between $\{a, b\}$ and $\{c, d\}$ as well as $\{a, c\}$ and $\{b, d\}$ are positive. But the OA value between $\{c\}$ and $\{d, e\}$ is negative.

14.3 Concept of Influence

Let X and Y be two itemsets in database D. We wish to find influence of X on Y in D. In the previous section, we have proposed overall association between two itemsets. The influence of X on Y seems to be different from the overall association between X and Y.

Let $X = \{x_1, x_2, \ldots x_p\}$ and $Y = \{y_1, y_2, \ldots y_q\}$ be two itemsets in database D. The influence of X on Y could be judged based on the following events: (i) Whether a customer purchases all the items of Y when he/she purchases all the items of X, and (ii) Whether a customer purchases all the items of Y when they do not purchase all the items of X. Such behaviors could be modeled using supports of $X \cap Y$ and $\neg X \cap Y$. The expression $supp(X \cap Y, D)/supp(X, D)$ measures the strength of positive association of X on Y. The expression $supp(\neg X \cap Y, D)/supp(\neg X, D)$ measures the strength of negative association of X on Y. Thus, the expressions $supp(X \cap Y, D)/supp(X, D)$ and $supp(\neg X \cap Y, D)/supp(\neg X, D)$ could be important in measuring an overall influence of X on Y.

14.3.1 Influence of an Itemset on Another Itemset

Let X and Y be the two itemsets in database D. The interestingness of an association rule $r_1: X \rightarrow Y$ could be expressed by its support and confidence (*conf*) measures (Agrawal et al. 1993). These measures are defined as follows. $supp(r_1, D) = supp$ $(X \cap Y, D)$, and $conf(r_1, D) = supp(X \cap Y, D)/supp(X, D)$. The measure $conf(r_1, D)$ could be interpreted as the fraction of transactions containing itemset Y among the transactions containing X in D. In other words, $conf(r_1, D)$ could be viewed as the *positive influence* (*PI*) of X on Y. Let us consider the negative association rule $r_2: \neg X \rightarrow Y$. Confidence of r_2 in D could be viewed as fractions of transactions containing Y among the transactions containing $\neg X$. In other words, confidence of r_2 in D could be viewed as *negative influence* (*NI*) of X on Y. Similarly to the overall association defined in (14.1), one could define *overall influence* (*OI*) of X on Y in a database as follows:

Definition 14.1 Let X and Y be two itemsets in database D such that $X \cap Y = \phi$. Then overall influence of X on Y in D is defined as follows (Adhikari and Rao 2010):

$$OI(X, Y, D) = supp(X \cap Y, D)/supp(X, D) - supp(\neg X \cap Y, D)/supp(\neg X, D)$$

$$(14.2)$$

$OI(X, Y, D)$ represents the difference of the influence on Y when X is present in a transaction and the influence on Y when X is not present in the transaction. Let y be user-defined level of interestingness. Then $OI(X, Y, D)$ is *interesting* if $OI(X, Y, D) \geq y$.

If $OI(X, Y, D) > 0$ then the itemset X has positive influence on itemset Y in D. In other words, all the items in X together help promoting itemset Y in D. If $OI(X, Y, D) < 0$ then X has negative influence on Y in D. In other words, all the items in X in D together do not help promoting together all the items in Y. If $OI(X, Y, D) = 0$ then X has no influence on Y in D. In Example 14.2, we illustrate the concept of overall influence.

Example 14.2 We continue our discussion that started in Example 14.1. We have $PI(\{a, b\}, \{c, d\}, D_1) = 3/4$, $NI(\{a, b\}, \{c, d\}, D_1) = 1/5$, and $OI(\{a, b\}, \{c, d\}, D_1) = 11/20$. We observe that $PI(\{a, b\}, \{c, d\}, D_1)$ is more than $PA(\{a, b\}, \{c, d\}, D_1)$. Also, $NA(\{a, b\}, \{c, d\}, D_1)$ is more than $NI(\{a, b\}, \{c, d\}, D_1)$. So, $OI(\{a, b\}, \{c, d\}, D_1)$ is more than $OA(\{a, b\}, \{c, d\}, D_1)$. In similar to overall association, overall influence could be negative as well. Let $X = \{c\}$ and $Y = \{d, e\}$. $PI(X, Y, D_1) = 2/5$, $NI(X, Y, D_1) = 1/2$, and $OI(X, Y, D_1) = -1/10$. Thus, overall influence between two itemsets could be negative as well as positive.

In most cases, the overall influence between two itemsets in a large database is negative. In real-world databases, it might be possible that the overall influence between the two itemsets is positive. In Example 14.3, we consider some special cases to illustrate the measure of overall influence.

Example 14.3 Let database D_2 contain the following transactions: $\{a, b, e\}$, $\{a, e, g\}$, $\{b, e, g\}$, $\{a, b, d, e, g\}$, $\{b, d, e, g\}$ and $\{c, e, g\}$. We compute overall influence of an itemset X on another itemset Y considering various cases.

Case 1: $supp(X, D_2) > supp(Y, D_2)$
Let $X = \{e, g\}$, $Y = \{a, b\}$. $supp(X, D_2) = 5/6$, $supp(Y, D_2) = 2/6$ and $supp(X \cap Y, D_2) = 1/6$. We obtain $OI(X, Y, D_2) = -0.8$.
Case 2: $supp(X, D_2) < supp(Y, D_2)$
Let $X = \{a, b\}$, $Y = \{e, g\}$. $supp(X, D_2) = 2/6$, $supp(Y, D_2) = 5/6$ and $supp(X \cap Y, D_2) = 1/6$. We have $OI(X, Y, D_2) = -0.5$.

Though the values of overall influence are negative for the above cases, the influence might turn out to be positive for some databases. Let us consider another database $D_3 = \{\{a, b, c, d, g\}, \{b, c, g\}, \{c, d, g\}, \{a, b, c, d, e\}, \{b, c, e, g\}, \{a, b, c, d, e, g\}\}$

Case 1: $supp(X, D_3) > supp(Y, D_3)$
Let $X = \{c, d\}$, $Y = \{a, b\}$. $supp(X, D_3) = 4/6$, $supp(Y, D_3) = 3/6$ and $supp(X \cap Y, D_3) = 3/6$. We have $OI(X, Y, D_3) = 0.5$.
Case 2: $supp(X, D_3) < supp(Y, D_3)$. Let $X = \{a, b\}$, $Y = \{c, d\}$. $supp(X, D_3) = 3/6$, $supp(Y, D_3) = 4/6$ and $supp(X \cap Y, D_3) = 3/6$. Here we have $OI(X, Y, D_3) = 0.667$.

14.3.2 Properties of Influence Measures

For the purpose of computing influence of an itemset on another itemset, one needs to express OI in terms of supports of relevant itemsets. From (14.2), we get OI as follows:

$$OI(X, Y, D) = supp(X \cap Y, D)/supp(X, D)$$
$$- (supp(Y, D)/supp(X \cap Y, D))/(1 - supp(X, D))$$

Finally, we compute OI as follows:

$$OI(X, Y, D) = \frac{supp(X \cap Y, D) - supp(X, D) \times supp(Y, D)}{supp(X, D)[1 - supp(X, D)]}, \quad \text{if } supp(X, D) \neq 1 \text{ and } supp(Y, D) \neq 1$$
$$OI(X, Y, D) = 0, \qquad\qquad\qquad\qquad\qquad\qquad\qquad\qquad \text{otherwise}$$

$$(14.3)$$

From the above formula one could observe that if support of itemset X in D is 1 then influence of other itemsets on X will be zero. On the other hand, if $supp(Y, D) = 1$ then $supp(X \cap Y, D) = supp(X, D)$ and $supp(X, D) \times supp(Y, D) = supp(X, D)$. Therefore, the numerator of formula (14.3) will result to be zero and overall influence becomes zero. In the following lemma, we mention some properties of *PI* and *NI*. *OI* *(X, X, D)* = 1 at *X* = *Y*. Thus, *OI(X, X, D)* at *X* = *Y* could be termed as *trivial influence*.

Lemma 14.1 For itemsets *X*, *Y* in *D*, the following properties are satisfied: (i) $0 \leq PI(X, Y, D) \leq 1$, (ii) $0 \leq NI(X, Y, D) \leq 1$, (iii) $-1 \leq OI\ (X, Y, D) \leq 1$.

Lemma 14.2

$$OI(X, Y, D) = \frac{supp(Y)[Corr(X, Y, D) - 1]}{1 - supp(X)},$$

where *Corr(X, Y, D)* is the correlation coefficient between itemsets *X* and *Y* in database *D*. If *Corr(X, Y, D)* = 1 then *X* and *Y* are independent in database *D*. In other words, if *OI(X, Y, D)* = 0 then *X* and *Y* are independent in *D*. If *Corr(X, Y, D)* < 1 then *X* and *Y* are negatively correlated in database *D*. In other words, if *OI(X, Y, D)* < 0 then *X* and *Y* are negatively correlated. If *Corr(X, Y, D)* > 1 then *X* and *Y* are positively correlated in database *D*. If *OI(X, Y, D)* > 0 then *X* and *Y* are positively correlated.

14.3.3 Influence of an Item on a Set of Specific Items

Let $I = \{i_1, i_2, \dots, i_m\}$ be the set of items in database *D*. Also, let $SI = \{s_1, s_2, \dots, s_p\}$ be the set of specific items in database *D*. We would like to analyze the overall influence of each item on *SI*. The influence of an item on *SI* could be computed based on $OI(i_j, s_k, D)$, for $j = 1, 2, \dots, m$ and $k = 1, 2, \dots, p$. We say that the influence of i_j on s_k is interesting if $OI(i_j, s_k, D) \geq \gamma$, for $j = 1, 2, \dots, m$ and $k = 1, 2, \dots, p$. The value of γ depends on the level of data analysis to be performed. If the data analysis is performed in-depth then the value of γ is expected to be low. Also, the value of γ is dependent on the data to be analyzed. Normally, when the data are sparse the user needs to provide a low value of γ. On the other hand, γ could be given a reasonably high value for analyzing dense data. The procedure of determining influence of an item on a set of specific items could be explained using the following steps.

(i) Generate influence matrix (*IM*) of order $p \times m$ using $OI(i_j, s_k, D)$, for $j = 1, 2, \dots, m$ and $k = 1, 2, \dots, p$. (ii) An influence is counted when it is interesting. (iii) For each item, count the number of interesting influences on each of the specific items. (iv) The items in database *D* are sorted based on primary key as the number of interesting influences on the specific items, and secondary key as the support of an item. We explain steps (i)–(iv) using Example 14.4.

Example 14.4 Consider the database D_1 given in Example 14.1. Let $I = \{a, b, c, d, e, f, g\}$ and $SI = \{a, c, d\}$ (Table 14.3).

Table 14.3 Supports of each items in D_1

Items (x)	a	b	c	d	e	f	g
$supp(\{x\}, D_1)$	5/9	6/9	5/9	6/9	6/9	3/9	5/9

In this case, the influence matrix of size 3×7 is given below.

$$IM = \begin{bmatrix} item & a & b & c & d & e & f & g \\ a & 1 & 0.333 & 0.100 & 0.333 & -0.167 & -0.333 & 0.350 \\ c & 0.100 & 0.333 & 1 & 0.333 & -0.667 & 0.167 & 0.350 \\ d & 0.300 & -0.500 & 0.300 & 1 & 0 & 0 & -0.150 \end{bmatrix}$$

Let γ be set to 0.2. Also let $x(\eta)$ denote η number of interesting influences of item x on different specific items. The numbers of interesting influences of different items in D_1 are given as follows. $a(2)$, $b(2)$, $c(2)$, $d(3)$, $e(0)$, $f(0)$, $g(2)$. The items being sorted using step (iv) are given as follows: $d(3)$, $b(2)$, $a(2)$, $c(2)$, $g(2)$, $e(0)$, $f(0)$. Given the set of specific items $\{a, c, d\}$, one could conclude that the item d has the maximum and the item f has a minimum influence on the specific items.

14.3.4 Motivation

The concept of influence might not be new in the literature of data mining. For example, $conf(X \rightarrow Y, D)$ refers to positive influence of X on Y. In other words, it implies how likely a customer purchases the items of Y when he has already purchased all the items of X. In addition, the concept of negative influence is present in the literature on data mining. $conf(\neg X \rightarrow Y, D)$ refers to the amount of negative influence of items of X in purchasing the items of Y. In other words, it implies how likely a customer purchases the items of Y when the customer has not purchased all the items of X. In many data analyses it might be required to consider the overall influence of a set of items on another set of items. Our work introduces the notion of overall influence that could be useful in dealing with many real life problems. In the following paragraph, we justify that an existing measure might not be appropriate to study the overall influence of an itemset on another itemset.

The analysis of relationships among variables is a fundamental task being at the heart of many data mining problems. For example, metrics such as support, confidence, lift, correlation, and collective strength have been used extensively to evaluate the interestingness of association patterns. These metrics are defined in terms of the frequency counts tabulated in a 2×2 contingency table as shown in Table 14.4. To illustrate this, let us consider ten example contingency tables, E_1 to E_{10}, given in Table 14.5. Tan et al. (2003) presented an overview of twenty one interestingness measures proposed in the statistics, machine learning, and data mining literature.

In the following discussion, we observe why these measures fail to compute overall influence of an itemset on another itemset. In Examples 14.2 and 14.3, we have observed that the overall influence of an itemset on another itemset could be

Table 14.4 A 2 × 2
contingency table for
variables x and y

	Y	$\neg Y$	Total
X	f_{11}	f_{10}	f_{1+}
$\neg X$	f_{01}	f_{00}	f_{0+}
Total	$f_{.+1}$	$f_{.+0}$	N

Table 14.5 Examples of
contingency tables

Example	f_{11}	f_{10}	f_{01}	f_{00}
E1	8,123	83	424	1,370
E2	8,330	2	622	1,046
E3	9,481	94	127	298
E4	3,954	3,080	5	2,961
E5	2,886	1,363	1320	4,431
E6	1,500	2,000	500	6,000
E7	4,000	2,000	1000	3,000
E8	4,000	2,000	2000	2,000
E9	1,720	7,121	5	1,154
E10	61	2,483	4	7,452

positive as well as negative. Thus, overall influence of an itemset on another itemset in a database lies in $[-1, 1]$. In a large database, where items are sparsely distributed over the transactions might result in negative overall influence of an itemset on another itemset. Based on these observations, one could consider the following five out of twenty one interestingness measures since overall influence of an itemset on another itemset lies in $[-1, 1]$. These measures are presented in Table 14.6.

Based on each formula present in Table 14.6, the above contingency tables have been ranked as shown in Table 14.7. A contingency table that gives the maximum value is ranked as number 1 based on the interestingness measure. For example, contingency tables $E1$ and $E2$ give maximum and the second maximum values based on ϕ.

Also, we rank the contingency tables based on the concept of overall influence explained in Example 14.1. In Table 14.8, we present the ranking of contingency tables when using the overall influence as stated in (14.3).

None of the five measures ranks contingency tables like the ranks given in Table 14.7. Thus, none of the above five measures serves as a measure of overall influence between two itemsets.

14.4 Problem Statement

Let D be a database of customer transactions grown over a period of time. In this chapter, we are interested in making an influence analysis of a set of specific items. We will see how each of the specific items becomes influenced by different items in the database. One could view the entire database as a sequence of time-based (temporal) databases. For instance, such databases may concern consecutive years. To provide an

Table 14.6 Relevant interestingness measures for association patterns

Symbol	Measure	Formula
ϕ	ϕ-coefficient	$\dfrac{P(\{x\}\cup\{y\})-P(\{x\})\times P(\{y\})}{\sqrt{P(\{x\})\times P(\{y\})\times(1-P(\{x\}))\times(1-P(\{y\}))}}$
Q	Yule's Q	$\dfrac{P(\{x\}\cup\{y\})\times P(\neg(\{x\}\cap\{y\}))-P(\{x\}\cup\neg\{y\})\times P(\neg\{x\}\cup\{y\})}{P(\{x\}\cup\{y\})\times P(\neg(\{x\}\cap\{y\}))-P(\{x\}\cup\neg\{y\})\times P(\neg\{x\}\cup\{y\})}$
Y	Yule's Y	$\dfrac{\sqrt{P(\{x\}\cup\{y\})\times P(\neg(\{x\}\cap\{y\}))}-\sqrt{P(\{x\}\cup\neg\{y\})\times P(\neg\{x\}\cup\{y\})}}{\sqrt{P(\{x\}\cup\{y\})\times P(\neg(\{x\}\cap\{y\}))}-\sqrt{P(\{x\}\cup\neg\{y\})\times P(\neg\{x\}\cup\{y\})}}$
κ	Cohen's	$\dfrac{P(\{x\}\cup\{y\})+P(\neg\{x\}\cup\neg\{y\})-P(\{x\})\times P(\{y\})-P(\neg\{x\})\times P(\neg\{y\})}{1-P(\{x\})\times P(\{y\})-P(\neg\{x\})\times P(\neg\{y\})}$
F	Certainty factor	$\max\left(\dfrac{P(\{y\}|\{x\})-P(\{y\})}{1-P(\{y\})},\ \dfrac{P(\{x\}|\{y\})-P(\{x\})}{1-P(\{x\})}\right)$

Table 14.7 Ranking of contingency tables using above interestingness measures

Example	ϕ	Q	Y	κ	F
E1	1	3	3	1	4
E2	2	1	1	2	1
E3	3	4	4	3	6
E4	4	2	2	5	2
E5	5	8	8	4	9
E6	6	7	7	7	7
E7	7	9	9	6	8
E8	8	10	10	8	10
E9	9	5	5	9	3
E10	10	6	6	10	5

Table 14.8 Ranking of contingency tables using overall influence

Example	Overall influence	Rank
E1	0.754	1
E2	0.627	3
E3	0.691	2
E4	0.560	4
E5	0.450	5
E6	0.352	7
E7	0.417	6
E8	0.167	9
E9	0.190	8
E10	0.023	10

incremental solution to this problem, one might need to mine only the current time database and combine the mining result with the previous mining results. Thus, one needs to mine only the current database for the purpose of making an analysis based on entire database. As a result one can obtain cost-effective and faster analysis based on the entire database. Since the database grows over time, an incremental solution to influence analysis of specific items becomes natural and desirable.

Each time database corresponds to the set of transactions made for a specific period of time. In this regard, the choice of time period corresponding to a database is an important issue. One could observe that the sales of items might vary over different seasons in a year. Instead of processing all the data together, we process data on a yearly basis. Then, the result of processing for the current year could be combined with that of previous years. Such incremental analysis might be appropriate since a season re-appears on a yearly basis. Otherwise, the processed result might be biased due to seasonal variations.

Our goal is to make an influence analysis of a set of items in a database. Let D_t be the database for the tth period of time, $t = 1, 2, \ldots, n$. For computing overall influence between two items in a database, one needs to mine supports of itemsets of sizes one and two. The *size* of an itemset refers to the number of items in the itemset. Let $D_{1, k}$ be the collection of databases $D_1, D_2, ..., D_k$. For computing $OI(x, y, D_{1,k+1})$, we assume that $OI(x, y, D_{1,k})$ is available to us for items x, y in $D_{1,k}$. In other words, for computing $OI(x, y, D_{1,k+1})$, we have $supp(x, D_{1,k})$, $supp(y, D_{1,k})$, and $supp(x \cap y, D_{1,k})$. Thus, our incremental procedure needs to compute $supp(x, D_{1,k+1})$, $supp(y, D_{1,k+1})$, and $supp(x \cap y, D_{1,k+1})$ using (i) $supp(x, D_{1,k})$, $supp(y, D_{1,k})$, and $supp(x \cap y, D_{1,k})$, (ii) $supp(x, D_{k+1})$, $supp(y, D_{k+1})$, and $supp(x \cap y, D_{k+1})$. In general, for an itemset X in $D_{1,k}$, $supp(X, D_{1,k+1})$ could be obtained incrementally as follows.

$$supp(X, D_{1,k+1}) = \frac{size(D_{k+1}) \times supp(X, D_{k+1}) + size(D_{1,k}) \times supp(X, D_{1,k})}{size(D_{k+1}) + size(D_{1,k})}$$

$$(14.4)$$

The $size(D)$ refers to the number of transactions in database D.

14.5   Related Work

For analyzing positive association between itemsets in a database, support-confidence framework was established by Agrawal et al. (1993). In Sect. 14.3.4, we have discussed why a confidence measure alone is not sufficient in determining an overall influence of an itemset on another itemset. Also, interestingness measures such as support, collective strength (Aggarwal and Yu 1998) and Jaccard (Tan et al. 2003) are not relevant in this context, since they are single-argument measures.

The χ^2 test (Greenwood and Nikulin 1996) only tells us whether two or more items are dependent. Such a test provides answers either "yes" or "no" to the question of whether the association is meaningful, and hence it might not be suitable for the specific requirement of our problem.

The interestingness measures such as lift (Tan et al. 2003), correlation (Tan et al. 2003), conviction (Brin et al. 1997), and odds-ratio (Tan et al. 2003) are semantically different from the measure of overall influence. Moreover, the values of each of these measures lie in $[0, \infty)$.

Shapiro (1991) has proposed leverage measure developed in the context of mining strong rules in a database. However, it might not be suitable for the specific requirement of our problem.

14.6   Design of Algorithms

Based on the discussion held in previous section, we design two algorithms for measuring influence of an item on another item and influence of an item on a set of specific items.

14.6.1 Designing Algorithm for Measuring Overall Influence of an Item on Another Item

In this algorithm (Adhikari and Rao 2010), we measure influence of an item on each of the items incrementally. We have expressed influence of an itemset on another itemset using supports of the relevant itemsets. Each itemset could be described by its *itemset* and *support*. We maintain arrays *IS1* and *IS2* for storing itemsets in $D_{1,k}$ of size one and two, respectively. *Itemset* attribute of ith itemset in *IS1* could be accessed using the notation *IS1(i).itemset*. Similar notation is used to access *support* attribute of an itemset. Also, we maintain arrays $\Delta IS1$ and $\Delta IS2$ for storing itemsets in D_{k+1} of size one and two, respectively. We merge *IS1* and $\Delta IS1$ to obtain supports of itemsets of size one in $D_{1,k+1}$ and are stored in array *OIS1*. Similarly, we merge *IS2* and $\Delta IS2$ to obtain supports of itemsets of size two in $D_{1,k+1}$ and are stored in array *OIS2*. Using *OIS1* and *OIS2*, we compute an overall influence between items in $D_{1,k+1}$. The overall influence between items is computed using formula (14.3) and stored in array *IOI*. The overall influence (*oi*) corresponding to j-th pair of items is accessed by *IOI(j).oi*.

Algorithm 14.1. Find top q overall influences in the given database over time.
procedure *Top-q-OI(q, IS1, IS2, $\Delta IS1$, $\Delta IS2$, IOI)*
Inputs:
q: an integer representing the number of top influences
IS1: array of supports of itemsets of size one in $D_{1,k}$
IS2: array of supports of itemsets of size two in $D_{1,k}$
$\Delta IS1$: array of supports of itemsets of size one in D_{k+1}
$\Delta IS2$: array of supports of itemsets of size two in D_{k+1}
Outputs:
IOI: array of overall influences in $D_{1,k+1}$
01: sort array $\Delta IS1$ on *itemset* attribute in non-decreasing order;
02: sort array $\Delta IS2$ on *itemset* attribute in non-decreasing order;
03: call *Merge (IS1, $\Delta IS1$, OIS1)*;
04: call *Merge (IS2, $\Delta IS2$, OIS2)*;
05: **let** $j = 1$;
06: **for** $i = 1$ to |OIS2| **do**
07:     search *OIS2(i).item1* in *OIS1*;
08:     search *OIS2(i).item2* in *OIS1*;
09:     *IOI(j).oi = OI(OIS2(i).item1, OIS2(i).item2, D)*;
10:     *IOI(j).item1 = OIS2(i).item1; IOI(j).item2= OIS2(i).item2*;
11:     increase j by 1;
12:     *IOI(j).oi = OI(OIS2(i).item2, OIS2(i).item1, D)*;
13:     *IOI(j).item1 = OIS2(i).item2; IOI (j).item2 = OIS2(i).item1*;
14:     increase j by 1;
15: **end for**
16: sort array *IOI* in non-increasing order on *oi* attribute;
17: return first q influences;
18: **end procedure**

The procedure *Merge* (A, B, C) merges sorted arrays A and B and generates output array C. In this context, sorting is based on support of an itemset. The time complexity of procedure *Merge* is $O(|A| + |B|)$ (Knuth 1998). Now, *OIS1* contains the supports of items in $D_{1,k+1}$. Also, *OIS2* contains the supports of itemsets of size two in $D_{1,k+1}$. The information contained in *OIS1* and *OIS2* is used to compute overall influence of an item on another item in $D_{1,k+1}$. Using line 09 we have computed influence of a singleton itemset on another singleton itemset. Suppose $\{6, 8\}$ be a frequent 2-itemset in $D_{1,k+1}$ stored in 4th cell of *OIS2*. Then *OI(OIS2*(4). *item1*, *OIS2*(4).*item2*, $D_{1,k+1}$) refers to overall influence of $\{6\}$on$\{8\}$in $D_{1,k+1}$. In lines 6–15, we have computed and stored overall influences of a singleton itemset on another singleton itemset in $D_{1,k+1}$. In line 16, we have sorted overall influences in non-increasing order. Finally, we display first q overall influences.

Let *IS1* and *IS2* contain M and N itemsets respectively. Let $\Delta IS1$ and $\Delta IS2$ contain m and n elements respectively. Lines 1 and 2 take $O(m \times \log(m))$ and $O(n \times \log(n))$ time, respectively. Also, lines 3 and 4 take $O(M + m)$ and $O(N + n)$ time, respectively. Each of the search statements in lines 7 and 8 take $O(\log(M + m))$ time, since *OIS1* is sorted. The sort statement in line 16 takes $O((N + n) \times \log(N + n))$. The time complexity of lines 6–15 is $O((N + n) \times \log(M + m))$. Thus, the time complexity of algorithm *Top-q-OI* is *maximum* $\{O(M + m), O((N + n) \times \log(N + n)), O((N + n) \times \log(M + m))\}$.

14.6.2 Designing Algorithm for Measuring Overall Influence of an Item on Each of the Specific Items

One could store specific items in an array. The proposed algorithm seems to be the same as Algorithm 14.1 except that every time it measures an overall influence of an item on a specific item.

14.6.3 Designing Algorithm for Identifying Top Influential Items on a Set of Specific Items

In Algorithm 14.2 (Adhikari and Rao 2010), we find influence of an item on a set of specific items in a database. We construct influence matrix (*IM*) from the arrays of specific items (*SI*) and overall influence between items (*IOI*). The algorithm scans *IM* for each item to count the number of interesting influences, which are stored in array called *count*. Finally, we sort *count* in descending order based on primary key count value and secondary key support.

Algorithm 14.2. Find influence of an item on a set of specific items in the database over time.
procedure *Top-q-items(q, SI, IS1, IS2, ΔIS1, ΔIS2, OIS1, OIS2, IOI)*
Inputs:
q: an integer representing the number of top influences
SI: array of specific items
IS1, IS2, ΔIS1, ΔIS2, OIS1, OIS2, IOI: as specified in Algorithm 14.1
Outputs:
count: array of number of interesting influences
01: **for** i = 1 to $|SI|$ **do**
02:    **for** j = 1 to $|IOI|$ **do**
03:       **if** $(SI(i)= IOI(j).item1)$ **then**
04:          $IM(i)(j) = IOI(j).oi$;
05:       **end if**
06:    **end for**
07: **end for**
08: **for** j = 1 to $|IOI|$ **do**
09:    **let** $count(j) = 0$;
10:    **for** i = 1 to $|SI|$ **do**
11:       **if** $(IM(j)(i) \geq \gamma)$ **then**
12:          increase $count(j)$ by 1;
13:       **end if**
14:    **end for**
15: **end for**
16: sort *count* on non-increasing order on primary key count value and secondary key support;
17: return first q items;
end procedure

Let array *SI* contains p items. Line 1 repeats for p times. Line 2 repeats O $(M + m)$ times. So, lines 1–7 take $O(p \times (M + m))$ time. Line 8 repeats $O(M + m)$ times. Line 10 repeats p times. Thus, line 8–15 take $O(p \times (M + m))$ time. Therefore, the time complexity of the above algorithm is $O(p \times (M + m))$, where $M > m$. Also, sorting statement at line 16 takes $O((M + m) \times \log(M + m))$. The time complexity of algorithm *Top-q-items* is $maximum\{O(p \times (M + m)), O((M + m) \times \log(M + m))\}$.

14.7   Experiments

We present the experimental results using three real-world databases and one synthetic database. The databases *mushroom*, *retail* (Frequent itemset mining dataset repository)[1] and *ecoli* are real-world databases. Database *ecoli* is a subset of *ecoli database* (UCI ML repository)[2] and it has been processed for the purpose of conducting experiments. *Random-68* is a synthetic database. The symbols used in different tables are explained as follows. Let *D, NT, ALT, AFI,* and *NI* denote database, the number of transactions, average length of a transaction, average frequency of an item, and number of items, respectively. The details of these databases are given in Table 14.9.

---

[1] Frequent itemset mining dataset repository, http://fimi.cs.helsinki.fi/data.

[2] UCI ML repository, http://www.ics.uci.edu/~mlearn/MLSummary.html.

Table 14.9 Database characteristics

Database	NT	ALT	AFI	NI
mushroom (M)	8124	24.000	1624.800	120
ecoli (E)	336	7.000	25.835	91
random-68 (R)	3000	5.460	280.985	68
retail (Rt)	88,162	11.306	99.674	10,000

Each database has been divided into 10 databases, called input databases, for the purpose of conducting experiments on multiple time databases. The input databases obtained from *mushroom, ecoli, random-68* and *retail* are named as M_i, E_i, R_i, and Rt_i, $i = 0, 1, ..., 9$. We present some characteristics of the input databases, in Table 14.10. Top 10 overall influences in different databases are shown in Table 14.11.

We have studied the execution time with respect to the number of data sources. We observe in Figs. 14.1, 14.2, 14.3 and 14.4 that this time increases as the number of data sources gets higher.

The size of each input database generated from *mushroom* and *retail* are significantly larger than an input database generated from *ecoli*. As a result, we observe a steeper relationship in Figs. 14.1 and 14.4. The number of frequent itemsets decreases as the minimum support increases.

Table 14.10 Time database characteristics

D	NT	ALT	AFI	NI	D	NT	ALT	AFI	NI
M_0	812	24.000	295.273	66	M_5	812	24.000	221.454	88
M_1	812	24.000	286.588	68	M_6	812	24.000	216.533	90
M_2	812	24.000	249.846	78	M_7	812	24.000	191.059	102
M_3	812	24.000	282.435	69	M_8	812	24.000	229.271	85
M_4	812	24.000	259.840	75	M_9	816	24.000	227.721	86
E_0	33	7.000	4.620	50	E_5	33	7.000	3.915	59
E_1	33	7.000	5.133	45	E_6	33	7.000	3.500	66
E_2	33	7.000	5.500	42	E_7	33	7.000	3.915	59
E_3	33	7.000	4.813	48	E_8	33	7.000	3.397	68
E_4	33	7.000	3.397	68	E_9	39	7.000	4.550	60
R_0	300	5.590	28.676	68	R_5	300	5.140	26.676	68
R_1	300	5.417	28.000	68	R_6	300	5.510	28.353	68
R_2	300	5.360	27.647	68	R_7	300	5.497	28.338	68
R_3	300	5.543	28.456	68	R_8	300	5.537	28.471	68
R_4	300	5.533	28.382	68	R_9	300	5.477	28.235	68
Rt_0	9,000	11.244	12.070	8384	Rt_5	9,000	10.856	16.710	5,847
Rt_1	9,000	11.209	12.265	8225	Rt_6	9,000	11.200	17.416	5,788
Rt_2	9,000	11.337	14.597	6990	Rt_7	9,000	11.155	17.346	5,788
Rt_3	9,000	11.490	16.663	6206	Rt_8	9,000	11.997	18.690	5,777
Rt_4	9,000	10.957	16.039	6148	Rt_9	7,162	11.692	15.348	5,456

Table 14.11 Top 10 overall influences in different databases

M (supp = 0.15)			E (supp = 0.12)			R (supp = 0.03)			Rt (supp = 0.12)		
{x}	{y}	OI	{x}	{y}	OI	{x}	{y}	OI	{x}	{y}	OI
86	34	0.997	24	48	0.946	19	29	−0.017	41	39	0.200
34	86	0.992	89	50	0.913	29	19	−0.020	39	48	0.180
58	24	0.991	53	48	0.693	8	56	−0.023	48	39	0.175
67	76	0.986	63	50	0.665	56	8	−0.023	41	48	0.129
76	67	0.986	87	50	0.660	15	14	−0.031	39	41	0.114
24	58	0.963	56	50	0.621	14	15	−0.032	48	41	0.071
93	59	0.895	61	50	0.618	18	52	−0.035	48	7	−0.234
93	76	0.884	27	48	0.618	52	18	−0.036	39	7	−0.292
93	67	0.881	83	50	0.540	54	58	−0.044	48	2	−0.293
102	24	0.875	56	48	0.488	58	54	−0.047	48	1	−0.316

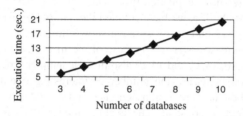

Fig. 14.1 Execution time versus number of databases at *supp* = 0.2 (*mushroom*)

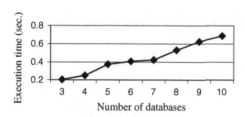

Fig. 14.2 Execution time versus number of databases at *supp* = 0.12 (*ecoli*)

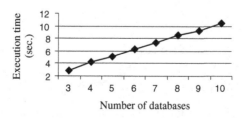

Fig. 14.3 Execution time versus number of databases at *supp* = 0.03 (*random-68*)

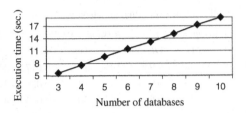

Fig. 14.4 Execution time versus number of databases at *supp* = 0.2 (*retail*)

In Figs. 14.5, 14.6, 14.7 and 14.8 it is shown how the execution time decreases over the increase of the minimum support value.

By comparing Figs. 14.1, 14.2, 14.3, 14.4, one notes that the steepness of a graph increases as the size of branch databases increases. Similar observation holds true on Figs. 14.5, 14.6, 14.7, 14.8.

In Sect. 14.3.1 we have explained the concept of interesting overall influence. Given a threshold value of γ, we have counted the number of overall influences. In Figs. 14.9, 14.10, 14.11 and 14.12 we have shown how the number of interesting overall influence decreases over the increase of the minimum influence level.

Figures 14.9, 14.10, 14.11 and 14.12 also provide another type of insight. As the size of a transaction increases, the number of interesting overall influences also increases, provided the number of transactions in a branch database and the level of overall influence remain constant. The average transaction length of mushroom

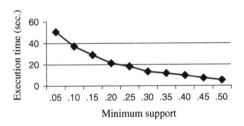

Fig. 14.5 Execution time versus minimum support (*mushroom*)

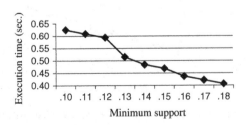

Fig. 14.6 Execution time versus minimum support (*ecoli*)

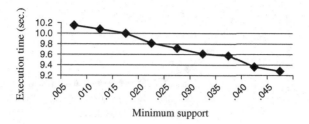

Fig. 14.7 Execution time versus minimum support (*random-68*)

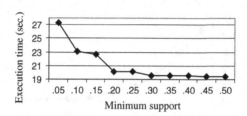

Fig. 14.8 Execution time versus minimum support (*retail*)

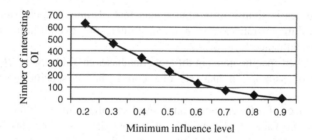

Fig. 14.9 Number of interesting *OI* values versus γ at *supp* = 0.2 (*mushroom*)

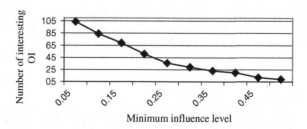

Fig. 14.10 Number of interesting *OI* values versus γ at *supp* = 0.12 (*ecoli*)

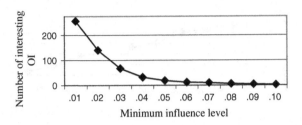

Fig. 14.11 Number of interesting *OI* values versus γ at *supp* = 0.015 (*random-68*)

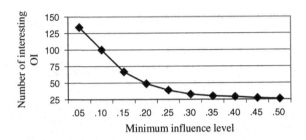

Fig. 14.12 Number of interesting *OI* values versus γ at *supp* = 0.02 (*retail*)

branch databases is significantly higher than that of other branch databases. The mining algorithm generates a large number of interesting overall influences even at the minimum influence level of 0.2.

We have taken specific items in different databases in Table 14.12. Based on the requirement of association analysis one could choose specific items in time databases.

The influences of different items on a set of specific items in different databases are presented in Table 14.13. In the *mushroom* database, item 86 is the most influential item because 3 specific items are influenced by it. Item 48 in *ecoli* database exhibits a significant influence on the set of specific items. It shows that item 48 has high influence on 5 out of 10 specific items. In the same way one could conclude that item 18 in *random-68* is the most influential item with respect to the given set of specific items. 5 out of 10 specific items are influenced significantly by item 18. Item 413 influences 8 out of 10 specific items significantly in *retail* database. Therefore, it is the most influential item in *retail*.

Table 14.12 Specific items in different databases

M	E	R	Rt
SI = {1, 2, 3, 6, 9, 10, 11, 13, 16, 23}	*SI* = {37, 39, 40, 41, 42, 44, 48, 49, 50, 51}	*SI* = {1, 2, 3, 4, 5, 6, 7, 8, 9, 10}	*SI* = {0, 1, 2, 3, 4, 5, 6, 7, 8, 9}

Table 14.13 Influences of different items on a set of specific items in different databases

M (supp = 0.2)		E (supp = 0.12)		R (supp = 0.015)		Rt (supp = 0.03)	
γ	$x(\eta)$	γ	$x(\eta)$	γ	$x(\eta)$	γ	$x(\eta)$
0.3	86(3), 34(3), 36 (3), 39(2), 59 (2), 63(2), 2(2), 93(2), 36(2), 23 (2), 90(1), 24(1)	0.07	48(5), 37(2), 50 (1), 42(1), 44 (1), 39(1), 40 (1), 49(1), 41(1)	0.05	18(5), 15(3), 65(2), 55(2), 61(2), 7(1), 54(1), 27(1), 35(1), 66(1), 22(1)	0.05	413(8), 310(2), 0 (1), 1(1), 8(1), 2(1), 3(1),5(1), 9(1), 4(1)

14.8 Conclusions

The concept of positive influence might not be sufficient in many data analyses. One could perform an effective data analysis by using the measure of overall influence. Measuring influence over time becomes an important issue, since many companies possess data over a long period of time so that they could be exploited in an efficient manner. In this chapter, we have designed two algorithms using the measure of overall influence. The first algorithm reports all the significant influences in a database. In the second algorithm, we have sorted items based on their influences on a set of specific items. Such analyses might be interesting since the proposed measure of influence considers both positive and negative influence of an itemset on another itemset.

References

Adhikari A, Ramachandrarao P, Pedrycz W (2011) Study of select items in different data sources by grouping. Knowl Inf Syst 27(1):23–43

Adhikari A, Rao PR (2008a) Mining conditional patterns in a database. Pattern Recogn Lett 29 (10):1515–1523

Adhikari A, Rao PR (2008b) Synthesizing heavy association rules in different real data sources. Pattern Recogn Lett 29(1):59–71

Adhikari A, Rao PR (2007) A framework for mining arbitrary Boolean expressions induced by frequent itemsets. In: Proceedings of the international conference on artificial intelligence, pp 5–23

Adhikari J, Rao PR (2010) Measuring influence of an item in a database over time. Pattern Recogn Lett 31(3):179–187

Adhikari J, Rao PR, Adhikari A (2009) Clustering items in different data sources induced by stability. Int Arab J Inf Technol 6(4):66–74

Aggarwal C, Yu P (1998) A new framework for itemset generation. In: Proceedings of PODS, pp 18–24

Agrawal R, Imielinski T, Swami A (1993) Mining association rules between sets of items in large databases. In: Proceedings of the ACM SIGMOD conference management of data, pp 207–216

Agrawal R, Srikant R (1994) Fast algorithms for mining association rules. In: Proceedings of the international conference on very large databases, pp 487–499

Brin S, Motwani R, Ullman JD, Tsur S (1997) Dynamic itemset counting and implication rules for market basket data. In: Proceedings of the ACM SIGMOD international conference on management of data, pp 255–264

Greenwood PE, Nikulin MS (1996) A Guide to Chi-Squared testing. Wiley-Interscience

Han J, Pei J, Yiwen Y (2000) Mining frequent patterns without candidate generation. In: Proceedings of the ACM-SIGMOD international conference management of data, pp 1–12

Knuth DE (1998) The art of computer programming, sorting and searching. Vol 3, Addison-Wesley

Savasere A, Omiecinski E, Navathe S (1995) An efficient algorithm for mining association rules in large databases. In: Proceedings of the international conference on very large data bases, pp 432–443

Shapiro P (1991) Discovery, analysis, and presentation of strong rules. In: Proceedings of knowledge discovery in databases, pp 229–248

Tan PN, Kumar V, Srivastava J (2003) Selecting the right interestingness measure for association patterns. In: Proceedings of SIGKDD conference, pp 32–41

Wu X, Zhang C, Zhang S (2004) Efficient mining of both positive and negative association rules. ACM Trans Inf Sys 22(3):381–405

Chapter 15
Clustering Multiple Databases Induced by Local Patterns

In view of answering queries provided in multiple large databases, it might be required to mine relevant databases *en block*. In this chapter, we present an effective solution to clustering multiple large databases. Two measures of similarity between a pair of databases are presented and study their main properties. In the sequel, we design an algorithm for clustering multiple databases based on an introduced similarity measure. Also, we present a coding, referred to as IS coding, to represent itemsets space efficiently. The coding of this nature enables more frequent itemsets to participate in the determination of the similarity between two databases. Thus the invoked clustering process becomes more accurate. We also show that the IS coding attains maximum efficiency in most of the cases of the mining processes. The clustering algorithm becomes improved (in terms of its time complexity) when contrasted with the existing clustering algorithms. The efficiency of the clustering process has been improved using several strategies that is by reducing execution time of the clustering algorithm, using more suitable similarity measure, and storing frequent itemsets space efficiently.

15.1 Introduction

Effective data analysis using a traditional data mining technique on multi-gigabyte repositories has proven difficult. A quick discovery of approximate knowledge from large databases would be adequate for many decision support applications.

As before, let us consider a company that deals with multiple large databases. The company might need to carry out association analysis involving non-profit making items (products). The ultimate objective is to identify the items that neither make much profit nor help promoting other products. An association analysis involving non-profit making items might identify such items. The company could then stop dealing with them. The analysis of this nature might require identifying similar databases. Let us note that two databases are deemed similar if they contain many similar transactions. Again, two transactions are similar if they have many common items. We observe later that two databases containing many common

A. Adhikari and J. Adhikari, *Advances in Knowledge Discovery in Databases*,
Intelligent Systems Reference Library 79, DOI 10.1007/978-3-319-13212-9_15

items are not necessarily very similar. First, let us define a few concepts used frequently in this chapter.

Let $I(D)$ be the set of items in database D. An *itemset* is a set of items in a database. An itemset X in D is associated with a statistical measure called support (Agrawal et al. 1993), denoted by $supp(X, D)$, for $X \subseteq I(D)$. *Support* of an itemset X in D is the fraction of transactions in D containing X. The importance of an itemset could be judged by quantifying its support. X is called a *frequent itemset* (FIS) in D if $supp(X, D) \geq \alpha$, where α is the user-defined *minimum support*. A frequent itemset possesses higher support. Thus the collection of frequent itemsets determines major characteristics of a database. One could define similarity between a pair databases in terms of their frequent itemsets. We may observe that two databases are similar if they have many common frequent itemsets.

Based on the similarity between two databases, one could cluster branch databases. Once the clustering process has been completed, one could mine all the databases in a class together to make an approximate association analysis involving frequent items. An approximate association analysis could be performed using the frequent itemsets in the union of all the databases in a class. In this manner, clustering of databases helps reducing data for analyzing the items. In what follows, we study the problem of clustering transactional databases using the local frequent itemsets.

For clustering transactional databases, Wu et al. (2005b) have proposed two similarity measures, denoted as sim_1, and sim_2. Let $D = \{D_1, D_2, ..., D_n\}$, where D_i is the database corresponding to the ith branch of a multi-branch company, $i = 1, 2, ..., n$. sim_1 is based on the items present in the databases, and becomes defined as follows:

$$sim_1(D_1, D_2) = |I(D_1) \cap I(D_2)| / |I(D_1) \cup I(D_2)|$$

Let S_i be the set of association rules present in D_i, $i = 1, 2, ..., n$. The measure sim_2 is based on the items generated from S_i, $i = 1, 2, ..., n$. Let $I(S_i)$ be the set of items generated from S_i, $i = 1, 2, ..., n$. The similarity measure sim_2 is expressed in the form:

$$sim_2(D_1, D_2) = |I(S_1) \cap I(S_2)| / |I(S_1) \cup I(S_2)|$$

$I(S_i) \subseteq I(D_i)$, for $i = 1, 2, ..., n$. sim_1 estimates similarity between two databases more correctly than sim_2, since the number of items which participate in determining the value of the similarity between two databases under sim_1 is higher than that of sim_2. A database may not extract any association rule for given values of (α, β), where β is the user-defined *minimum confidence level*. In such situations, the accuracy of sim_2 is low. In the following example, we discuss a situation where the accuracy of sim_1 and sim_2 are low.

Example 15.1 Consider a multi-branch company that possesses following three databases:

$$DB_1 = \{\{a,b,c,e\}, \{a,b,d,f\}, \{b,c,g\}, \{b,d,g\}\}$$
$$DB_2 = \{\{a,g\}, \{b,e\}, \{c,f\}, \{d,g\}\} \text{ and}$$
$$DB_3 = \{\{a,b,c\}, \{a,b,d\}, \{b,c\}, \{b,d,g\}\}$$

Here, $I(DB_1) = \{a, b, c, d, e, f, g\}$, $I(DB_2) = \{a, b, c, d, e, f, g\}$, $I(DB_3) = \{a, b, c, d, g\}$. Thus, $sim_1(DB_1, DB_2) = 1.0$ (maximum), and $sim_1(DB_1, DB_3) = 0.71429$. Ground realities are as follows: (i) The similarity between DB_1 and DB_2 is low, since they contain dissimilar transactions. (ii) The similarity between DB_1 and DB_3 is higher than the similarity observed between DB_1 and DB_2, since DB_1 and DB_3 contain similar transactions. Hence the similarity measures sim_1 produces low accuracy in finding the similarity between two databases. There are no frequent itemsets in DB_2, if $\alpha > 0.25$. Thus, $I(S_2) = \phi$, if $\alpha > 0.25$. Hence, the accuracy of sim_2 is low in finding the similarity between DB_1 and DB_2 if $\alpha > 0.25$.

We have observed that the similarity measures based on items in databases might not be appropriate in finding similarity between two databases. A more suitable similarity measure could be designed based on frequent itemsets present in both the databases. The frequent itemsets in two databases could find better the similarity among transactions in two databases. Thus, frequent itemsets in two databases could find similarity between two databases correctly.

Wu et al. (2005a) have proposed a solution of inverse frequent itemset mining. They argued that one could efficiently generate a synthetic market basket database from the frequent itemsets and their supports. Thus, the similarity between two databases could be estimated correctly by involving supports of the frequent itemsets. We propose two measures of similarity based on the frequent itemsets and their supports. A new algorithm for clustering databases is designed based on a proposed measure of similarity.

The existing industry practice is to refresh a data warehouse on a periodic basis. Let λ be the frequency of this process of data warehouse refreshing. In this situation, an incremental mining algorithm (Lee et al. 2001) could be used to obtain updated supports of the existing frequent itemsets in a database. But, there could be addition or, deletion of frequent itemsets over time. We need to mine the databases individually and again this is being done in a periodic manner. Let Λ be the periodicity of data warehouse mining. The values of λ and Λ could be chosen in such way that $\Lambda > \lambda$. Based on the updated local frequent itemsets, one could cluster the databases afresh.

Another alternative for taming multi-gigabyte data could be sampling. Let us note that a commonly used technique for approximate query answering is sampling (Babcock et al. 2003). If an itemset is frequent in a large database then it is likely that this itemset is frequent in a sample database. Thus, one could analyze approximately a database by analyzing the frequent itemsets present in a sample database.

The chapter is organized so that it reflects the main objectives identified above. We formulate the problem in Sect. 15.2. In Sect. 15.3, we discuss some related work. In Sect. 15.4, we show how to cluster all the branch databases. The experimental results are presented in Sect. 15.5.

15.2 Problem Statement

Let there are n branch databases. Also, let $FIS(D_i, \alpha)$ be the set of frequent itemsets corresponding to database D_i at a given value of α, $i = 1, 2, ..., n$. The problem is stated succinctly as follows:

Find the best non-trivial partition (if it exists) of $\{D_1, D_2, ..., D_n\}$ using $FIS(D_i, \alpha)$, $i = 1, 2, ..., n$.

A partition (Liu 1985) is a specific type of clustering. A formal definition of a non-trivial partition will be given in Sect. 15.4.

15.3 Related Work

Jain et al. (1999) have presented an overview of clustering methods from a statistical pattern recognition perspective, with a goal of providing a useful advice and references to fundamental concepts accessible to the broad community of clustering practitioners. A traditional clustering technique (Zhang et al. 1997) is based on *metric* attributes. A *metric* attribute is one whose values can be represented by explicit coordinates in a Euclidean space. Thus a traditional clustering technique might not work in this case, since we are interested in clustering databases. Ali et al. (1997) have proposed a partial classification technique using association rules. The clustering of databases using local association rules might not be a good idea. The number of frequent itemsets obtained from a set of association rules might be much less than the number of frequent itemsets extracted using the apriori algorithm (Agrawal et al. 1993). In this way, the efficiency of the clustering process could be low. Liu et al. (2001) have proposed a multi-database mining technique that searches only the relevant databases. Identifying relevant databases is based on selecting the relevant tables (relations) that contain specific, reliable and statistically significant information pertaining to the query. Our study involves clustering transactional databases. Yin and Han (2005) have proposed a new strategy for relational heterogeneous database classification. This strategy might not be suitable for clustering transactional databases. Yin et al. (2006) have proposed two scalable methods for multi-relational classification: CrossMine-Rule, a rule-based method and CrossMine-Tree, a decision-tree-based method. Bandyopadhyay et al. (2006) have proposed a technique for clustering homogeneously distributed data in a peer-to-peer environment like sensor networks. It is based on the idea of the K-Means clustering. It works in a localized asynchronous manner by realizing a communication with the neighboring nodes.

In the context of similarity measures, Tan et al. (2002) have presented an overview of twenty one interestingness measures available in statistics, machine learning, and data mining literature. Support and confidence measures (Agrawal et al. 1993) are used to identify frequently occurring association rules between two sets of items in large databases. Our first measure, $simi_1$, is similar to the Jaccard measure (Tan et al. 2002). Measures such as support, interest (Tan et al. 2002), cosine (Tan et al. 2002) are expressed as a ratio of two quantities. Their numerators represent a kind of closeness between two objects. But their denominators are not appropriate to make these ratios as measures of association. As a result they do not serve as sound measures of similarity.

Zhang et al. (2003) designed a local pattern analysis for mining multiple databases. Zhang (2002) studied various strategies for mining multiple databases. For utilizing low-cost information and knowledge on the internet, Su et al. (2006) have proposed a logical framework for identifying knowledge of sound quality coming from different data sources. It helps working towards the development of a generally acceptable ontology.

Privacy concerns over sensitive data have become important in knowledge discovery. Usually, data owners have different levels of concerns over different data attributes, which adds complexity to data privacy. Moreover, collusion among malicious adversaries poses a severe threat to data security. Yang and Huang (2008) have proposed an efficient clustering method for distributed multi-party data sets using the orthogonal transformation and perturbation techniques. It allows data owners to set up different levels of privacy for different attributes.

In many large e-commerce organizations, multiple data sources are often used to describe the same customers, thus it is important to consolidate data of multiple sources for intelligent business decision making. Ling and Yang (2006) have proposed a method that predicts the classification of data from multiple sources without class labels in each source.

15.4 Clustering Databases

The approach of finding the best partition of a set of databases can be explained through a sequence of the following steps:

(i) Find $FIS(D_i, \alpha)$, for $i = 1, 2, ..., n$.
(ii) Determine the similarity between each pair of databases using the proposed measure of similarity $simi_2$.
(iii) Check for the existence of partitions at the required similarity levels (as mentioned in Theorem 15.5).
(iv) Calculate the goodness values for all the non-trivial partitions.
(v) Report the non-trivial partition for which the goodness value attains its maximum.

The steps (i)–(v) will be followed and explained with the help of a running example. We start with an example of a multi-branch company that has multiple databases.

Example 15.2 A multi-branch company has seven branches. The branch databases are given below.

$$D_1 = \{(a, b, c), \ (a, c), \ (a, c, d)\}$$
$$D_2 = \{(a, c), \ (a, b), \ (a, c, e)\}$$
$$D_3 = \{(a, e), \ (a, c, e), \ (a, b, c)\}$$
$$D_4 = \{(f, d), \ (f, d, h), \ (e, f, d), \ (e, f, h)\}$$
$$D_5 = \{(g, h, i), \ (i, j), \ (h, i), \ (i, j, g)\}$$
$$D_6 = \{(g, h, i), \ (i, j, h), \ (i, j)\}$$
$$D_7 = \{(a, b), \ (g, h), \ (h, i), \ (h, i, j)\}$$

The sets of frequent itemsets are shown below:

$FIS(D_1, 0.35) = \{(a, 1.0), \ (c, 1.0), \ (ac, 1.0)\}$
$FIS(D_2, 0.35) = \{(a, 1.0), \ (c, 0.67), \ (ac, 0.67)\}$
$FIS(D_3, 0.35) = \{(a, 1.0), \ (c, 0.67), \ (e, 0.67), \ (ac, 0.67), \ (ae, 0.67)\}$
$FIS(D_4, 0.35) = \{(d, 0.75), \ (e, 0.5), \ (f, 1.0), \ (h, 0.5), \ (df, 0.75), \ (ef, 0.5), \ (fh, 0.5)\}$
$FIS(D_5, 0.35) = \{(g, 0.5), \ (h, 0.5), \ (i, 1.0)\}, \ (j, 0.5), \ (gi, 0.5), \ (hi, 0.5), \ (ij, 0.5)\}$
$FIS(D_6, 0.35) = \{(i, 1.0), \ (j, 0.67), \ (h, 0.67), \ (hi, 0.67), \ (ij, 0.67)\}$
$FIS(D_7, 0.35) = \{(h, 0.75), \ (i, 0.5), \ (hi, 0.5)\}.$

Based on the sets of frequent itemsets in a pair of databases, one could define many measures of similarity between them. The two measures of similarity between a pair of databases are suitable for dealing with the problem at hand. The first measure $simi_1$ (Adhikari and Rao 2008) is defined as follows:

Definition 15.1 The measure of similarity $simi_1$ between databases D_1 and D_2 is defined as the following ratio:

$$simi_1(D_1, D_2, \alpha) = \frac{|FIS(D_1, \alpha) \cap FIS(D_2, \alpha)|}{|FIS(D_1, \alpha) \cup FIS(D_2, \alpha)|},$$

where the symbols $\cap$ and $\cup$ stand for the intersection and union operations used in set theory, respectively.

The similarity measure $simi_1$ is the ratio of the number frequent itemsets common to D_1 and D_2, and the total number of distinct frequent itemsets in D_1 and D_2. Frequent itemsets are the dominant patterns that determine major characteristics of a database. There are many implementations of mining frequent itemsets in a database (FIMI 2004). Let X and Y be two frequent itemsets in database DB. The itemset X is more dominant than the itemset Y in DB if $supp(X, DB) > supp(Y, DB)$. Therefore the characteristics of DB are revealed more by the pair $(X, supp(X, DB))$ than that of $(Y, supp(Y, DB))$. In other words, a sound measure of similarity between two databases is a function of the supports of the frequent itemsets in the databases.

The second measure of similarity $simi_2$ (Adhikari and Rao 2008) comes in the form:

Definition 15.2 The measure of similarity $simi_2$ between databases D_1 and D_2 is defined as follows:

$$simi_2(D_1, D_2, \alpha) = \frac{\sum_{X \in \{FIS(D_1, \alpha) \cap FIS(D_2, \alpha)\}} minimum\{supp(X, D_1), supp(X, D_2)\}}{\sum_{X \in \{FIS(D_1, \alpha) \cup FIS(D_2, \alpha)\}} maximum\{supp(X, D_1), supp(X, D_2)\}},$$

Here we assume that $supp(X, D_i) = 0$, if $X \notin FIS(D_i, \alpha)$, for $i = 1, 2$.

With reference to Example 15.1, the frequent itemsets in different databases are given as follows:

$FIS(DB_1, 0.3) = \{a(0.5), b(1.0), c(0.5), d(0.5), g(0.5), ab(0.5), bc(0.5), bd(0.5)\}$

$FIS(DB_2, 0.3) = \{g(0.5)\}$

$FIS(DB_3, 0.3) = \{a(0.5), b(1.0), c(0.5), d(0.5), ab(0.5), bc(0.5), bd(0.5)\}$

We obtain $simi_1(DB_1, DB_2, 0.3) = 0.125$, $simi_1(DB_1, DB_3, 0.3) = 0.875$, $simi_2(DB_1, DB_2, 0.3) = 0.111$, and $simi_2(DB_1, DB_3, 0.3) = 0.889$. Thus, the proposed measures $simi_1$ and $simi_2$ are more suitable than the existing measures mentioned in Example 15.1. Theorem 15.1 justifies the fact that $simi_2$ is more appropriate measure than $simi_1$.

Theorem 15.1 *The similarity measure $simi_2$ exhibits higher discriminatory power than that of the similarity measure $simi_1$.*

Proof The support of a frequent itemset could be considered as its weight in the database. We attach weight 1.0 to itemset X in database D_i, under the similarity measure $simi_1$, if $X \in FIS(D_i, \alpha)$, $i = 1, 2$. We attach an weight $supp(X, D_i)$ to the itemset X in database D_i, under the similarity measure $simi_2$, if $X \in FIS(D_i, \alpha)$, $i = 1, 2$. The similarity measures sim_1 and sim_2 are defined as a ratio of two quantities. If $X \in FIS(D_i, \alpha)$, and $X \in FIS(D_j, \alpha)$, then it is more justifiable to add $minimum\{supp(X, D_i), supp(X, D_j)\}$ (instead of 1.0) in the numerator and $maximum\{supp(X, D_i), supp(X, D_j)\}$ (instead of 1.0) in the denominator for the itemset X, $i, j \in \{1, 2\}$. If $X \in FIS(D_i, \alpha)$, and $X \notin FIS(D_j, \alpha)$, then it is more justifiable to add 0 in the numerator and $supp(X, D_i)$ (instead of 1.0) in the denominator for itemset X, $i, j \in \{1, 2\}$. Hence, the theorem has been proved. ☐

In Example 15.3, we verify that $simi_2$ is more appropriate measure than $simi_1$.

Example 15.3 With reference to Example 15.2, $supp(\{a\}, D_1) = supp(\{c\},$ $D_1) = supp(\{a, c\}, D_1) = 1.0$, $supp(\{a\}, D_2) = 1.0$, and $supp(\{c\}, D_2) = supp(\{a, c\},$ $D_2) = 0.67$. $simi_2(D_1, D_2, 0.35) = 0.78$, and $simi_1(D_1, D_2, 0.35) = 1.0$. We observe that the databases D_1 and D_2 are highly similar, but they are not the same. Thus, the similarity computed by $simi_2$ is more suitable.

We highlight some interesting properties of $simi_1$ and $simi_2$ by presenting Theorems 15.2, 15.3 and 15.4.

Theorem 15.2 *The similarity measure $simi_k$ satisfies the following properties $(k = 1, 2)$, (i) $0 \leq simi_k(D_i, D_j, \alpha) \leq 1$, (ii) $simi_k(D_i, D_j, \alpha) = simi_k(D_j, D_i, \alpha)$, (iii) $simi_k(D_i, D_i, \alpha) = 1$, for $i, j = 1, 2, ..., n$.*

Proof The properties follow from the definition of $simi_k$, $(k = 1, 2)$. $\qquad\square$

Now we express the distance between two databases in term of their similarity.

Definition 15.3 The distance measure $dist_k$ between two databases D_1 and D_2 based on the similarity measure $simi_k$ is defined as $dist_k(D_1, D_2, \alpha) = 1 - simi_k(D_1,$ $D_2, \alpha)$, $(k = 1, 2)$.

A "meaningful" distance satisfies the metric properties (Barte 1976). The higher the distance between two databases, the lower is the similarity between them. For the purpose of concise presentation, we use the notation I_i in place of $FIS(D_i, \alpha)$ used so far in Theorems 15.3 and 15.4, for $i = 1, 2$.

Theorem 15.3 $dist_1$ *is a metric over* $[0, 1]$.

Proof We show that $dist_1$ satisfies the triangular inequality. Other properties of a metric follow from Theorem 15.2.

$$dist_1(D_1, D_2, \alpha) = 1 - \frac{|I_1 \cap I_2|}{|I_1 \cup I_2|} \geq \frac{|I_1 - I_2| + |I_2 - I_1|}{|I_1 \cup I_2 \cup I_3|} \qquad (15.1)$$

Thus,

$$dist_1(D_1, D_2, \alpha) + dist_1(D_2, D_3, \alpha) \geq \frac{|I_1 - I_2| + |I_2 - I_1| + |I_2 - I_3| + |I_3 - I_2|}{|I_1 \cup I_2 \cup I_3|}$$

$$(15.2)$$

$$= \frac{|I_1 \cup I_2 \cup I_3| - |I_1 \cap I_2 \cap I_3| + |I_1 \cap I_3| + |I_2| - |I_1 \cap I_2| - |I_2 \cap I_3|}{|I_1 \cup I_2 \cup I_3|}$$

$$(15.3)$$

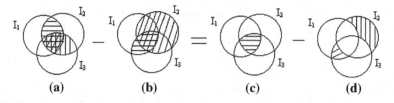

Fig. 15.1 Simplification of the expression (15.5) using Venn diagram

$$= 1 - \frac{|I_1 \cap I_2 \cap I_3| \;-\; |I_1 \cap I_3| \;-\; |I_2| \;+\; |I_1 \cap I_2| \;+\; |I_2 \cap I_3|}{|I_1 \cup I_2 \cup I_3|} \qquad (15.4)$$

$$= 1 - \frac{\{|I_1 \cap I_2 \cap I_3| \;+\; |I_1 \cap I_2| \;+\; |I_2 \cap I_3|\} - \{|I_1 \cap I_3| \;+\; |I_2|\}}{|I_1 \cup I_2 \cup I_3|} \qquad (15.5)$$

Let the number of elements in the shaded regions of Fig. 15.1c, d be N_1 and N_2, respectively. Then the expression (15.5) becomes

$$1 - \frac{N_1 - N_2}{|I_1 \cup I_2 \cup I_3|} \geq \begin{cases} 1 - \frac{N_1 - N_2}{|I_1 \cup I_2 \cup I_3|}, & \text{if } N_1 \geq N_2 \quad \text{(case 1)} \\ 1 - \frac{|I_1 \cap I_3|}{|I_1 \cup I_2 \cup I_3|}, & \text{if } N_1 < N_2 \quad \text{(case 2)} \end{cases} \qquad (15.6)$$

In case 1, the expression remains the same. In case 2, a positive quantity $|I_1 \cap I_3|$ has been put in place of a negative quantity $N_1 - N_2$. Thus, the expression (6.6) reads as

$$\geq \begin{cases} 1 - \frac{N_1 - N_2}{|I_1 \cup I_3|}, & \text{if } N_1 \geq N_2 \\ 1 - \frac{|I_1 \cap I_3|}{|I_1 \cup I_3|}, & \text{if } N_1 < N_2 \end{cases} \geq \begin{cases} 1 - \frac{N_1}{|I_1 \cap I_3|}, & \text{if } N_1 \geq N_2 \\ 1 - \frac{|I_1 \cap I_3|}{|I_1 \cup I_3|}, & \text{if } N_1 < N_2 \end{cases} \qquad (15.7)$$

$$\geq \begin{cases} 1 - \frac{|I_1 \cap I_3|}{|I_1 \cup I_3|}, & \text{if } N_1 \geq N_2 \\ 1 - \frac{|I_1 \cap I_3|}{|I_1 \cup I_3|}, & \text{if } N_1 < N_2 \end{cases} , \quad \text{where, } N_1 = |I_1 \cap I_2 \cap I_3| \leq |I_1 \cap I_3| \quad (15.8)$$

Therefore, irrespective of the relationship between N_1 and N_2, $dist_1(D_1, D_2, \alpha) + dist_1(D_2, D_3, \alpha) \geq dist_1(D_1, D_3, \alpha)$. Thus, $dist_1$ satisfies the triangular inequality. □

We also show that $dist_2$ satisfies the metric properties.

Theorem 15.4 $dist_2$ *is a metric over* [0, 1].

Proof We show that $dist_2$ satisfies the triangular inequality. The remaining properties of a metric follow from Theorem 15.2.

$$dist_2(D_1, D_2, \alpha) = 1 - \frac{\sum_{x \in I_1 \cap I_2} minimum \ \{supp(x, D_1), \ supp(x, D_2)\}}{\sum_{x \in I_1 \cup I_2} maximum \ \{supp(x, D_1), \ supp(x, D_2)\}}$$

$$= 1 - \frac{\sum_{x \in I_1 \cap I_2} \min_{12}(x)}{\sum_{x \in I_1 \cup I_2} \max_{12}(x)} \qquad (15.9)$$

where, $\max_{ij}(x) = maximum\{supp(x, D_i), supp(x, D_j)\}$, and $\min_{ij}(x) = minimum$ $\{supp(x, D_i), supp(x, D_j)\}$, for $i \neq j$. Also, let $\max_{123}(x) = maximum\{supp(x, D_1),$ $supp(x, D_2), supp(x, D_3)\}$, and $\min_{123}(x) = minimum\{supp(x, D_1), supp(x, D_2), supp(x, D_3)\}$.

Thus, $dist_2(D_1, D_2, \alpha) + dist_2(D_2, D_3, \alpha)$

$$= \frac{\sum_{x \in I_1 \cup I_2} \max_{12}(x) - \sum_{x \in I_1 \cap I_2} \min_{12}(x)}{\sum_{x \in I_1 \cup I_2} \max_{12}(x)}$$

$$+ \frac{\sum_{x \in I_2 \cup I_3} \max_{23}(x) - \sum_{x \in I_2 \cap I_3} \min_{23}(x)}{\sum_{x \in I_2 \cup I_3} \max_{23}(x)} \qquad (15.10)$$

$$\geq \frac{\sum_{x \in I_1 - I_2} \max_{12}(x) + \sum_{x \in I_2 - I_1} \max_{12}(x)}{\sum_{x \in I_1 \cup I_2} \max_{12}(x)}$$

$$+ \frac{\sum_{x \in I_2 - I_3} \max_{23}(x) + \sum_{x \in I_3 - I_2} \max_{23}(x)}{\sum_{x \in I_2 \cup I_3} \max_{23}(x)} \qquad (15.11)$$

$$\geq \frac{\sum_{x \in I_1 - I_2} \max_{12}(x) + \sum_{x \in I_2 - I_1} \max_{12}(x) + \sum_{x \in I_2 - I_3} \max_{23}(x) + \sum_{x \in I_3 - I_2} \max_{23}(x)}{\sum_{x \in I_1 \cup I_2 \cup I_3} \max_{123}(x)}$$

$$\qquad (15.12)$$

Using the simplification visualized graphically in Fig. 15.2, the expression (15.12) becomes

$$\frac{\sum_{x \in I_1 \cup I_2 \cup I_3} \max_{123}(x) - N_1 + N_2}{\sum_{x \in I_1 \cup I_2 \cup I_3} \max_{123}(x)} \qquad (15.13)$$

where N_1 and N_2 are the value of $\sum_x \max_{123}(x)$ over the shaded regions of Fig. 15.2c, d, respectively. The expression (15.13) is equal to

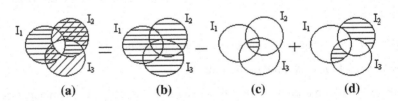

        (a)   (b)   (c)   (d)

Fig. 15.2 Simplification of the expression (15.12) using Venn diagram

$$1 - \frac{N_1 - N_2}{\sum_{x \in I_1 \cup I_2 \cup I_3} \max_{123}(x)} \geq \begin{cases} 1 - \frac{N_1}{\sum_{x \in I_1 \cup I_2 \cup I_3} \max_{123}(x)}, & \text{if } N_1 \geq N_2 \\ 1 - \frac{N_1 - N_2}{\sum_{x \in I_1 \cup I_2 \cup I_3} \max_{123}(x)}, & \text{if } N_1 < N_2 \end{cases}$$

$$\geq \begin{cases} 1 - \frac{\sum_{x \in I_1 \cap I_3} \max_{13}(x)}{\sum_{x \in I_1 \cup I_2 \cup I_3} \max_{123}(x)}, & \text{if } N_1 \geq N_2 \\ 1 - \frac{\sum_{x \in I_1 \cap I_3} \max_{13}(x)}{\sum_{x \in I_1 \cup I_2 \cup I_3} \max_{123}(x)}, & \text{if } N_1 < N_2 \end{cases} \qquad (15.14)$$

Therefore, irrespective of the relationship between N_1 and N_2, $dist_2(D_1, D_2, \alpha) + dist_2(D_2, D_3, \alpha) \geq dist_2(D_1, D_3, \alpha)$. Thus, $dist_2$ satisfies the triangular inequality. □

Given a set of databases, the similarity between a collection of pairs of databases could be expressed by a square matrix, called *database similarity matrix* (DSM). We define DSM of a set of databases as follows:

Definition 15.4 Let $D = \{D_1, D_2, ..., D_n\}$ be the set of all databases. The database similarity matrix DSM_k of D expressed by the measure of similarity $simi_k$, is a symmetric square matrix of size n by n, whose (i, j)th element $DSM_k^{i,j}(D, \alpha) = simi_k(D_i, D_j, \alpha)$; for $D_i, D_j \in D$, and $i, j = 1, 2, ..., n$, $(k = 1, 2)$.

For n databases, there are $^nC_2$ pairs of databases. For each pair of databases, we determine the calculations of similarity between them. If the similarity is high then the databases may be placed in the same class. We define this class as follows:

Definition 15.5 Let $D = \{D_1, D_2, ..., D_n\}$. A class $class_k^\delta$ formed at the level of similarity δ under the measure of similarity $simi_k$, is defined as

$$class_k^\delta(D, \alpha)$$
$$= \begin{cases} P : P \subseteq D, |P| \geq 2, \text{ and } simi_k(A, B, \alpha) \geq \delta, \text{ for } A, B \in P \\ P : P \subseteq D, |P| = 1 \end{cases}, (k = 1, 2).$$

A *DSM* could be viewed as a complete weighted graph. Each database forms a vertex of the graph. A weight of the edge is the similarity between the pair of the corresponding databases. During the process of clustering, we assume that the databases $D_1, D_2, ..., D_r$ have been included in some classes, and the remaining databases are yet to be clustered. Then the clustering process forms the next class by finding a maximal complete sub-graph of the complete weighted graph containing vertices $D_{r+1}, D_{r+2}, ..., D_n$. A maximal complete sub-graph is defined as follows:

Definition 15.6 A weighted complete sub-graph g of a complete weighted graph G is maximal at the similarity level δ if the following conditions are satisfied: (i) The weight of every edge of g is greater than or equal to δ. (ii) The addition of one more vertex (i.e., a database) to g leads to the addition of at least one edge to g having weight less than δ.

We need to find out a maximal weighted complete sub-graph of the complete weighted graph of the remaining vertices to form the next class. This process continues till all the vertices have been clustered. A clustering of databases is defined as follows:

Definition 15.7 Let D be a set of databases. Let $\pi_k^\delta(D, \alpha)$ be a clustering of databases in D at the similarity level δ under the similarity measure $simi_k$. Then, $\pi_k^\delta(D, \alpha) = \{X: X \in \rho(D),$ and X is a $class_k^\delta (D, \alpha)\}$, where $\rho(D)$ is the power set of D, $(k = 1, 2)$.

During the clustering process we may like to impose a restriction that each database belongs to at least one class. This restriction makes a clustering complete. We define a complete clustering as follows:

Definition 15.8 Let D be a set of databases. Let $\pi_k^\delta(D, \alpha) = \{C_{k,1}^\delta(D, \alpha),$ $C_{k,2}^\delta(D, \alpha), \ldots, C_{k,m}^\delta(D, \alpha)\}$, where $C_{k,i}^\delta(D, \alpha)$ is the ith class of $\pi_k^\delta, i = 1, 2, \ldots, m$. π_k^δ is complete, if $\cup_{i=1}^m C_{k,i}^\delta(D, \alpha) = D, (k = 1, 2)$.

In complete clustering, two classes may have a common database. We may be interested in forming clustering of mutually exclusive classes. We define mutually exclusive clustering as follows:

Definition 15.9 Let D be a set of databases. Let $\pi_k^\delta(D, \alpha) = \{C_{k,1}^\delta(D, \alpha),$ $C_{k,2}^\delta(D, \alpha), \ldots, C_{k,m}^\delta(D, \alpha)\}$, where $C_{k,i}^\delta(D, \alpha)$ is the ith class of $\pi_k^\delta, i = 1, 2, \ldots, m$. π_k^δ is mutually exclusive if $C_{k,i}^\delta(D, \alpha) \cap C_{k,j}^\delta(D, \alpha) = \phi$, for $i \neq j, 1 \leq i, j \leq m, (k = 1, 2)$.

We may be interested in realizing such a mutually exclusive and complete clustering. Here we have

Definition 15.10 Let D be a set of databases. Also, let $\pi_k^\delta(D, \alpha)$ be a clustering of databases in D at the similarity level δ under the similarity measure $simi_k$. If $\pi_k^\delta(D, \alpha)$ is a mutually exclusive and complete clustering then it is called a partition $(k = 1, 2)$.

Definition 15.11 Let D be a set of databases. Also let $\pi_k^\delta(D, \alpha)$ be a partition of D at the similarity level δ under the similarity measure $simi_k$. $\pi_k^\delta(D, \alpha)$ is called a non-trivial partition if $1 < |\pi_k^\delta| < n(k = 1, 2)$.

A clustering does not necessarily lead to a partition. In the following example, we wish to find partitions (if they exist) of a set of databases.

Example 15.4 With reference to Example 15.2, consider the set of databases $D = \{D_1, D_2, ..., D_7\}$. The corresponding DSM_2 is given as follows.

$$DSM_2(D, 0.35) = \begin{bmatrix} 1.0 & 0.780 & 0.539 & 0.0 & 0.0 & 0.0 & 0.0 \\ 0.780 & 1.00 & 0.636 & 0.0 & 0.0 & 0.0 & 0.0 \\ 0.539 & 0.636 & 1.0 & 0.061 & 0.0 & 0.0 & 0.0 \\ 0.0 & 0.0 & 0.061 & 1.0 & 0.063 & 0.065 & 0.087 \\ 0.0 & 0.0 & 0.0 & 0.063 & 1.0 & 0.641 & 0.353 \\ 0.0 & 0.0 & 0.0 & 0.065 & 0.641 & 1.0 & 0.444 \\ 0.0 & 0.0 & 0.0 & 0.087 & 0.353 & 0.444 & 1.0 \end{bmatrix},$$

We arrange all non-zero and distinct $DSM_2^{i,j}(D, 0.35)$ values in non-increasing order, for $1 \leq i < j \leq 7$. The arranged similarity values are given as follows: 0.780, 0.641, 0.636, 0.539, 0.444, 0.353, 0.087, 0.065, 0.063, 0.061. We obtain many non-trivial partitions formed at different similarity levels. At the similarity levels equal to 0.780, 0.641, 0.539, and 0.353, we get non-trivial partitions as

$$\pi_2^{0.780} = \{\{D_1, D_2\}, \{D_3\}, \{D_4\}, \{D_5\}, \{D_6\}, \{D_7\}\},$$
$$\pi_2^{0.641} = \{\{D_1, D_2\}, \{D_3\}, \{D_4\}, \{D_5, D_6\}, \{D_7\}\},$$
$$\pi_2^{0.539} = \{\{D_1, D_2, D_3\}, \{D_4\}, \{D_5, D_6\}, \{D_7\}\}, \text{ and }$$
$$\pi_2^{0.353} = \{\{D_1, D_2, D_3\}, \{D_4\}, \{D_5, D_6, D_7\}\}, \text{ respectively.}$$

Our *BestDatabasePartition* algorithm (as presented in Sect. 15.4.1) is based on binary similarity matrix (*BSM*). We derive binary similarity matrix BSM_k from the corresponding DSM_k ($k = 1, 2$). BSM_k is defined as follows:

Definition 15.12 The (i, j)th element of the binary similarity matrix BSM_k at the similarity level δ using the similarity measure $simi_k$ is defined as follows.

$$BSM_k^{i,j}(D, \alpha, \delta) = \begin{cases} 1, & \text{if } simi_k(D_i, D_j, \alpha) \geq \delta \\ 0, & \text{otherwise} \end{cases}, \text{ for } i, j = 1, 2, ..., n(k = 1, 2).$$

We take an example of BSM_2 and observe the distribution of 0s and 1s.

Example 15.5 With reference to Example 15.4, the BSM_2 at the similarity level 0.353 is given below.

$$BSM_2(D, 0.35, 0.353) = \begin{bmatrix} 1 & 1 & 1 & 0 & 0 & 0 & 0 \\ 1 & 1 & 1 & 0 & 0 & 0 & 0 \\ 1 & 1 & 1 & 0 & 0 & 0 & 0 \\ 0 & 0 & 0 & 1 & 0 & 0 & 0 \\ 0 & 0 & 0 & 0 & 1 & 1 & 1 \\ 0 & 0 & 0 & 0 & 1 & 1 & 1 \\ 0 & 0 & 0 & 0 & 1 & 1 & 1 \end{bmatrix}, \text{ where}$$

$$D = \{D_1, D_2, \ldots, D_7\}.$$

There may exist two the same partitions at two distinct similarity levels. Two partitions are distinct if they are not the same. In the following, we define two same partitions at two distinct similarity levels.

Definition 15.13 Let D be a set of databases. Let $C \subseteq D$, and $C \neq \phi$. Two partitions $\pi_k^{\delta_1}(D, \alpha)$ and $\pi_k^{\delta_2}(D, \alpha)$ are the same, if the following statement is true: $C \in \pi_k^{\delta_1}$ if and only if $C \in \pi_k^{\delta_2}$, for $\delta_1 \neq \delta_2$.

We would like to enumerate the maximum number of possible distinct partitions. In Theorem 15.5, we find the maximum number of possible distinct partitions of a set of databases (Adhikari and Rao 2008).

Theorem 15.5 *Let D be a set of databases. Let m be the number of distinct non-zero similarity values in the upper triangle of DSM_2. Then the number of distinct partitions is less than or equal to m.*

Proof We arrange the non-zero similarity values of the upper triangle of DSM_2 in non-increasing order. Let $\delta_1, \delta_2, \ldots, \delta_m$ be m non-zero ordered similarity values. Let δ_i, δ_{i+1} be two consecutive similarity values in the sequence of non-increasing similarity values. Let $x, y \in [\delta_i, \delta_{i+1})$, for some $i = 1, 2, \ldots, m$, where $\delta_{m+1} = 0$. Then $BSM_2(D, \alpha, x) = BSM_2(D, \alpha, y)$. Thus, there exists at the most one distinct non-trivial partition in the interval $[\delta_i, \delta_{i+1})$, for $i = 1, 2, \ldots, m$. We have m such semi-closed intervals $[\delta_i, \delta_{i+1})$, $i = 1, 2, \ldots, m$. The theorem follows. □

For the purpose of finding partitions of the input databases, we first design a simple algorithm that uses the apriori property (Agrawal et al. 1993). The similarity values considered here are based on the similarity measure $simi_2$. Initially, we have n database classes, where n is the number of databases. At this time, each class contains a single database object. These classes are assumed at level 1. Based on the classes at level 1, we construct database classes at level 2. At level 1, we assume that the ith class contains database D_i, $i = 1, 2, \ldots, n$. ith class and jth class of level 1 could be merged if $simi_2(D_i, D_j) \geq \delta$, where δ is the user defined level of similarity.

We proceed further until no more classes could be generated and no more levels could be generated. The algorithm (Adhikari and Rao 2008) is presented below.

Algorithm 15.1. Find partitions (if they exist) of a set of databases using apriori property.
procedure *AprioriDatabaseClustering* (n, DSM_2)
Input: n, DSM_2
n: number of databases
DSM_2: database similarity matrix
Output: Partitions (if they exist) of input databases
01: sort all the non-zero values that exist in the upper triangle of DSM_2 in non-increasing
02: order into an array called *simValues*; **let** the number of non-zero values be m;
03: **let** $k = 1$; let $simValues(m+1) = 0$; **let** $delta = simValues(k)$;
04: **while** $(delta > 0)$ **do**
05:    construct n classes, where each class contains a single database;   // level: 1
06:    **repeat** line 7 **until** no more level could be generated;
07:       construct all possible classes at level $(i+1)$ using lines 8-10;   // level: $i+1$
08:          **let** A and B be two classes at the i-th level such that $|A \cap B| = i-1$;
09:          **let** $a \in (A\text{-}B)$, and $b \in (B\text{-}A)$;
10:          **if** $DSM_2^{a,b} \geq \delta$ then construct a new class $A \cup B$; **end if**
11:    **repeat** line 12 from top level to level 1;
12:       **for** each class at the current level **do**
13:          **if** all databases of the current class are not included a class generated earlier **then**
14:             generate the current class;
15:          **end if**
16:       **end for**
17:       **if** the current clustering is a partition then store it; **end if**
18:    increase k by 1; **let** $delta = simValues(k)$;
19: **end while**
20: display all the partitions;
end procedure

Lines 1–2 take $O(m \times \log(m))$ time to sort m data. *While*-loop at line 4 executes m times. Line 5 takes $O(n)$ time. Initially (at line 5), n classes are constructed. At the first iteration of line 6, the maximum number of classes generated is $^nC_2$. At the second iteration, the maximum number of classes generated is $^nC_3$. Lastly, at the $(n-1)$th iteration, the maximum number of classes generated is ^nC_n. Thus, the maximum number of possible classes is $O\left(\sum_{i=1}^{n} n\ C_i\right)$, i.e., $O(2^n)$. Let p be the average size of a class. Line 8 takes $O(p)$ time. Also, line 11 takes $O(2^n)$ time, since the maximum number of possible classes is $O(2^n)$. Thus, the time complexity of lines 4–19 is $O(m \times p \times 2^n)$. The line 20 takes time $O(m \times n)$, since the maximum number of partitions is m. Thus, the time complexity of the procedure *Apriori-DatabaseClustering* is maximum$\{O(m \times \log(m)), O(m \times p \times 2^n), O(m \times n)\}$, i.e., $O(m \times p \times 2^n)$, since $p \times 2^n > 2^n > n^2 > m > \log_2(m)$, for $p > 1$ and $n > 4$. The *AprioriDatabaseClustering* algorithm generates all possible classes level-wise. It is a simple but not an efficient clustering technique, since the time-complexity of the algorithm is an exponential function of n.

In Theorems 15.6–15.9, we discuss some properties of BSM_2.

Theorem 15.6 *Let* $D = \{D_1, D_2, ..., D_n\}$. *Let* $\pi_2^\delta(D, \alpha)$ *be a clustering of databases in D at the similarity level* δ. π_2^δ *is a partition if and only if the corresponding* BSM_2 *gets transformed into the following form by inter-changing jointly a row and the corresponding column with another row and the corresponding column.*

$$
\begin{bmatrix}
U_1 & 0 & \cdots & 0 \\
0 & U_2 & \cdots & 0 \\
\cdots & \cdots & \cdots & \cdots \\
0 & 0 & \cdots & U_m
\end{bmatrix},
$$

U_i *is a matrix of size* $n_i \times n_i$, *containing all elements as* 1, *where* $\sum_{i=1}^m n_i = n$, $\left|\pi_2^\delta\right| = m$.

Proof Let $\left\{D_1^i, D_2^i, ..., D_{n_i}^i\right\}$ be the *i*th database class of the partition at the similarity level δ, $i = 1, 2, ..., m$. The row referring to D_j^i of BSM_2 corresponds to a unique combination of 0s and 1s, $j = 1, 2, ..., n_i$. Similarly, the column corresponding to D_j^i of BSM_2 results as a unique combination of 0s and 1s, $j = 1, 2, ..., n_i$. All such n_i rows and columns may not be initially consecutive, $i = 1, 2, ..., m$. We keep these n_i rows and columns consecutive, $i = 1, 2, ..., m$. Initially, we keep n_1 rows and the corresponding columns of the first database class to be consecutive. Then, we keep n_2 rows and the corresponding columns of the second database class consecutively and so on. In general, to fix the matrix U_i at the proper position, we interchange jointly $\left(\sum_1^{i-1} n_j + k\right)$th row and $\left(\sum_1^{i-1} n_j + k\right)$th column with D_k^ith row and D_k^ith column of BSM_2, $1 \leq k \leq n_i$, $i = 1, 2, ..., m$. □

Referring to BSM_2 in Example 15.5, we apply Theorem 15.6, and conclude that a partition exists at similarity level of 0.353.

Theorem 15.7 *Let D be a set of databases. Let* $\pi_2^\delta(D, \alpha)$ *be a clustering of databases in D at the similarity level* δ. *Let* $\{D_1^i, D_2^i, ..., D_{n_i}^i\}$ *be the ith database class of* $\pi_2^\delta(D, \alpha)$. *Then* D_k^ith *row (or,* D_k^ith *column) of* BSM_2 *contains* n_i 1s, $k = 1, 2, ..., n_i$, $i = 1, 2, ..., \left|\pi_2^\delta\right|$.

Proof If possible, let D_k^ith row or, D_k^ith column has $(n_i + 1)$ 1s. Then D_k^ith database would belong to two database classes. It contradicts the mutual exclusiveness of classes of a partition. If possible, let D_k^ith row or, D_k^ith column contains $(n_i - 1)$ 1s. It contradicts the fact that $BSM_2^{D_j^i, D_k^i} = 1$, $j = 1, 2, ..., n_i$ and $j \neq k$. □

Theorem 15.8 *Let D be a set of databases. Let $\pi_2^\delta(D, \alpha)$ be a clustering of databases in D at the similarity level δ. Then, the rank of the corresponding BSM_2 is $\left|\pi_2^\delta\right|$.*

Proof Let $\left\{D_1^i, D_2^i, \ldots, D_{n_i}^i\right\}$ be the *i*th database class of π_2^δ. Then,

$$BSM_2^{D_j^i, D_k^i}(D, \alpha) = \begin{cases} 1, & \text{for } D_j^i, D_k^i \in \{D_1^i, D_2^i, \ldots, D_{n_i}^i\} \\ 0, & \text{for } D_j^i \in \{D_1^i, D_2^i, \ldots, D_{n_i}^i\} \text{ and } D_k^i \notin \{D_1^i, D_2^i, \ldots, D_{n_i}^i\} \\ 0, & \text{for } D_j^i \notin \{D_1^i, D_2^i, \ldots, D_{n_i}^i\} \text{ and } D_k^i \in \{D_1^i, D_2^i, \ldots, D_{n_i}^i\} \end{cases}$$

The row corresponding to D_j^i of BSM_2 corresponds to a unique combination of 0s and 1s, for $j = 1, 2, \ldots, n_i$. So, all the rows of BSM_2 are divided into $\left|\pi_2^\delta\right|$ groups such that all the rows in a group correspond to a unique combination of 0s and 1s. Thus, BSM_2 has $\left|\pi_2^\delta\right|$ independent rows. □

Theorem 15.9 *Let $D = \{D_1, D_2, \ldots, D_n\}$. At a given value of the triplet (D, α, δ), there exists at the most one partition of D.*

Proof At a given value of the pair (D, α), the element $DSM_2^{i, j}$ is unique, $i, j = 1, 2, \ldots, n$. Thus at a given value of the tuple (D, α, δ) the element $BSM_2^{i, j}$ is unique, for $i, j = 1, 2, \ldots, n$. There exists a partition if the BSM_2 gets transformed into a specific form (as outlined in Theorem 15.6), by jointly interchanging a row and the corresponding column with another row and the corresponding column. Hence, the theorem follows. □

15.4.1 Finding the Best Non-trivial Partition

Now we get back to Example 15.4. We observed that at different similarity levels there may exist different partitions. We have observed the existence of four non-trivial partitions. We would like to find the best partition among these partitions. The best partition is based on the principle of maximizing the intra-class similarity and maximizing the inter-class distance. The intra-class similarity and inter-class distance are defined as follows.

Definition 15.14 The intra-class similarity *intra-sim* of a partition π at the similarity level δ using the similarity measure $simi_2$ is defined as follows:

$$Intra\text{-}sim(\pi_2^\delta) = \sum_{C \in \pi_2^\delta} \sum_{D_i, D_j \in C; \, i < j} simi_2(D_i, D_j, \alpha).$$

Definition 15.15 The inter-class distance *inter-dist* of a partition π at the similarity level δ using the similarity measure $simi_2$ is defined as follows:

$$Inter\text{-}dist(\pi_2^\delta) = \sum_{C_p, C_q \in \pi_2^\delta;\, p<q} \sum_{D_i \in C_p;\, D_j \in C_q;\, i<j} dist_2(D_i, D_j, \alpha).$$

The best partition among a set of partitions is selected on the basis of goodness value of a partition. The goodness measure itself, *goodness*, of a partition is defined as follows:

Definition 15.16 The goodness of a partition π at similarity level δ using the similarity measure $simi_2$ is expressed as follows:

$$goodness(\pi_2^\delta) = intra\text{-}sim(\pi_2^\delta) + inter\text{-}dist(\pi_2^\delta) - \left|\pi_2^\delta\right|,$$

where $\left|\pi_2^\delta\right|$ is the number classes in π.

Note that we have subtracted the term $\left|\pi_2^\delta\right|$ from the sum of intra-class similarity and inter-class distance to remove the bias of goodness value of a partition. The higher the value of *goodness*, the better is the corresponding partition. Now, we partition the set of databases D using the proposed goodness measure.

Example 15.6 Continuing Example 15.4, we calculate the goodness value of each of the non-trivial partitions using $simi_2$ as follows:

$$intra\text{-}sim(\pi_2^{0.353}) = 3.185,\ inter\text{-}dist(\pi_2^{0.353}) = 15.276,$$
$$\left|\pi_2^{0.353}\right| = 3,\ \text{and}\ goodness(\pi_2^{0.353}) = 15.461.$$
$$intra\text{-}sim(\pi_2^{0.539}) = 2.596,\ inter\text{-}dist(\pi_2^{0.539}) = 16.666,$$
$$\left|\pi_2^{0.539}\right| = 4,\ \text{and}\ goodness(\pi_2^{0.539}) = 15.262.$$
$$intra\text{-}sim(\pi_2^{0.641}) = 1.421,\ inter\text{-}dist(\pi_2^{0.641}) = 17.491,$$
$$\left|\pi_2^{0.641}\right| = 5,\ \text{and}\ goodness(\pi_2^{0.641}) = 13.912.$$
$$intra\text{-}sim(\pi_2^{0.780}) = 0.780,\ inter\text{-}dist(\pi_2^{0.780}) = 17.118,$$
$$\left|\pi_2^{0.780}\right| = 6,\ \text{and}\ goodness(\pi_2^{0.780}) = 11.898.$$

The goodness value corresponding to the partition $\pi_2^{0.353}$ attains the maximal value. The partition $\pi_2^{0.353} = \{\{D_1, D_2, D_3\},\ \{D_4\},\ \{D_5, D_6, D_7\}\}$ is the best among all the non-trivial partitions. Let us look back into the databases presented in Example 15.2. We find that the partition $\pi_2^{0.353}$ matches the best the ground reality among the partitions reported.

We present an algorithm (Adhikari and Rao 2008) for finding the best non-trivial partition of a set of databases.

Algorithm 15.2. Find the best non-trivial partition (if it exists) of a set of databases.
procedure *BestDatabasePartition* (*n, DSM₂*)

Input: *n, DSM₂*

n: number of databases

DSM₂: database similarity matrix

Output: The best partition (if it exists) of input databases

01: sort all the non-zero values that exist in the upper triangle of *DSM₂* in non-increas-
02: ing order into an array called *simValues*; **let** the number of non-zero values be *m*;
03: **let** *k* = 1; **let** *simValues*(*m* + 1) = 0; **let** *delta* = *simValues*(*k*);
04: **while** (*delta* > 0) **do**
05:      **for** *i* = 1 to *n* **do** *class*(*i*) = 0; **end for**
06:      construct the *BSM₂* at current level of the similarity *delta*;
07:      **let** *currentClass* = 1; **let** *currentRow* = 1; **let** *class*(1) = *currentClass*;
08:      **for** *col* = (*currentRow* + 1) to *n* **do**
09:          **if** ($BSM_2^{currentRow, col}$ = 1) **then**
10:              **if** (*class*(*col*) = 0) **then** *class*(*col*) = *currentClass*;
11:              **else if** (*class*(*col*) ≠ *currentClass*) **then** go to line 24; **end if**
12:          **end if**
13:      **end if**
14:      **end for**
15:      **let** *i* = 1; **let** *class*(*n* +1) = 0;
16:      **while** (class(*i*) ≠ 0) **do** increase *i* by 1; **end while**
17:      **if** (*i* = *n* +1) **then**
18:          store the content of array *class* and current similarity level *delta*;
19:      **else**
20:          increase *currentRow* by 1;
21:          **if** (*class*(*currentRow*) = 0) **then** increase *currentClass* by 1; **end if**
22:          go to line 8;
23:      **end if**
24:      increase *k* by 1; **let** *delta* = *simValues*(*k*);
25: **end while**
26: **for** each non-trivial partition **do**
27:      calculate the goodness value of the current partition;
28: **end for**
29: **return** the partition whose goodness value is the maximum;
end procedure

Initially, we have sorted all non-zero values in the upper triangle of *DSM₂* in non-increasing order. The algorithm checks the existence of a partition starting with the maximum of all the similarity values. At line 5, we initialize the class label of each database to 0. The algorithm starts forming a class with D_1 (the first database) as the variable *currentRow* is initialized with 1. Also, class label starts with 1 as the variable *currentClass* is initialized with 1. Lines 8–14 are used to check the similarity of $D_{currentRow}$ with other databases. If the condition at line 9 is true then databases $D_{currentRow}$ and D_{col} are similar. At line 10, D_{col} is put in the *currentClass* if it is still unlabelled. If D_{col} is already labeled with a class label not equal to current class label then D_{col} get another label. Thus, partition does not exist at the current similarity level.

Operations in Line 1 take $O(m \times \log(m))$ time. Line 3 repeats m times. Line 6 constructs BSM_2 in $O(n^2)$ time as the order of BSM_2 is $n \times n$. Each of lines 5 and 16 takes $O(n)$ time. For-loop positioned at line 8, repeats maximum n times. Line 18 takes $O(n)$ time, since the time required to store a partition is $O(n)$. Thus, the time complexity of lines 4–25 is $O(m \times n^2)$. Therefore, the time complexity of the procedure *best-database-partition* is maximum $\{O(m \times \log(m)), O(m \times n^2)\}$, i.e., $O(m \times n^2)$, since $n^2 > m > \log_2(m)$.

The drawback of *BestClassification* (Wu et al. 2005b) algorithm is that the step value for assigning the next similarity level has to be user-defined. Thus, the method might fail to find the exact similarity level at which a partition exists. *BestDatabasePartition* algorithm reports the exact similarity level at which a partition exists. Also, the algorithm works faster, since it is required to check for the existence of partitions only at m similarity levels. Li et al. (2009) have recently proposed *BestCompleteClass* algorithm for partitioning a set of databases. But, the *BestCompleteClass* algorithm has followed the strategies that we have already reported in the *BestDatabasePartition* algorithm. In Theorem 15.10, we prove the correctness of the proposed algorithm.

Theorem 15.10 *Algorithm BestDatabasePartition works correctly.*

Proof Let $D = \{D_1, D_2, \ldots, D_n\}$. Let there are m distinct non-zero similarity values in the upper triangle of DSM_2. Using Theorem 15.5, one could conclude that the maximum number of partitions of D is m at a given value of pair (D, α). *While*-loop at line 4 checks for the existence of partitions at m similarity levels. At each similarity level, we get a new BSM_2. The existence of a partition is determined from the BSM_2. We have an array *class* that stores the class label given to each database under the current level of similarity. In a partition, each database has a unique class label. The existence of a partition is checked based on the principle that every database receives a unique class label. As soon as we find that a labeled database receives another class label, we conclude that a partition does not exist at the current level of similarity *delta* (line 11). Initially, we put the class label 0 to all databases using line 5. Then, we start from the row 1 of BSM_2 that corresponds to database D_1. Thus, D_1 is kept in the first database class. If there is a 1 in the jth column of BSM_2, then we put class label of D_j as 1 using line 10. We find a database D_i that has not been clustered yet using lines 15–16. Then, we start at row i of BSM_2. If there is a 1 in the jth column of row i, then we put database D_j in the current class. Thus, the algorithm *BestDatabasePartition* works correctly.   □

15.4.2 Efficiency of Clustering Technique

The proposed clustering algorithm is based on the similarity measure $simi_2$. The same similarity measure $simi_2$ is based on the supports of the frequent itemsets in databases. If we vary the value of α then the number of frequent itemsets in a database changes. The accuracy of similarity between two databases increases as the number of frequent itemsets increases. Therefore, a clustering process would be more accurate for lower values of α. The frequent itemsets participate in the clustering process is limited by main memory. If we can store more frequent itemsets in main memory then $simi_2$ could determine similarity between two databases more accurately. Thus, the clustering process would be more accurate. This limitation begs for a space efficient representation of the frequent itemsets in main memory. For this purpose, we propose a coding that efficiently represents frequent itemsets. The coding allows more frequent itemsets to participate in determining the similarity between two databases.

15.4.2.1 Space Efficient Representation of Frequent Itemsets in Different Databases

In this technique, we represent each frequent itemset using a bit vector. Each frequent itemset has three components: database identification, frequent itemset, and support. Let the number of databases be n. There exists an integer p such that $2^{p-1} < n \leq 2^p$. Then p bits are enough to represent a database. Let k be the number of digits after the decimal point to represent support. Support value 1.0 could be represented as 0.99999, for $k = 5$. If we represent the support s as an integer d containing of k digits then $s = d \times 10^{-k}$. The number digits required to represent a decimal number could be obtained by Theorem 15.3. The proposed coding is described with the help of Example 15.7.

Example 15.7 We refer again to Example 15.2. The frequent itemsets sorted in non-increasing order with regard to the number of extractions are given as follows:

$$(h, 4), (a, 3), (ac, 3), (c, 3), (hi, 3), (i, 3), (j, 2), (e, 2), (ij, 2),$$
$$(ae, 1), (d, 1), (df, 1), (ef, 1), (f, 1), (fh, 1), (g, 1), (gi, 1).$$

(X, μ) denotes itemset X having number of extractions equal to μ. We code the frequent itemsets of the above table from left to right. The frequent itemsets are coded using a technique similar to Huffman coding (Huffman 1952). We attach code 0 to itemset h, 1 to itemset a, 00 to itemset ac, 01 to itemset c, etc. Itemset h gets a code of minimal length, since it has been extracted maximum number of times. We call this coding as itemset (IS) coding. It is a lossless coding (Sayood 2000). IS coding and Huffman coding are not the same, in the sense that an IS code may be a prefix of another IS code. Coded itemsets are given as follows:

$(h, 0), (a, 1), (ac, 00), (c, 01), (hi, 10), (i, 11), (j, 000), (e, 001), (ij, 010), (ae, 011), (d, 100), (df, 101), (ef, 110), (f, 111), (fh, 0000), (g, 0010), (gi, 0011)$. Here (X, v) denotes itemset X having IS code v.

15.4.2.2 Efficiency of IS Coding

Using the above representation of the frequent itemsets, one could store more frequent itemsets in the main memory during the clustering process. This enhances the efficiency of the clustering process.

Definition 15.17 Let there are n databases $D_1, D_2, ..., D_n$. Let $S^T\left(\cup_{i=1}^n FIS(D_i)\right)$ be the amount of storage space (in bits) required to represent $\cup_{i=1}^n FIS(D_i)$ by a technique T. Let $S_{\min}\left(\cup_{i=1}^n FIS(D_i)\right)$ be the minimum amount of storage space (in bits) required to represent $\cup_{i=1}^n FIS(D_i)$. Let τ, κ, and λ denote a clustering algorithm, similarity measure, and computing resource under consideration, respectively. Let Γ be the set of all frequent itemset representation techniques. We define efficiency of a frequent itemset representation technique T at a given value of triplet (τ, κ, λ) as follows:

$$\varepsilon(T|\tau, \kappa, \lambda) = S_{\min}\left(\cup_{i=1}^n FIS(D_i)\right) / S^T\left(\cup_{i=1}^n FIS(D_i)\right), \quad \text{for } T \in \Gamma.$$

One could store an itemset conveniently using the following components: database identification, items in the itemset, and support. Database identification, an item and a support could be stored as a short integer, an integer and a real type data, respectively. A typical compiler represents a short integer, an integer and a real number using 2, 4 and 8 bytes, respectively. Thus, a frequent itemset of size 2 could consume $(2 + 2 \times 4 + 8) \times 8$ bits, i.e. 144 bits. An itemset representation may have an overhead of indexing frequent itemsets. Let $OI(T)$ be the overhead of indexing coded frequent itemsets using technique T.

Theorem 15.11 *IS coding stores a set of frequent itemsets using minimum storage space, if $OI(IS\ coding) \leq OI(T)$, for $T \in \Gamma$.*

Proof A frequent itemset has three components, viz., database identification, itemset, and support. Let the number of databases be n. Then $2^{p-1} < n \leq 2^p$, for an integer p. We need minimum p bits to represent a database identifier. The representation of database identification is independent of the corresponding frequent itemsets. If we keep k digits to store a support then $\lceil k \times \log_2 10 \rceil$ binary digits are needed to represent a support (as mentioned in Lemma 16.3). Thus, the representation of support becomes independent of the other components of the frequent itemset. Also, the sum of all IS codes is the minimum because of the way they are constructed. Thus, the space used by the IS coding for representing a set of frequent itemsets attains the minimum. □

Thus, the efficiency of a frequent itemset representation technique T could be expressed as follows:

$$\varepsilon(T|\tau, \kappa, \lambda) = S^{\text{IS coding}} \left(\bigcup_{i=1}^{n} FIS(D_i) \right) / S^{\text{T}} \bigcup_{i=1}^{n} FIS(D_i), \tag{15.15}$$

provided $OI(\text{IS coding}) \leq OI(T)$, for $T \in \Gamma$.

If the condition in (15.15) is satisfied, then the IS coding performs better than any other techniques. If the condition in (15.15) is not satisfied, then the IS coding performs better than any other techniques in almost all cases. The following corollary is derived from Theorem 15.11.

Corollary 15.1. *Efficiency of IS coding attains maximum, if $OI(\text{IS coding}) \leq OI(T)$, for $T \in \Gamma$.*

Proof $\varepsilon(\text{IS coding}|\tau, \kappa, \lambda) = 1.0$.   □

The IS coding maintains an index table to decode/search a frequent itemset. In the following example, we compute the amount of space required to represent the frequent itemsets using an ordinary method and the IS coding.

Example 15.8 With reference to Example 15.7, there are 33 frequent itemsets in different databases. Among them, there are 20 itemsets of size 1 and 13 itemsets of size 2. An ordinary method could use $(112 \times 20 + 144 \times 13) = 4{,}112$ bits. The amount of space required to represent frequent itemsets in seven databases using IS coding is equal to $P + Q$ bits, where P is the amount of space required to store frequent itemsets, and Q is the amount of space required to maintain the index table. Since there are seven databases, we need 3 bits to identify a database. The amount of memory required to represent the database identification for 33 frequent itemsets is equal to 33×3 bits $= 99$ bits. Suppose we keep 5 digits after the decimal point for a support. Thus, $\lceil 5 \times \log_2(10) \rceil$ bits, i.e., 17 bits are required to represent a support. The amount of memory required to represent the supports of 33 frequent itemsets is equal to 33×17 bits $= 561$ bits. Let the number of items be 10,000. Therefore, 14 bits are required to identify an item. The amount of storage space would require for itemsets h and ac are 14 bits and 28 bits respectively. To represent 33 frequent itemsets, we need $(20 \times 14 + 13 \times 28)$ bits $= 644$ bits. Thus, $P = (99 + 561 + 644)$ bits $= 1{,}304$ bits. There are 17 frequent itemsets in the index table. Using IS coding, 17 frequent itemsets consume 46 bits. To represent 17 frequent itemsets, we need $14 \times 9 + 28 \times 8$ bits $= 350$ bits. Thus, $Q = 350 + 46$ bits $= 396$ bits. The total amount of memory space required (including the overhead of indexing) to represent frequent itemsets in 7 databases using IS coding is equal to $P + Q$ bits, i.e., 1,700 bits. The amount of space saving in compared to an ordinary method is equal to 2,412 bits, i.e., 58.66 % approximately. A technique without optimization (TWO) may not maintain index table separately. In this case, $OI(\text{TWO}) = 0$. In spite of that, IS coding performs better than a TWO in most of the cases.

Finally, we claim that our clustering technique is more accurate. There are two reasons for this claim: (i) We propose more appropriate measures of similarity than the existing ones. We have observed that the similarity between two databases based on items might not be appropriate. The proposed measures are based on the similarity between transactions of two databases. As a consequence the similarity between two databases is estimated more accurately. (ii) Also, the proposed IS coding enables us to mine local databases further at a lower level of α to accommodate more frequent itemsets in main memory. As a result, more frequent itemsets could participate in the clustering process.

15.5 Experiments

We have carried out a number of experiments to study the effectiveness of our approach. We present experimental results using two synthetic databases, and one real database. The synthetic databases *T10I4D100K* (Frequent Itemset Mining Dataset Repository 2004) and *T40I10D100K* (Frequent Itemset Mining Dataset Repository 2004) have been generated using synthetic database generator from IBM Almaden Quest research group. The real database *BMS-Web-Wiew-1* could be found at the KDD CUP 2000 repository (KDD CUP 2000). The characteristics of these databases are presented in Table 15.1. Let *NT*, *ALT*, *AFI*, and *NI* denote the number of transactions, the average length of a transaction, the average frequency of an item, and the number of items in the database (*DB*), respectively.

Each of the above databases is divided into 10 databases for the purpose of carrying out experiments. The databases obtained from *T10I4D100K*, and *T40I10D100K* are named T_{1j}, and T_{4j}, respectively, $j = 0, 1, ..., 9$. The databases obtained from *BMS-Web-Wiew-1* are named B_{1j}, $j = 0, 1, ..., 9$. The databases T_{ij} and B_{1j} are called input databases, for $i = 1, 4$, and $j = 0, 1, ..., 9$. Some characteristics of these input databases are presented in the Table 15.2.

At a given value of α, there may exist many partitions. Partitions of the set of input databases are presented in Table 15.3. If we vary the value of α, the set of frequent itemsets in a database varies. Apparently, the similarity between a pair of databases changes over the change of α.

At a lower value of α, more frequent itemsets are reported from a database and hence the database is represented more correctly by its frequent itemsets. We obtain a more accurate value of similarity between a pair of databases. Thus, the partition

Table 15.1 Dataset characteristics

Dataset	*NT*	*ALT*	*AFI*	*NI*
T10I4D100K (T1)	1,00,000	11.102280	1276.124138	870
T40I10D100K (T4)	1,00,000	40.605070	4310.516985	942
BMS-Web-Wiew-1 (B1)	1,49,639	2.000000	155.711759	1,922

Table 15.2 Input database characteristics

DB	NT	ALT	AFI	NI	DB	NT	ALT	AFI	NI
T_{10}	10,000	11.06	127.66	866	T_{15}	10,000	11.14	128.63	866
T_{11}	10,000	11.13	128.41	867	T_{16}	10,000	11.11	128.56	864
T_{12}	10,000	11.07	127.65	867	T_{17}	10,000	11.10	128.45	864
T_{13}	10,000	11.12	128.44	866	T_{18}	10,000	11.08	128.56	862
T_{14}	10,000	11.14	128.75	865	T_{19}	10,000	11.08	128.11	865
T_{40}	10,000	40.57	431.57	940	T_{45}	10,000	40.51	430.46	941
T_{41}	10,000	40.58	432.19	939	T_{46}	10,000	40.74	433.44	940
T_{42}	10,000	40.63	431.79	941	T_{47}	10,000	40.62	431.71	941
T_{43}	10,000	40.63	431.74	941	T_{48}	10,000	40.53	431.15	940
T_{44}	10,000	40.66	432.56	940	T_{49}	10,000	40.58	432.16	939
B_{10}	14,000	2.00	14.94	1,874	B_{15}	14,000	2.00	280.00	100
B_{11}	14,000	2.00	280.00	100	B_{16}	14,000	2.00	280.00	100
B_{12}	14,000	2.00	280.00	100	B_{17}	14,000	2.00	280.00	100
B_{13}	14,000	2.00	280.00	100	B_{18}	14,000	2.00	280.00	100
B_{14}	14,000	2.00	280.00	100	B_{19}	23,639	2.00	472.78	100

Table 15.3 Partitions of the input databases for a given value of α

Databases	α	Non-trivial distinct partition (π)	δ	Goodness (π)
$\{T_{10}, ..., T_{19}\}$	0.03	$\{\{T_{10}\},\{T_{11}\},\{T_{12}\},\{T_{13}\},\{T_{14},T_{18}\},\{T_{15}\},$ $\{T_{16}\},\{T_{17}\},\{T_{19}\}\}$	0.881	0.01
$\{T_{40}, ..., T_{49}\}$	0.1	$\{\{T_{40}\},\{T_{41}, T_{45}\},\{T_{42}\},\{T_{43}\},\{T_{44}\},\{T_{46}\},$ $\{T_{47}\},\{T_{48}\},\{T_{49}\}\}$	0.950	−3.98
		$\{\{T_{40}\},\{T_{41}, T_{45}\},\{T_{42}\},\{T_{43}\},\{T_{44}\},\{T_{46}\},$ $\{T_{47}\},\{T_{48}, T_{49}\}\}$	0.943	11.72
		$\{\{T_{40}\}, \{T_{41}, T_{43}, T_{45}\}, \{T_{42}\}, \{T_{44}\},\{T_{46}\},$ $\{T_{47}\}, \{T_{48}, T_{49}\}\}$	0.942	24.21
$\{B_{10}, ..., B_{19}\}$	0.009	$\{\{B_{10}\},\{B_{11}\},\{B_{12}, B_{14}\},\{B_{13}\},\{B_{15}\},\{B_{16}\},$ $\{B_{17}\},\{B_{18}\},\{B_{19}\}\}$	0.727	11.70
		$\{\{B_{10}\},\{B_{11}\},\{B_{12}, B_{14}\},\{B_{13}\},\{B_{15}\},\{B_{16}, B_{19}\},\{B_{17}\},\{B_{18}\}\}$	0.699	27.69
		$\{\{B_{10}\},\{B_{11}\},\{B_{12}, B_{13}, B_{14}\}, \{B_{15}\},\{B_{16}, B_{19}\},\{B_{17}\}, \{B_{18}\}\}$	0.684	36.97
		$\{\{B_{10}\}, \{B_{11}\},\{B_{12}, B_{13}, B_{14}, B_{15}, B_{16}, B_{19}, B_{17}, B_{18}\}\}$	0.582	55.98
		$\{\{B_{10}, B_{11}\}, \{B_{12}, B_{13}, B_{14}, B_{15}, B_{16},B_{17}, B_{18}, B_{19}\}\}$	0.536	81.03

generated at a smaller value of α would be more correct. In Tables 15.4 and 15.5, we have presented best partitions of a set of databases obtained for different values of α. So, the best partition of a set of databases may change over the change of α.

Table 15.4 Best partitions of $\{T_{10}, T_{11}, ..., T_{19}\}$

α	Best partition (π)	δ	Goodness (π)
0.07	$\{\{T_{10},T_{13},T_{14},T_{16},T_{17}\}, \{T_{11}\}, \{T_{12},T_{15}\}, \{T_{18},T_{19}\}\}$	0.725	85.59
0.06	$\{\{T_{10},T_{11},T_{15},T_{16},T_{17},T_{18}\}, \{T_{12}\}, \{T_{13},T_{14},T_{19}\}\}$	0.733	81.08
0.05	$\{\{T_{10}\}, \{T_{11}\}, \{T_{12}\},\{T_{13}\}, \{T_{14},T_{16}\}, \{T_{15}\}, \{T_{17},T_{19}\}, \{T_{18}\}\}$	0.890	13.35
0.04	$\{\{T_{10}\}, \{T_{11},T_{13}\}, \{T_{12}\}, \{T_{14}\}, \{T_{15}\}, \{T_{16}\}, \{T_{17}\}, \{T_{18}\}, \{T_{19}\}\}$	0.950	−2.07
0.03	$\{\{T_{10}\}, \{T_{11}\}, \{T_{12}\}, \{T_{13}\}, \{T_{14},T_{18}\}, \{T_{15}\}, \{T_{16}\}, \{T_{17}\}, \{T_{19}\}\}$	0.881	0.01

Table 15.5 Best partitions of $\{B_{10}, B_{11}, ..., B_{19}\}$

α	Best partition (π)	δ	Goodness (π)
0.020	$\{\{B_{10}\}, \{B_{11},B_{12},B_{13},B_{14},B_{15},B_{16},B_{17},B_{18},B_{19}\}\}$	0.668	51.90
0.017	$\{\{B_{10}\}, \{B_{11},B_{12},B_{13},B_{14},B_{15},B_{16},B_{17},B_{18},B_{19}\}\}$	0.665	66.10
0.014	$\{\{B_{10}\}, \{B_{11},B_{12},B_{13},B_{14},B_{15},B_{16},B_{17},B_{18},B_{19}\}\}$	0.581	72.15
0.010	$\{\{B_{10},B_{11}\}, \{B_{12},B_{13},B_{14},B_{15},B_{16},B_{17},B_{18},B_{19}\}\}$	0.560	63.67
0.009	$\{\{B_{10},B_{11}\}, \{B_{12},B_{13},B_{14}\}, \{B_{15}\}, \{B_{16},B_{19}\}, \{B_{17}\}, \{B_{18}\}\}$	0.536	81.03

Thus, a partition may not remain the same over the change of α. But, we have observed a general tendency that the databases show more similarity over larger values of α. As the value of α becomes smaller, more frequent itemsets are reported from a database, and databases become more dissimilar.

In Fig. 15.3, we have shown how the execution time of an experiment increases as the number databases increases. The execution time increases faster as we increase input databases from database T_1. The reason is that the size of each local database obtained from T_1 is larger than that of T_4 and B_1.

The number of frequent itemsets decreases as the value of α increases. Thus, the execution time of an experiment decreases as α increases. We observe this phenomenon in Figs. 15.4 and 15.5.

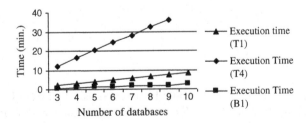

Fig. 15.3 Execution time versus the number of databases

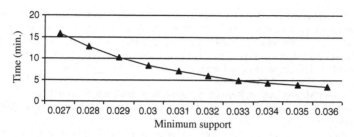

Fig. 15.4 Execution time versus α for experiment with $\{T_{10}, T_{11}, \ldots, T_{19}\}$

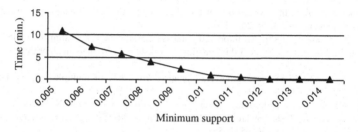

Fig. 15.5 Execution time versus α for experiment with $\{B_{10}, B_{11}, \ldots, B_{19}\}$

15.6 Conclusions

Clustering a set of databases is an important activity. It reduces cost of searching relevant information required for many problems. We provided an efficient solution to this problem in three ways. Firstly, we proposed more suitable measures of similarity between two databases. Secondly, we showed that there is a need to figure out the existence of the best clustering only at a few similarity levels. Thus, the proposed clustering algorithm executes faster. Lastly, we introduce IS coding for storing frequent itemsets in the main memory. It allows more frequent itemsets to participate in the clustering process. The IS coding enhances the accuracy of the clustering process. Thus, the proposed clustering technique is efficient in finding clusters in a set of databases.

References

Adhikari A, Rao PR (2008) Efficient clustering of databases induced by local patterns. Decis Support Syst 44(4):925–943

Agrawal R, Imielinski T, Swami A (1993) Mining association rules between sets of items in large databases. In: Proceedings of ACM SIGMOD conference, pp 207–216

Ali K, Manganaris S, Srikant R (1997) Partial classification using association rules. In: Proceedings of the 3rd international conference on knowledge discovery and data mining, pp 115–118

Babcock B, Chaudhury S, Das G (2003) Dynamic sample selection for approximate query processing. In: Proceedings of ACM SIGMOD conference management of data, pp 539–550

Bandyopadhyay S, Giannella C, Maulik U, Kargupta H, Liu K, Datta S (2006) Clustering distributed data streams in peer-to-peer environments. Inf Sci 176(14):1952–1985

Barte RG (1976) The elements of real analysis, 2nd edn. Wiley, New York

FIMI (2004) http://fimi.cs.helsinki.fi/src/

Frequent Itemset Mining Dataset Repository (2004) http://fimi.cs.helsinki.fi/data

Huffman DA (1952) A method for the construction of minimum redundancy codes. Proc IRE 40 (9):1098–1101

Jain AK, Murty MN, Flynn PJ (1999) Data clustering: a review. ACM Comput Surv 31(3):264–323

KDD CUP (2000) http://www.ecn.purdue.edu/KDDCUP

Lee C-H, Lin C-R, Chen M-S (2001) Sliding-window filtering: an efficient algorithm for incremental mining. In: Proceedings of the 10th international conference on information and knowledge management, pp 263–270

Li H, Hu X, Zhang Y (2009) An improved database classification algorithm for multi-database mining. In: Proceedings of the 3rd international workshop on frontiers in algorithmics, pp 346–357

Ling CX, Yang Q (2006) Discovering classification from data of multiple sources. Data Min Knowl Disc 12(2–3):181–201

Liu CL (1985) Elements of discrete mathematics, 2nd edn. McGraw-Hill, New York

Liu H, Lu H, Yao J (2001) Toward multi-database mining: identifying relevant databases. IEEE Trans Knowl Data Eng 13(4):541–553

Sayood K (2000) Introduction to data compression. Morgan Kaufmann, Los Altos

Su K, Huang H, Wu X, Zhang S (2006) A logical framework for identifying quality knowledge from different data sources. Decis Support Syst 42(3):1673–1683

Tan P-N, Kumar V, Srivastava J (2002) Selecting the right interestingness measure for association patterns. In: Proceedings of SIGKDD conference, pp 32–41

Wu X, Wu Y, Wang Y, Li Y (2005a) Privacy-aware market basket data set generation: a feasible approach for inverse frequent set mining. In: Proceedings of SIAM international conference on data mining, pp 103–114

Wu X, Zhang C, Zhang S (2005b) Database classification for multi-database mining. Inf Syst 30 (1):71–88

Yang W, Huang S (2008) Data privacy protection in multi-party clustering. Data Knowl Eng 67 (1):185–199

Yin X, Han J (2005) Efficient classification from multiple heterogeneous databases. In: Proceedings of 9th European conference on principles and practice of knowledge discovery in databases, pp 404–416

Yin X, Yang J, Yu PS, Han J (2006) Efficient classification across multiple database relations: a crossmine approach. IEEE Trans Knowl Data Eng 18(6):770–783

Zhang S (2002) Knowledge discovery in multi-databases by analyzing local instances, Ph.D. thesis, Deakin University

Zhang S, Wu X, Zhang C (2003) Multi-database mining. IEEE Comput Intell Bull 2(1):5–13

Zhang T, Ramakrishnan R, Livny M (1997) BIRCH: a new data clustering algorithm and its applications. Data Min Knowl Disc 1(2):141–182

Chapter 16
Enhancing Quality of Patterns in Multiple Related Databases

Multi-database mining using local pattern analysis could be considered as an approximate method of mining multiple large databases. Assuming this point of view, it might be required to enhance the quality of knowledge synthesized from multiple databases. Also, many decision-making applications are directly based on the available local patterns present in different databases. The quality of synthesized knowledge/decision based on local patterns present in different databases could be enhanced by incorporating more local patterns in the knowledge synthesizing/ processing activities. Thus, the available local patterns play a crucial role in building efficient multi-database mining applications. We represent patterns in a condensed form by employing a so-called ACP (antecedent-consequent pair) coding. It allows one to consider more local patterns by lowering further the user-defined characteristics of discovered patterns, like minimum support and minimum confidence. The ACP coding enables more local patterns participate in the knowledge synthesizing/processing activities and thus the quality of synthesized knowledge based on local patterns becomes enhanced significantly with regard to the synthesizing algorithm and required computing resources. To secure a convenient access to association rule, we introduce an index structure. We demonstrate that ACP coding represents rulebases by making use of the least amount of storage space in comparison to any other rulebase representation technique. Furthermore a technique for storing rulebases in the secondary storage is presented.

16.1 Introduction

In Chaps. 6, 10 and 12, we have discussed how to improve multi-database mining by adopting different mining techniques. Also, we have learnt that a single multi-database mining technique might not be sufficient in all situations. Chapters 9 and 10 present different variations of multi-database mining using local pattern analysis. Multi-database mining using local pattern analysis could be considered as an approximate method of mining multiple large databases. In this chapter, we employ a coding, referred to as antecedent-consequent pair (ACP) coding, to improve the

© Springer International Publishing Switzerland 2015   333
A. Adhikari and J. Adhikari, *Advances in Knowledge Discovery in Databases*,
Intelligent Systems Reference Library 79, DOI 10.1007/978-3-319-13212-9_16

quality of synthesized knowledge coming from multi-database mining. The ACP coding enables an efficient storage for association rules in multiple databases space. One could extract knowledge of better quality by storing more association rules in the main memory. In this way, applications dealing with association rules in multiple databases become more efficient.

Consider a multi-branch company that operates at different locations. Each branch generates a large database and subsequently we have to deal with multiple large databases. In particular, the company might be interested in identifying the global association rules in the union of all databases. Let $X \rightarrow Y$ be an association rule extracted from a few databases. Then local pattern analysis might return approximate association rule $X \rightarrow Y$ in the union of all databases, since the association rule might not get extracted from all the databases. As the higher number of data sources report the association rule, the quality of synthesized association rule gets elevated. We discuss how to enhance the quality of synthesized association rules in multiple databases.

Many multi-database mining applications often handle a large number of patterns. In multi-database mining applications, local patterns could be used in two ways. In the first category of applications, global patterns are synthesized from local patterns (Wu and Zhang 2003; Zhang et al. 2004). Synthesized global patterns could be used in various decision-making problems. In the second category of applications, various decisions are taken based on the local patterns present in different databases (Adhikari and Rao 2008; Wu et al. 2005). Thus, the available local patterns could play an important role in finding a solution to a problem. For a problem positioned in the first category, the quality of a global pattern is influenced by the pattern synthesizing algorithm and the locally available patterns. Also, we observe that a global pattern synthesized from local patterns might be approximate. For a given pattern synthesizing algorithm, one could enhance the quality of synthesized patterns by increasing the number of local patterns in a process of knowledge synthesis. For the problems pertinent to the second category, the quality of the resulting decision is implied by the quality of measure used in the decision-making process. Again, the quality of measure is based on the correctness of the measure itself and the available local patterns. For the purpose of database clustering, Wu et al. (2005) have proposed two such measures expressing similarity between two databases. For a given measure of decision-making, one could enhance the quality of decision by increasing the number of local patterns in the decision making process. In other words, the number of available local patterns plays a crucial role in building efficient multi-database mining applications. One could increase the number of local patterns by lowering the user-defined inputs, such as minimum support and minimum confidence. More patterns could be stored in main memory by applying a space efficient pattern base representation technique. In this chapter, we present the ACP coding (Adhikari and Rao 2007) to represent a set of association rules present in different databases space.

As before, let D_i be the database corresponding to the ith branch of the company, $i = 1, 2, \ldots, n$, while D stands for the union of these databases. The data mining model adopted in this chapter for association rule is the support (supp)-confidence (conf) framework established by Agrawal et al. (1993). The set of association rules extracted from a database is called a *rulebase*. Before proceeding with the algorithmic details, let us introduce some useful notations. Let RB_i be the rulebase corresponding to database D_i at the *minimum support level* α and *minimum confidence level* β, $i = 1, 2, ..., n$. Also, let RB be the union of rulebases corresponding to different databases. Many interesting algorithms have been reported on mining association rules in a database (Agrawal and Srikant 1994; Han et al. 2000; Savasere et al. 1995). Let T be a technique for representing RB in main memory. Let φ and ψ denote the pattern synthesizing algorithm and computing resource used for a data mining application, respectively. Also, let $\xi(RB \mid T, \alpha, \beta, \varphi, \psi)$ denote the collection of synthesized patterns over RB at a given tuple $(T, \alpha, \beta, \varphi, \psi)$. The quality of synthesized patterns could be enhanced if the number of local patterns increases. Thus, the quality of $\xi(RB \mid T, \alpha_1, \beta_1, \varphi, \psi)$ is lower than the quality of $\xi(RB \mid T, \alpha_2, \beta_2, \varphi, \psi)$, if $\alpha_2 < \alpha_1$ and $\beta_2 < \beta_1$. Thus, the problem of enhancing the quality of synthesized patterns translates to the problem of designing a space-efficient technique for representing rulebases corresponding to different databases.

As the frequent itemsets are the natural form of compression for association rules, the following reasons motivate us to compress association rules rather than frequent itemsets. Firstly, applications dealing with the association rules could be developed efficiently. Secondly, a frequent itemset might not generate any association rule at a given minimum confidence.

In this chapter, we present a space efficient technique to represent RB in a main memory. Let $SP^T(RB \mid \alpha, \beta, \psi)$ and $SP^T_{min}(RB \mid \alpha, \beta, \psi)$ describe the amount of space (expressed in bits) and minimum amount of space (expressed in bits) consumed by RB using a rulebase representation technique T, respectively. We observe that a rulebase representation technique might not represent RB at its minimum level because of the random nature of the set of transactions contained in the database. In other words, a frequent itemset might not generate all the association rules. For example, the association rule $X \rightarrow Y$ might not get extracted from any one of the given databases, even if the itemset $\{X, Y\}$ is frequent in some databases. Thus $SP^T_{min}(RB \mid \alpha, \beta, \psi) \leq SP^T(RB \mid \alpha, \beta, \psi)$, for a given tuple (α, β, ψ), where $0 < \alpha \leq \beta \leq 1$. Let Γ be the set of all techniques for representing a set of association rules. We are interested in finding a technique $T_1 \in \Gamma$ for representing RB, such that $SP^{T_1}(RB \mid \alpha, \beta, \psi) \leq SP^T(RB \mid \alpha, \beta, \psi)$, for all $T \in \Gamma$. Let $SP_{min}(RB \mid \alpha, \beta, \psi) = $ minimum $\{SP^T_{min}(RB \mid \alpha, \beta, \psi): T \in \Gamma\}$. The efficiency of T for representing RB is evaluated by comparing $SP^T(RB \mid \alpha, \beta, \psi)$ with $SP_{min}(RB \mid \alpha, \beta, \psi)$. We would like to design an efficient rulebase representation technique T_1 such that $SP^{T_1}(RB \mid \alpha, \beta, \psi) \leq SP^T(RB \mid \alpha, \beta, \psi)$, for $T \in \Gamma$.

The study presented in this chapter is based on a collection of rulebases RB_i, $i = 1, 2, ..., n$. One could lower α and β further so that each RB_i represents the corresponding database reasonably well. The work is not concerned with mining branch databases. The coding presented in this chapter reduces RB significantly, so that the coded RB becomes available in the main memory during the execution of pattern processing/synthesizing algorithm. The benefits of coding RB are given as follows. Firstly, the quality of processed/synthesized knowledge gets enhanced, since the number of local association rules participate in the pattern processing/ synthesizing algorithm is higher. Secondly, the pattern processing/synthesizing algorithm could access all the local association rules conveniently, since coded *RB* becomes available in the main memory. This arrangement might be possible, since the coded *RB* is reasonably small. For the purpose of achieving latter benefit, we present an index structure to access the coded association rules. Finally, the coded RB and the corresponding index table could be stored in the secondary storage for the usage of different multi-database mining applications. The following issues are discussed here:

- We present the ACP coding, for representing rulebases corresponding to different databases space efficiently. It enables us to incorporate more association rules for synthesizing global patterns or decision-making activities.
- We present an index structure to access the coded association rules.
- We prove that the ACP coding represents *RB* using the least amount of storage space in comparison to any other rulebase representation technique.
- We present a technique for storing rulebases corresponding to different databases in the secondary storage.
- We conduct experiments to express the effectiveness of the proposed approach.

The chapter is organized as follows. In Sect. 16.2, we discuss related work. A simple coding, called SBV coding, for representing different rulebases is presented in Sect. 16.3. In Sect. 16.4, we present the ACP coding for representing rulebases space. Experimental results are covered in Sect. 16.5.

16.2 Related Work

Our objective is to enhance the quality of decisions induced by local association rules. To achieve this objective, we present ACP coding for reducing the storage space of rulebases corresponding to different databases. There are three approaches to reducing the amount of storage space of different rulebases. Firstly, one could devise a mining technique for reducing the number of association rules extracted from a database. Secondly, one could adopt a suitable data structure for reducing the storage space for representing association rules in main memory. Thirdly, one could devise a post-mining technique along with a suitable data structure for reducing storage space required for association rules in the main memory. The first and second approaches to reducing the storage space are normally followed during a data mining task. Here, we concentrate on the third approach to reduce the storage space of different rulebases.

While mining association rules, we observe that there may exist many redundant association rules in a database. Using the semantics based on the closure of the Galois connection (Fraleigh 1982), one could define a condensed representation of association rules (Pasquier et al. 2005). This representation is characterised by frequent closed itemsets and their generators (Zaki and Ogihara 1998). It contains the non-redundant association rules having minimal antecedent and maximal consequent. These rules are the most relevant since they are the most general non-redundant association rules. Mining association rule is iterative and interactive nature. The user has to refine his/her mining queries until he/she is satisfied with the discovered patterns. To support such an interactive process, an optimized sequence of queries is proposed by means of a cache that stores information from previous queries (Jeudy and Boulicaut 2002). The technique uses condensed representations like free and closed itemsets for both data mining and caching. A condensed representation of the frequent patterns called disjunction-free sets (Bykowski and Rigotti 2003), could be used to regenerate all the frequent patterns and their exact frequencies without any access to the original data. In what follows, we discuss work related to the second approach to reducing the storage space of different rulebases.

Shenoy et al. (2000) have proposed a vertical mining algorithm that applies some optimization techniques for mining frequent itemsets in a database. Coenen et al. (2004) have proposed two new structures for association rule mining, the so-called T-tree, and P-tree, together with associated algorithms. The T-tree offers significant advantages in terms of generation time, and storage requirements compared to hash tree structures (Agrawal and Srikant 1994). The P-tree offers significant pre-processing advantages in terms of generation time and storage requirements compared to the FP-tree (Han et al. 2000). The T-tree and P-tree data structures are useful during the mining of a database. At the top level, T-tree stores supports for 1-itemsets, the second level for 2-itemsets, and so on. In T-tree, each node is an object containing support and a reference to an array of child T-tree nodes. The implementation of this data structure could be optimised by storing levels in the tree in the form of arrays, thus reducing the number of links needed and providing indexing. P-tree is different from T-tree in some ways. The idea behind the construction of P-tree can be outlined as follows. At the first pass of scanning input data, the entire database is copied into a data structure, which maintains all the relevant aspects of the input, and then mines this structure. P-tree offers two advantages: (i) it merges the duplicated records and records with common leading substrings, thus reducing the storage and processing requirements, and (ii) it allows partial counts of the support for individual nodes within the tree to be accumulated effectively as the tree is constructed. The top level is comprised of an array of nodes, each index describing a 1-itemset, with child references to body P-tree nodes. Each node at the top level contains the following fields: (i) a field for the support, and (ii) a link to a body P-tree node. A body P-tree node contains the following fields: (i) a support field, (ii) an array of short integers for the itemset that the node represents, and (iii) child and sibling links to further P-tree nodes. T-tree and P-tree structures are not suitable for storing and accessing association rules. These structures do not provide explicit provisions for storing confidence and database identification of association rules in different databases. It is difficult to handle effectively association rules in different databases during post-mining of rulebases corresponding to different databases. Ananthanarayana et al. (2003) have proposed PC-tree to represent data completely and minimally in main memory. It is built by scanning database only once. It could be used to represent dynamic databases with the help of knowledge that is either static or dynamic. It is not suitable for storing and accessing association rules. Furthermore PC-tree lacks the capability of handling association rules in different databases during post-mining of rulebases corresponding to different databases.

The proposed work falls under the third category of solutions to reducing storage of different rulebases. It is useful for handling association rules effectively during post-mining of association rules in different databases. No work has been reported so far under this category.

In the context of mining good quality of knowledge from different data sources, Su et al. (2006) have proposed a framework for identifying trustworthy knowledge from external data sources. Such framework might not be useful in this context.

Zhang and Zaki (2002) have edited a study on various problems related to multi-database mining. Zhang (2002) studied various strategies for mining multiple databases. Kum et al. (2006) have presented an algorithm, ApproxMAP, to mine approximate sequential patterns, called *consensus patterns*, from large sequence databases in two steps. First, sequences are clustered by similarity. Then, consensus patterns are mined directly from each cluster through multiple alignments.

16.3 Simple Bit Vector (SBV) Coding

We need to process all the association rules in different local databases for synthesizing patterns, or decision-making applications. We use a tuple (ant, con, s, c) to represent an association rule in a symbolic manner, where $ant, con, s,$ and c represent antecedent, consequent, support and confidence of the association rule $ant \rightarrow con$, respectively. The following example serves as a pertinent illustration of this representation.

Example 16.1 A multi-branch company has four branches. Let $\alpha = 0.35$, and $\beta = 0.45$. The rulebases corresponding to these different databases are given below:

$RB_1 = \{(A, C, 1.0, 1.0), (C, A, 1.0, 1.0), (A, B, 0.42, 0.42), (B, A, 0.42, 0.74),$
$\quad\quad (B, C, 0.40, 0.71), (C, B, 0.40, 0.40), (A, BC, 0.36, 0.36),$
$\quad\quad (B, AC, 0.36, 0.64), (C, AB, 0.36, 0.36), (AB, C, 0.36, 0.74),$
$\quad\quad (AC, B, 0.36, 0.36), (BC, A, 0.36, 0.90)\}$
$RB_2 = \{(A, C, 0.67, 0.67), (C, A, 0.67, 1.0)\}$
$RB_3 = \{(A, C, 0.67, 0.67), (C, A, 0.67, 1.0), (A, E, 0.67, 0.67), (E, A, 0.67, 1.0)\}$
$RB_4 = \{(F, D, 0.75, 0.75), (D, F, 0.75, 1.0), (F, E, 0.50, 0.50), (E, F, 0.50, 1.0),$
$\quad\quad (F, H, 0.50, 0.50), (H, F, 0.50, 1.0)\}.$

One could conveniently represent an association rule using an object (or a record). A typical object representing an association rule consists of following attributes: database identification, number of items in the antecedent, items in the antecedent, number of items in the consequent, items in the consequent, support and confidence. We further calculate the space requirement for such an object by continuing Example 16.1.

Example 16.2 A typical compiler represents an integer and a real number using 4 and 8 bytes, respectively. An item could be considered as an integer. Consider the association rule $(A, BC, 0.36, 0.36)$ of RB_1. Each of the following components of an association rule could consume 4 bytes: database identification, number of items in the antecedent, item A, number of items in the consequent, item B, and item C. Support and confidence of an association rule could consume 8 bytes each. The association rule $(A, BC, 0.36, 0.36)$ of RB_1 thus consumes 40 bytes. The association rule $(A, C, 1.0, 1.0)$ of RB_1 could consume 36 bytes. Thus, the amount of space required to store four rulebases is equal to $(18 \times 36 + 6 \times 40)$ bytes, i.e. 7,104 bits. A technique without optimization (TWO) could consume 7,104 bits to represent these rulebases.

Let I be the set of all items in D. Let X, Y and Z be three itemsets such that Y, $Z \subseteq X$. Then $\{Y, Z\}$ forms a *2-itemset partition* of X if $Y \cup Z = X$, and $Y \cap Z = \phi$. We define *size* of itemset X as the number of items in X, denoted by $|X|$. Then, we have $2^{|X|}$ 2-itemset partitions of X. For example, $\{\{a\}, \{b, c\}\}$ is a 2-itemset partition of $\{a, b, c\}$. An association rule $Y \rightarrow Z$ corresponds to a 2-itemset partition of X, for $Y, Z \subseteq X$. The antecedent and consequent of an association rule are non-null. We arrive at the following lemma.

Lemma 16.1 *An itemset X can generate maximum $2^{|X|} - 2$ association rules for $|X| \geq 2$.*

Let there are 10 items. The number of itemsets using 10 items is 2^{10}. Thus, 10 bits would be enough to represent an itemset. The itemset ABC, i.e. $\{A, B, C\}$ could be represented by the bit combination 1110000000. 2-itemset partitions of ABC are $\{\phi, ABC\}$, $\{A, BC\}$, $\{B, AC\}$, $\{C, AB\}$, $\{AB, C\}$, $\{AC, B\}$, $\{BC, A\}$, and $\{ABC, \phi\}$. Number of 2-itemset partitions of a set containing 3 items is 2^3. Every 2-itemset partition corresponds to an association rule, except the partitions $\{\phi, ABC\}$ and $\{ABC, \phi\}$. For example, the partition $\{A, BC\}$ corresponds to the association rule $A \rightarrow BC$. Thus, 3 bits are sufficient to identify an association rule generated from ABC. If the number of items is large, then this method might take significant amount of memory space to represent itemsets and the association rules generated from the itemsets. Thus, this technique is not suitable to represent association rules in databases containing large number of items.

16.3.1 Dealing with Databases Containing Large Number of Items

We explain the crux of the SBV coding with the help of the following example.

Example 16.3 We continue here the discussion we started in Example 16.1. Let the number of items be 10,000. We need 14 bits to identify an item, since $2^{13} < 10{,}000 \le 2^{14}$. We assume that the support and confidence of an association rule are represented using 5 decimal digits. Thus support/confidence value 1.0 could be represented as 0.99999. We use 17-bit binary number to represent support/confidence, since $2^{16} < 99{,}999 \le 2^{17}$.

Let us consider the association rule $(A, BC, 0.36107, 0.\,36107)$ of RB_1. There are 4 databases viz., D_1, D_2, D_3, and D_4. We need 2 bits to identify a database, since $2^1 < 4 \le 2^2$. Also 4 bits are enough to represent the number of items in an association rule. We place bit 1 at the beginning of binary representation of an item, if it appears in the antecedent of the association rule. We use bit 0 at the beginning of binary representation of an item, if it appears in the consequent of the association rule. Using this arrangement, the lengths of the antecedent and consequent are not required to be stored. The following bit vector could represent the above association rule.

```
00001110000000000000001000000000000010
 1  2 3     4        5    6
00000000000011010001101010001011 01000110100001011
 7     8                9                    10
```

The components of above bit vector are explained below.

Component 1 represents the first database (i.e., D_1)
Component 2 represents the number of items in the association rule (i.e., 3)
Component 3 (i.e., bit 1) implies that the current item (i.e., item A) belongs to antecedent
Component 4 represents item A (i.e., item number 1)
Component 5 (i.e., bit 0) implies that the current item (i.e., item B) belongs to consequent
Component 6 represents item B (i.e., item number 2)
Component 7 (i.e., bit 0) implies that the current item (i.e., item C) belongs to consequent
Component 8 represents item C (i.e., item number 3)
Component 9 represents support of association rule
Component 10 represents confidence of association rule

The storage space required for an association rule containing two items and three items are 70 and 85 bits, respectively. Therefore, the amount of storage space required to represent different rulebases is equal to $(18 \times 70 + 6 \times 85)$ bits, i.e., 1,770 bits. A technique without optimization could consume 7,104 bits (as mentioned in Example 16.2) to represent the same structure. We note that SBV coding significantly reduces the amount of storage space for representing different rulebases.

In the following section, we consider a special case of bit vector coding. It optimizes the storage space for representing different rulebases which is based on the fact that many association rules have the same antecedent-consequent pair. Before we move on to the next section, we consider the following lemma.

Lemma 16.2 *Let there are p items. Let m be the minimum number of bits required to represent an item. Then, $m = \lceil \log_2(p) \rceil$.*

Proof We have $2^{m-1} < p \le 2^m$, for an integer m. Thus we get $m < \log_2(p) + 1$, and $\log_2(p) \le m$, since $\log_k(x)$ is a monotonically increasing function of x, for $k > 1$. Combining these two inequalities we obtain $\log_2(p) \le m < \log_2(p) + 1$.   □

16.4   Antecedent-Consequent Pair (ACP) Coding

The central office generates sets of frequent itemsets from different rulebases. Let FIS_i be the set of frequent itemsets generated from RB_i, $i = 1, 2, ..., n$. Also let FIS denote the union of all frequent itemsets being reported from different databases. In a symbolic way, we denote a frequent itemset as a pair (itemset, support). The association rules $(F, D, 0.75, 0.75)$ and $(D, F, 0.75, 1.0)$ of RB_4 generate the following frequent itemsets: $(D, 0.75)$, $(F, 1.0)$ and $(DF, 0.75)$. In the following example, we generate FIS_i, for $i = 1, 2, ..., n$.

Example 16.4 Continuing Example 16.1, the sets of frequent itemsets generated by the central office comes as follows:

$$FIS_1 = \{(A, 1.0), (C, 1.0), (B, 0.57), (AC, 1.0), (AB, 0.42),$$
$$(BC, 0.40), (ABC, 0.36)\}$$
$$FIS_2 = \{(A, 1.0), (C, 0.67), (AC, 0.67)\}$$
$$FIS_3 = \{(A, 1.0), (C, 0.67), (E, 0.67), (AC, 0.67), (AE, 0.67)\}$$
$$FIS_4 = \{(D, 0.75), (E, 0.50), (F, 1.0), (H, 0.50), (DF, 0.75),$$
$$(EF, 0.50), (FH, 0.50)\}.$$

The ACP coding is a special case of bit vector coding, where antecedent-consequent pairs of the associations rules are coded in a specific order. The ACP coding is lossless (Sayood 2000) and similar to the Huffman coding (Huffman 1952). The ACP coding and the Huffman coding are not the same, in the sense that an ACP code may be a prefix of another ACP code. Then a question arises: how does a search procedure detect antecedent-consequent pair of an association rule correctly? We arrive at the answer to this question in Sect. 16.4.1.

Let X be a frequent itemset generated from an association rule. Also, let $f(X)$ be the number of rulebases that generate itemset X. Furthermore let $f_i(X) = 1$, if X is extracted from the ith database, and $f_i(X) = 0$, otherwise; for $i = 1, 2, ..., n$. Then,

Table 16.1 Sorted frequent itemsets of size greater than or equal to 2

X	AC	AB	AE	BC	DF	EF	FH	ABC
$f(X)$	3	1	1	1	1	1	1	1

$f(X) \leq \sum_{i=1}^{n} f_i(X)$. The central office sorts the frequent itemsets X using $|X|$ as the primary key and $f(X)$ as the secondary key, for $X \in FIS$ and $|X| \geq 2$. Initially, the itemsets are sorted on size in non-decreasing order. Then the itemsets of the same size are sorted on $f(X)$ in non-increasing order. If $f(X)$ is high then the number of association rules generated from X is expected to be high. Therefore, we represent antecedent-consequent pair of such an association rule using a code of smaller size. We explain the essence of this coding with the help of Example 16.5.

Example 16.5 We continue here the discussion of Example 16.4. We sort all the frequent itemsets of size greater than or equal to 2. Sorted frequent itemsets are presented in Table 16.1.

The coding process is described as follows. Find an itemset that has a maximal f-value. Itemset AC has the maximum f-value. We code AC as 0. The maximum number of association rules could be generated from AC is two. Thus we code association rules $A \rightarrow C$ and $C \rightarrow A$ as 0 and 1, respectively. Now, 1-digit codes are no more available. Then we find an itemset that has a second maximal f-value. We choose AB. We could have chosen any itemset from $\{AB, AE, BC, DF, EF, FH\}$, since every itemset in the set has the same size and the same f-value. We code AB as 00. The maximum number of association rules could be generated from AB is two. Thus we code the association rules $A \rightarrow B$ and $B \rightarrow A$ as 00 and 01, respectively. We follow in the same way and code the association rules $A \rightarrow E$ and $E \rightarrow A$ as 10 and 11, respectively. Now, we have constructed 2-digit codes. Finally, we choose ABC. We code ABC as 0000. The association rules $A \rightarrow BC$, $B \rightarrow AC$, $C \rightarrow AB$, $AB \rightarrow C$, $AC \rightarrow B$, and $BC \rightarrow A$ get coded as 0000, 0001, 0010, 0011, 0100, and 0101, respectively. Each frequent itemset receives a code. We call it an *itemset code*. Also, antecedent-consequent pair of an association rule is assigned a code. We call it a *rule code*.

Now an association rule could be represented in the main memory using the following components: database identification number, ACP code, support and confidence. Let n be the number of databases. Then we have $2^{k-1} < n \leq 2^k$, for an integer k. Thus, we need k bits to represent the database identification number. We represent support/confidence using p decimal digits. If we represent a fraction f using an integer d while f is given through the formula: $f = d \times 10^{-p}$. We represent support/confidence by storing the corresponding integer. The following lemma determines the minimum number of binary digits required to store a decimal number.

Lemma 16.3 *A p-digit decimal number can be represented by a$\lceil p \times \log_2 10 \rceil$-digit binary number.*

Proof Let t be the minimum number of binary digits required to represent a p-digit decimal number x. Then we have $x < 10^p < 2^t$. So, $t > p \times \log_2 10$, since $\log_k(y)$ is a monotonically increasing function of y, for $k > 1$. Thus the minimum integer t for which $x < 2^t$ is true is given as $\lceil p \times \log_2 10 \rceil$.   □

The following lemma specifies the minimum amount of storage space required to represent *RB* under some conditions.

Lemma 16.4 *Let M be the number of association rules having distinct antecedent-consequent pairs among N association rules extracted from n databases, where $2^{m-1} < M \leq 2^m$, and $2^{p-1} < n \leq 2^p$, for some positive integers m and p. Suppose the support and confidence of an association rule are represented by a fractions containing k digits after the decimal point. Assume that a frequent itemset X generates all possible association rules, for $X \in FIS$, and $|X| \geq 2$. Then the minimum amount of storage space required to represent RB in the main memory is given as follows.*

$$SP^{ACP\ coding}_{min,\ main}(RB\,|\alpha,\ \beta,\ \psi) = M \times (m-1) - 2 \times (2^{m-1} - m)$$
$$+ N \times (p + 2 \times \lceil k \times \log_2 10 \rceil)\ bits,\quad if\ M < 2^m - 2;\ and$$
$$SP^{ACP\ coding}_{min,\ main}(RB\,|\alpha,\ \beta,\ \psi) = M \times m - 2 \times (2^m - m - 1)$$
$$+ N \times (p + 2 \times \lceil k \times \log_2 10 \rceil)\ bits,\quad otherwise.$$

Proof p bits are required to identify a database. The amount of memory required to represent database identifiers of N association rules is equal to $P = N \times p$ bits. The minimum amount of memory required to represent both support and confidence of N association rules is equal to $Q = N \times 2\lceil k \times \log_2 10 \rceil R$ bits (as shown in Lemma 16.3). Let be the minimum amount of memory required to represent ACPs of M association rules. The expression R could be obtained from the fact that 2^1 ACPs are of length 1, 2^2 ACPs are of length 2, and so on. The expression of R is given as follows.

$$R = \sum_{i=1}^{m-2} i \times 2^i + (m-1) \times \left(M - \sum_{i=1}^{m-2} 2^i \right) bits,\ if\ \left(M - \sum_{i=1}^{m-2} 2^i \right) < 2^{m-1};\ and$$
$$R = \sum_{i=1}^{m-1} i \times 2^i + m \times \left(M - \sum_{i=1}^{m-1} 2^i \right) bits,\ if\ \left(M - \sum_{i=1}^{m-2} 2^i \right) \geq 2^{m-1}. \quad (16.1)$$

R assumes second form of expression for a few cases. For example, if ($M = 15$) then R assumes second form of expression. The ACP codes are given as follows: 0, 1, 00, 01, 10, 11, 000, 001, 010, 011, 100, 101, 110, 111, 0000. Then, the minimum amount of storage space required to represent *RB* is equal to $(P + Q + R)$ bits. Now,

$\sum_{i=1}^{m-2} 2^i = 2^{m-1} - 2$, and $\sum_{i=1}^{m-2} i \times 2^i = (m-3) \times 2^{m-1} + 2$. Thus the lemma follows. $\qquad\square$

In the following example, we calculate the amount of storage space for representing rulebases of Example 16.1.

Example 16.6 The discussion of Example 16.1 is continued here. The number of association rules in *RB* is 24. With reference to Lemma 16.4, we have $N = 24$, $M = 20$, and $n = 4$. Thus, $m = 5$, and $p = 2$. Assume that the support and confidence of an association rule are represented by fractions containing 5 decimal digits.

Thus $k = 5$. Then, the minimum amount of storage space required to represent *RB* is 922 bits.

The ACP coding may assign some codes for which there exists no associated rule. Let *ABC* be a frequent itemset extracted from some databases. Assume that the association rule $AC \rightarrow B$ is not extracted from any database that extracts *ABC*. Let the itemset code corresponding to *ABC* is 0000. Then the ACP code for $AC \rightarrow B$ is 0100, i.e., the 4th association rule generated from *ABC*. Therefore the ACP coding does not always store rulebases at the minimum level.

All rule codes are the ACP codes. But, the converse statement is not true. Some ACP codes do not have assigned association rules, since the assigned association rules are not extracted from any one of the given databases. An ACP code *X* is *empty* if *X* is not a rule code.

Lemma 16.5 *Let $X \in FIS$ such that $|X| \geq 2$. We assume that X generates at least one association rule. Let m (≥ 2) be the maximum size of a frequent itemset in FIS. Let n_i be the number of distinct frequent itemsets in FIS of size i, i = 2, 3, ..., m. Then the maximum number of empty ACP codes is equal to $\sum_{i=2}^{m} (2^i - 3) \times n_i$.*

Proof In the extreme case, only one association rule is generated for each frequent itemset *X* in *FIS*, such that $|X| \geq 2$. Using Lemma 16.1, one can note that a frequent itemset *X* could generate maximum $2^{|X|} - 2$ association rules. In such a situation, $2^{|X|} - 3$ ACP codes are empty for *X*. Thus, the maximum number of empty ACP codes for the frequent itemsets of size *i* is equal to $(2^i - 3) \times n_i$. Hence the result follows. $\qquad\square$

To search an association rule we maintain all the itemsets in an index table along with their itemset codes such that the size of an itemset is greater than one. We generate rule codes of the association rules from the corresponding itemset code. In Sect. 16.4.1, we discuss a procedure for constructing index table and accessing mechanism for the association rules.

Table 16.2 Index table for searching an association rule

Itemset	AC	AB	AE	BC	DF	EF	FH	ABC
Code	0	00	10	000	010	100	110	0000

16.4.1 Indexing Rule Codes

An index table contains the frequent itemsets of size greater than one and the corresponding itemset codes. These frequent itemsets are generated from different rulebases. Example 16.7, being a continuation of Example 16.5, illustrates the procedure of searching an association rule in the index table.

Example 16.7 Here we show how to construct an index table, Table 16.2.

The itemset code corresponding to AC is 0. The itemset code 0 corresponds to the set of association rules $\{A \rightarrow C, C \rightarrow A\}$. We would like to discuss the procedure for searching an association rule in the index table. Suppose we wish to search for the association rule corresponding to rule code 111. We apply binary search technique to find code 111. The binary search technique is based on the length of an itemset code. The search might end up at the fourth cell containing itemset code 000. Now, we apply sequential search towards the right side of the forth cell, since $value(000) < value(111)$. We find that 111 is not present in the index table. But, the code 111 is positioned in-between 110 and 0000, since $|111| < |0000|$ and $value(111) > value(110)$. We define $value$ of a code ω as the numerical value of the code, i.e. $value(\omega) = (\omega)_{10}$. For example, $value\,(010) = 2$. Thus, the sequential search stops at the cell containing itemset code 110. In general, for a rule code ω, we get a consecutive pair of itemset codes $(code_1, code_2)$ in the index table, such that $code_1 \leq \omega < code_2$. Then $code_1$ is the desired itemset code. Let Y be the desired itemset corresponding to the rule code ω. Then ω corresponds to an association rule generated by Y. Thus, the itemset code corresponding to the rule code 111 is 110. The frequent itemset corresponding to itemset code 110 is FH. Thus, the association rule corresponding rule code 111 is $H \rightarrow F$.

Initially, the binary search procedure finds an itemset code of desired length. Then it moves forward or backward sequentially till we get the desired itemset code. The algorithm for searching an itemset code is shown below.

Algorithm 16.1 Search for the itemset code corresponding to a rule code in the index table.
procedure *itemset-code-search* (ω, T, i, j)
Inputs:
ω: rule code (an ACP code)
T: index table
i: start index
j: end index
Outputs:
Index of the itemset code corresponding to ω
01: $x = |\omega|$;
02: $k = binary\text{-}search$ (x, T, i, j);
03: **if** ($value(\omega) \geq value((T(k).code))$ **then**
04:      $q = forward\text{-}sequential\text{-}search$ (ω, T, $k + 1$, j);
05: **else**
06:      $q = backward\text{-}sequential\text{-}search$ (ω, T, $k - 1$, i);
07: **end if**
08: **return**(q);
end procedure

The above algorithm is described as follows. The algorithm *itemset-code-search* (Adhikari and Rao 2007) searches index table T between the ith and jth cells and returns the index of the itemset code corresponding to the rule code ω. The procedure *binary-search* returns an integer k corresponding to rule code ω. If *value* (ω) $\geq$ $value(T(k).code)$ then we search sequentially in T from index ($k + 1$) to j. Otherwise, we search sequentially in T from index ($k - 1$) down to i. Let there are m cells in the index table. Then binary search requires maximum $\lfloor \log_2(m) \rfloor + 1$ comparisons (Knuth 1973). The sequential search makes $O(1)$ comparison, since codes ω and $T(k).code$ are close and the search is performed only once. Therefore, algorithm *itemset-code-search* takes $O(\log(m))$ time.

Now, we need to find the association rule generated from the itemset corresponding to the itemset code returned by algorithm *itemset-code-search*. We consider a certain example to illustrate the procedure for identifying association rule for a given rule code. Let us consider the rule code 0100. Using the above technique, we determine that 0000 is the itemset code corresponding to rule code 0100. The itemset corresponding to the itemset code 0000 is *ABC*. The association rules generated from itemset *ABC* could be numbered as follows: 0th association rule (i.e., $A \rightarrow BC$) has rule code 0000, 1th association rule (i.e., $B \rightarrow AC$) has rule code 0001, and so on. Proceeding in this way, we find that the 4th association rule (i.e., $AC \rightarrow B$) has rule code 0100.

We now find the association rule number corresponding to rule code ω. Let X be the itemset corresponding to rule code ω, and v be the itemset code corresponding to X. Let $RB(X)$ be the set of all possible association rules generated by X. From Lemma 16.1, we have $|RB(X)| = 2^{|X|} - 2$, for $|X| \geq 2$. If $|v| = |\omega|$ then ω corresponds to ($\omega_{10} - v_{10}$)th association rule generated from X, where Y_{10} denote the decimal value corresponding to binary code Y. If $|v| < |\omega|$ then ω corresponds to ($2^{|v|} - \omega_{10} + v_{10}$)th association rule generated from X. In this case, $v = 0000$, $\omega = 0100$, and $X = ABC$. Thus ω corresponds to 4th association rule generated from X.

Algorithm 16.2 Find itemset and association rule number corresponding to a rule code.
procedure *rule-generation* (k, T, C, X)
Input:
k: index
T: index table
C: rule code (an ACP code)
Output:
Itemset X corresponding to C
Association rule number corresponding to C
01: **let** $X = T(k)$.itemset;
02: **if** ($|T(k).\text{code}| = |C|$) **then**
03:     **return** $(C_{10} - (T(k).\text{code})_{10})$;
04: **else**
05:     **return** $(2^{|T(k).\text{code}|} - (C)_{10} + (T(k).\text{code})_{10})$;
06: **end if**
end procedure

We assume that the algorithm *itemset-code-search* returns k as the index of the itemset code corresponding to rule code C. Using index table T and k, the algorithm *rule-generation* returns the rule number and the itemset corresponding to rule code C. The itemset is returned through argument X, and rule number is returned by the procedure.

The ACP coding maintains an index table in main memory. We show an example to verify that the amount of space occupied by a rulebase (including the overhead of indexing) is significantly less than that of other techniques. We determine an overhead of maintaining index table in the following situation.

Example 16.8 We refer here to Example 16.7. We encounter 8 frequent itemsets in the index table. Let there are 10,000 items in the given databases. Therefore 14 bits are required to identify an item. Thus an amount of storage space would require for AC and ABC are equal to $2 \times 14 = 28$ bits, and $3 \times 14 = 42$ bits, respectively. The size of index file is the size of itemsets plus the size of itemset codes. In this case, the index table consumes $(28 \times 7 + 42 \times 1) + 21$ bits, i.e., we encounter 259 bits. The total space required (including the overhead of indexing) to represent RB is equal to $(259 + 922)$ bits (as mentioned in Example 16.6) $= 1{,}181$ bits. Based on the running example, we compare the amounts of storage space required to represent RB using different rulebase representation techniques (see Table 16.3).

We observe that the ACP coding consumes the least amount of space to represent RB. Let $OI(T)$ be the overhead of maintaining index table using technique T. A technique without optimization (TWO) might not maintain index table separately. In this case, $OI(\text{TWO}) = 0$ bit. But, the ACP coding performs better than the TWO because ACP coding optimizes storage spaces for representing components of an association rule.

Table 16.3 Amounts of storage space required for representing *RB* using different rulebase representation techniques

Technique for representing *RB*	TWO	SBV	ACP
Amount of space (bits)	7,104	1,770	1,181

We describe here the data structures used in the algorithm for representing rulebases using ACP coding. A frequent itemset could be described by the following attributes: database identification, itemset and support. The frequent itemsets generated from RB_i are stored into array FIS_i, i = 1, 2, ..., n. We keep all the generated frequent itemsets into array *FIS*. Also, we have calculated f-value for every distinct frequent itemset X in *FIS* such that $|X| \geq 2$. The frequent itemsets and their f-values are stored into array *IS_Table*. We present below an algorithm (Adhikari and Rao 2007) for representing different rulebases using the ACP coding.

Algorithm 16.3 Represent rulebases using ACP coding.
procedure *ACP-coding* (*n*, *RB*)
Input:
n: number of databases
RB: union of rulebases
Output:
Coded association rules
01: **let** *FIS* = ϕ;
02: **for** *i* = 1 to *n* **do**
03:      read RB_i from secondary storage;
04:      generate FIS_i from RB_i;
05:      *FIS* = *FIS* $\cup$ FIS_i;
06: **end for**
07: **let** *j* = 1;
08: **let** *i* = 1;
09: **while** (*i* ≤ |*FIS*|) **do**
10:      **if** (|*FIS*(*i*).itemset| ≥ 2) **then**
11:          compute $f(X)$;
12:          *IS_Table*(*j*).itemset = *X*;
13:          *IS_Table*(*j*).$f(X)$ = $f(X)$;
14:          increase index *j* by 1;
15:          update index *i* for processing the next frequent itemset in *FIS*;
16:      **end if**
17: **end for**
18: sort itemsets in *IS_Table* using |*X*| as the primary key and $f(X)$ as the secondary key;
19: **for** *i* = 1 to |*IS_Table*| **do**
20:      *C* = ACP code of *IS_Table* (*i*).itemset;
21:      *T*(*i*) .itemset = *IS_Table*(*i*).itemset;
22:      *T*(*i*).code = *C*;
23: **end for**
end procedure

In lines 1–6, we have generated frequent itemsets from different rulebases and are stored them into array *FIS*. We compute f-value for every frequent itemset X and store it into *IS_Table* (lines 7–17), for $|X| \geq 2$. At line 18, we sort frequent itemsets in *IS_Table* for the purpose of coding. Index table T is constructed by using lines 19–23.

We calculate time complexities of different statements except the shaded statement of the above algorithm, since it involves reading data from secondary storage. Let *maximum* $\{|FIS_i|: 1 \leq i \leq n\}$ be p. Then the total number of itemsets is $O(n \times p)$. Therefore, lines 7–17 take $O(n \times p)$ time. Line 18 takes $O(n \times p \times \log(n \times p))$ time to sort $O(n \times p)$ itemsets. Lines 19–23 take $O(n \times p)$ time to construct the index table.

16.4.2 Storing Rulebases in Secondary Memory

An association rule could be stored in main memory using the following components: database identification, rule code, support, and confidence. Database identification, support and confidence could be stored using the method described in Sect. 16.3. Furthermore we need to maintain an index table in main memory to code/decode an association rule.

The rulebases corresponding to different databases could be stored in secondary memory using a bit sequential file F. The first line of F contains the number of databases. The second line of F contains the number of association rules in the first rulebase. The following lines of F contain the association rules in the first rulebase. After keeping all the association rules in the first rulebase, we keep number of association rules in the second rulebase, and the association rules in the second rulebase thereafter. We illustrate the proposed file structure using the following example.

Example 16.9 Assume that there are 3 databases D_1, D_2, and D_3. Let the number of association rules extracted from these databases be 3, 4, and 2, respectively. The coded rulebases could be stored as follows:

$$\langle 3 \rangle \langle \backslash n \rangle$$
$$\langle 3 \rangle \langle \backslash n \rangle$$
$$\langle r_{11} \rangle \langle s_{11} \rangle \langle c_{11} \rangle \langle \backslash n \rangle$$
$$\langle r_{12} \rangle \langle s_{12} \rangle \langle c_{12} \rangle \langle \backslash n \rangle$$
$$\langle r_{13} \rangle \langle s_{13} \rangle \langle c_{13} \rangle \langle \backslash n \rangle$$
$$\langle 4 \rangle \langle \backslash n \rangle$$
$$\langle r_{21} \rangle \langle s_{21} \rangle \langle c_{21} \rangle \langle \backslash n \rangle$$
$$\langle r_{22} \rangle \langle s_{22} \rangle \langle c_{22} \rangle \langle \backslash n \rangle$$
$$\langle r_{23} \rangle \langle s_{23} \rangle \langle c_{23} \rangle \langle \backslash n \rangle$$
$$\langle r_{24} \rangle \langle s_{24} \rangle \langle c_{24} \rangle \langle \backslash n \rangle$$
$$\langle 2 \rangle \langle \backslash n \rangle$$
$$\langle r_{31} \rangle \langle s_{31} \rangle \langle c_{31} \rangle \langle \backslash n \rangle$$
$$\langle r_{32} \rangle \langle s_{32} \rangle \langle c_{32} \rangle \langle \backslash n \rangle$$

'\n' stands for the new line character. While storing an association rule in the secondary memory, if it contains a bit combination as that of '\n', then we need to insert one more '\n' after the occurrence of '\n'. We need not store the database identification along with an association rule, since the ith set of association rules corresponds to the ith database, $i = 1, 2, 3$. The terms r_{ij}, s_{ij}, and c_{ij} denote the rule code, support, and confidence of jth association rule reported from ith database, respectively, $j = 1, 2, \ldots, |RB_i|$, and $i = 1, 2, 3$.

Lemma 16.6 *Let M be the number of association rules with distinct antecedent-consequent pairs among N association rules reported from n databases, where $2^{m-1} < M \leq 2^m$, for an integer m. Suppose the support and confidence of an association rule are represented by fractions containing k digits after the decimal point. Assume that a frequent itemset X in FIS generates all possible association rules, for $|X| \geq 2$. Then the minimum amount of storage space required to represent RB in secondary memory is given as follows.*

$$SP^{ACP \text{ coding}}_{\min, \text{ secondary}}(RB\,|\alpha, \beta) = 12 \times n + M \times (m-1) + N \times (2 \times \lceil k \times \log_2 10 \rceil + 8) - 2$$
$$\times \left(2^{m-1} - m\right) + 12 \text{ bits, if } M < 2^m - 2; \text{ and}$$

$$SP^{ACP \text{ coding}}_{\min, \text{ secondary}}(RB\,|\alpha, \beta) = 12 \times n + M \times m + N \times (2 \times \lceil k \times \log_2 10 \rceil + 8) - 2$$
$$\times \left(2^m - m - 1\right) + 12 \text{ bits, otherwise.}$$

Proof We do not need to store the database identification in the secondary storage, as the rulebases are stored sequentially one after another. A typical compiler represents '\n' and an integer value using 1 and 4 bytes, respectively. The amount of memory required to represent the new line characters is equal to $P = 8 \times (N + n + 1)$ bits. The amount of memory required to store the number of databases and the number of association rules of each rulebase is equal to $Q = 4 \times (n + 1)$ bits. The amount of memory required to represent both the support and confidence of N rules is equal to $R = N \times 2 \times \lceil k \times \log_2 10 \rceil$ bits (as mentioned in Lemma 16.3). Let S be the minimum amount of memory required to represent the ACPs of M rules. Then, $S = M \times (m - 1) - 2 \times (2^{m-1} - m)$ bits, if $M < 2^m - 2$, and $S = M \times m - 2 \times (2^m - m - 1)$ bits, otherwise (as mentioned in Lemma 16.4). Thus the minimum amount of storage space required to represent *RB* in the secondary memory is equal to $(P + Q + R + S)$ bits. □

16.4.3   Space Efficiency of Our Approach

The effectiveness of a rulebase representation technique requires to be validated by its storage efficiency. There are many ways one could define the storage efficiency of a rulebase representation technique. We use the following definition to measure the storage efficiency of a rulebase representation technique.

Definition 16.1 Let RB_i be the rulebase corresponding to database D_i at a given pair (α, β), $i = 1, 2, ..., n$. Let RB be the union of rulebases corresponding to different databases. The space efficiency of technique T for representing RB is defined as follows:

$$\varepsilon(T, RB|\alpha, \beta, \psi) = \frac{SP_{\min}(RB|\alpha, \beta, \psi)}{SP^T(RB|\alpha, \beta, \psi)}, \quad for\ T \in \Gamma$$

The symbols and notation have been specified in Sect. 16.1.

We note that $0 < \varepsilon \leq 1$. We say that a rulebase representation technique is good if the value of ε is close to 1. We show that ACP coding stores rulebases at higher level of efficiency than that of any other representation technique.

Lemma 16.7 *Let RB_i be the set of association rules extracted from database D_i at a given pair (α, β), for $i = 1, 2, ..., n$. Let RB be the union of rulebases corresponding to different databases. Also, let Γ be the set of all rulebase representation techniques. Then ε (ACP coding, $RB \mid \alpha, \beta, \psi) \geq \varepsilon(T, RB \mid \alpha, \beta, \psi)$, for $T \in \Gamma$.*

Proof We show that ACP coding stores RB using minimum storage space at a given pair (α, β). A local association rule has the following components: database identification, antecedent, consequent, support, and confidence. We classify the above components into the following three groups: {database identification}, {antecedent, consequent}, and {support, confidence}. Among these three groups, the item of group 1 is independent of the items of other groups. If there are n databases, we need a minimum of $\lceil \log_2 n \rceil$ bits to represent the item of group 1 (as shown in Lemma 16.2). Many association rules may have the same antecedent-consequent pair. If an antecedent-consequent pair appears in many association rules, then it receives a shorter code. Therefore the antecedent-consequent pair of association rule having highest frequency is represented by a code of smallest size. ACP code starts from 0, and then follows the sequence 1, 00, 01, 10, 11, 000, 001, ... Therefore, no other technique would provide sizes of codes shorter than them. Therefore, the items of group 2 are expressed minimally using ACP codes. Again, the items of group 3 are related with the items of group 2. Suppose we keep p digits after the decimal point for representing an item of group 3. Then the representation an item of group 3 becomes independent of the one present for items of group 2. We need minimum $2 \times \lceil p \times \log_2 10 \rceil$ bits to represent support and confidence of an association rule (as mentioned in Lemma 16.3).

Thus, *minimum {representation of an association rule} = minimum {representation of items of group 1 + representation of items of group 2 + representation of items of group 3} = minimum {representation of items of group 1} + minimum {representation of items of group 2} + minimum {representation of items of group 3}.*

Also, there will be an entry in the index table for the itemset corresponding to an association rule for the coding/decoding process.

Thus we have *minimum {representation of index table} = minimum {representation of itemsets + representation of codes}.*

If there are p items then an itemset of size k could be represented by $k \times \lceil \log_2(p) \rceil$ bits (as mentioned in Lemma 16.2). Also, ACP codes consume minimum space because of the way they have been designed.

Thus *Minimum {representation of index table} = minimum {representation of itemsets} + minimum {representation of codes}.*

Therefore *minimum {representation of rulebases} = $\sum_r$ {representation of association rule r using ACP coding} + representation of index table used in ACP coding.*

Hence the lemma follows.   □

Lemma 16.8 *Let RB_i be the set of association rules extracted from database D_i at a given pair (α, β), i = 1, 2, ..., n. Let RB be the union of rulebases corresponding to different databases. Then, $SP_{min}(RB \mid \alpha, \beta, \psi) = SP_{min}^{ACP\ coding}(RB \mid \alpha, \beta, \psi)$.*

Proof From Lemma 16.7, we conclude that ACP coding represents rulebases using lesser amount of storage space than that of any other technique. Thus, $SP_{min}^{ACP\ coding}(RB \mid \alpha, \beta, \psi) \leq SP_{min}^{T}(RB \mid \alpha, \beta, \psi)$, for $T \in \Gamma$. We observe that a rulebase representation technique T might not represent rulebases at its minimum level because of the random nature of the set of transactions contained in a database. In other words, a frequent itemset may not generate all the association rules in a database. For example, the association rule $X \to Y$ may not get extracted from some of the given databases, even though the itemset $\{X, Y\}$ is frequent in the remaining databases. If the ACP coding represents RB using minimum storage space then it would be the minimum representation of RB at a given tuple (α, β, ψ).   □

There are many ways one could define the quality of synthesized patterns. We define the quality of synthesized patterns as follows.

Definition 16.2 Let RB_i be the rulebase extracted from database D_i at a given pair (α, β), i = 1, 2, ..., n. Let RB be the union of rulebases corresponding to different databases. We represent RB using a rulebase representation technique T. Let $\xi(RB \mid T, \alpha, \beta, \varphi, \psi)$ denote the collection of synthesized patterns over RB at a given tuple $(T, \alpha, \beta, \varphi, \psi)$. We define quality of $\xi(RB \mid T, \alpha, \beta, \varphi, \psi)$ as $\varepsilon(T, RB \mid \alpha, \beta, \varphi, \psi)$. The symbols and notation have been specified in Sect. 16.1.

Also, $\varepsilon(ACP\ coding, RB, \alpha, \beta) \geq \varepsilon(T, RB, \alpha, \beta)$, for $T \in \Gamma$ (as mentioned in Lemma 16.7). Thus the quality of $\xi(RB \mid ACP\ coding, \alpha, \beta, \varphi, \psi) \geq$ quality of $\xi(RB \mid T, \alpha, \beta, \varphi, \psi)$, for $T \in \Gamma$.

16.5   Experiments

We have carried out several experiments to study the effectiveness of ACP coding. The following experiments are based on the transactional databases T10I4D100K (T_1) (Frequent Itemset Mining Dataset Repository 2004), and T40I10D100K (T_2)

(Frequent Itemset Mining Dataset Repository 2004). These databases were generated using synthetic database generator from IBM Almaden Quest research group. We present some characteristics of these databases in Table 16.4.

For the purpose of conducting the experiments, we divide each of these databases into 10 databases. We call these two sets of 10 databases as the input databases. The database T_i has been divided into 10 databases T_{ij} of size 10,000 transactions each, for $j = 0, 1, 2, ..., 9$, and $i = 1, 4$. We present the characteristics of the input databases in Table 16.5.

The results of mining input databases are given in Table 16.6. The notations used in the above tables are explained as follows. *NT*, *ALT*, *AFI* and *NI* stand for number of transactions, average length of a transaction, average frequency of an item, and number of items in the data source, respectively.

Table 16.4 Database characteristics

Database	*NT*	*ALT*	*AFI*	*NI*
T_1	1,00,000	11.10	1276.12	870
T_4	1,00,000	40.41	4310.52	942

Table 16.5 Input database characteristics

Database	*ALT*	*AFI*	*NI*
T_{10}	11.06	127.66	866
T_{11}	11.13	128.41	867
T_{12}	11.07	127.65	867
T_{13}	11.12	128.44	866
T_{14}	11.14	128.75	865
T_{15}	11.14	128.63	866
T_{16}	11.11	128.56	864
T_{17}	11.10	128.45	864
T_{18}	11.08	128.56	862
T_{19}	11.081	128.11	865
T_{40}	40.57	431.56	940
T_{41}	40.58	432.19	939
T_{42}	40.63	431.79	941
T_{43}	40.63	431.74	941
T_{44}	40.66	432.56	940
T_{45}	40.51	430.46	941
T_{46}	40.74	433.44	940
T_{47}	40.62	431.71	941
T_{48}	40.53	431.15	940
T_{49}	40.58	432.16	939

Table 16.6 Results of data mining

Database	α	β	N2IR	N3IR	NkIR (k > 3)
$\bigcup_{i=1}^{10} T_{1i}$	0.01	0.2	136	29	0
$\bigcup_{i=1}^{10} T_{4i}$	0.05	0.2	262	0	0

Table 16.7 Different rulebase representation techniques-comparative analysis

Database	SP (TWO) (bits)	SP (SBV) (bits)	OI (bits)	SP (ACP) (bits)	MSO (bits)	AC (SBV)	AC (ACP)
$\bigcup_{i=1}^{10} T_{1i}$	48,448	10,879	619	7,121	7,051	1.79640	1.17586
$\bigcup_{i=1}^{10} T_{4i}$	75,456	16,768	549	10,681	10,661	1.77778	1.13242

Table 16.8 Comparison among different rulebase representation techniques (contd.)

Database	ε(TWO)	ε(SBV)	ε(ACP)
$\bigcup_{i=1}^{10} T_{1i}$	0.14554	0.64813	0.99017
$\bigcup_{i=1}^{10} T_{4i}$	0.14129	0.63579	0.99813

In the above table, *NkIR* stands for the number of *k*-item association rules resulting from different databases, for $k \geq 2$. We present a comparison among different rulebase representation techniques (see Tables 16.7 and 16.8).

In the above tables, we use the following abbreviations: *SP* stands for storage space (including overhead of indexing), *MSO* denotes the minimum storage space for representing rulebases including the overhead of indexing, and *AC(T)* stands for amount of compression (bits/byte) using technique *T*. In Fig. 16.1, we compare different rulebase representation techniques at different levels of minimum support .

We have fixed the value β at 0.2 for all the experiments. The results show that the ACP coding stores rulebases most efficiently among different rulebase representation techniques. Also, we find that the SBV coding reduces the size of a rulebase considerably, but stores less efficiently than the ACP coding. This coding achieves maximum efficiency when the following two conditions are satisfied: (i) All the databases are of similar type and extract an identical set of association rules, and (ii) Each of the frequent itemsets of size greater than one generates all possible association rules.

Nelson (1996) studied data compression with the Burrows-Wheeler Transformation (BWT) (Burrows and Wheeler 1994). Experiments were carried out on 18 different files and average compression obtained by techniques using BWT and PKZIP are 2.41 and 2.64 bits/byte, respectively.

The results of Fig. 16.1a, b are carried out at 11 different pairs of values of pairs of (α, β). Using the ACP coding, we have obtained average compression 1.15014 and 1.12190 bits/bytes for the experiments referring to Fig. 16.1a, b, respectively.

Fig. 16.1 Storage efficiency of different rulebase representation techniques. **a** For association rules extracted from T_{1i}, for $i = 0, 1, ..., 9$. **b** For association rules extracted from T_{4i}, for $i = 0, 1, ..., 9$

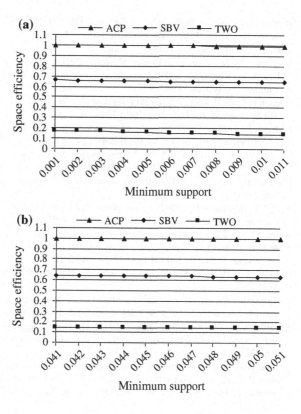

16.6 Conclusions

An efficient storage representation of a set of pattern bases could contribute to the foundations of a multi-database mining system. Based on them, many applications of data mining of global nature could be developed in an efficient manner as reported through experimental results presented in this chapter. Similar technique could be employed to store frequent itemsets in different databases.

References

Adhikari A, Rao PR (2007) Enhancing quality of knowledge synthesized from multi-database mining. Pattern Recogn Lett 28(16):2312–2324

Adhikari A, Rao PR (2008) Efficient clustering of databases induced by local patterns. Decis Support Syst 44(4):925–943

Agrawal R, Srikant R (1994) Fast algorithms for mining association rules. In: Proceedings of international conference on very large data bases, pp 487–499

Agrawal R, Imielinski T, Swami A (1993) Mining association rules between sets of items in large databases. In: Proceedings of ACM SIGMOD conference, pp 207–216

Ananthanarayana VS, Murty MN, Subramanian DK (2003) Tree *structure* for efficient *data mining* using rough sets. Pattern Recogn Lett 24(6):851–862

Burrows M, Wheeler DJ (1994) A block-sorting lossless data compression algorithm. DEC, Digital Systems Research Center, Research Report 124

Bykowski A, Rigotti C (2003) A condensed representation to find frequent patterns for efficient mining. Inf Syst 28(8):949–977

Coenen F, Leng P, Ahmed S (2004) Data structure for association rule mining: T-trees and P-trees. IEEE Trans Knowl Data Eng 16(6):774–778

Fraleigh JB (1982) A first course in abstract algebra, 3rd edn. Addison-Wesley, Reading, MA

Frequent Itemset Mining Dataset Repository (2004) http://fimi.cs.helsinki.fi/data

Han J, Pei J, Yiwen Y (2000) Mining frequent patterns without candidate generation. In: Proceedings of ACM SIGMOD conference on management of data, pp 1–12

Huffman DA (1952) A method for the construction of minimum redundancy codes. Proc IRE 40 (9):1098–1101

Jeudy B, Boulicaut JF (2002) Using condensed representations for interactive association rule mining. Proceedings of PKDD, LNAI 2431:225–236

Knuth DE (1973) The art of computer programming, vol 3. Addision-Wesley, Reading, MA

Kum H-C, Chang HC, Wang W (2006) Sequential pattern mining in multi-databases via multiple alignment. Data Min Knowl Disc 12(2–3):151–180

Nelson MR (1996) Data compression with the Burrows-Wheeler transformation. Dr. Dobb's J 9:46–50

Pasquier N, Taouil R, Bastide Y, Stumme G, Lakhal L (2005) Generating a condensed representation for association rules. J Intell Inf Syst 24(1):29–60

Savasere A, Omiecinski E, Navathe S (1995) An efficient algorithm for mining association rules in large databases. In: Proceedings of the 21st international conference on very large data bases, pp 432–443

Sayood K (2000) Introduction to data compression. Morgan Kaufmann, Los Altos

Shenoy P, Haritsa JR, Sudarshan S, Bhalotia G, Bawa M, Shah D (2000) Turbo-charging vertical mining of large databases. In: Proceedings of ACM SIGMOD conference on management of data, pp 22–33

Su K, Huang H, Wu X, Zhang S (2006) A logical framework for identifying quality knowledge from different data sources. Decis Support Syst 42(3):1673–1683

Wu X, Zhang S (2003) Synthesizing high-frequency rules from different data sources. IEEE Trans Knowl Data Eng 14(2):353–367

Wu X, Zhang C, Zhang S (2005) Database classification for multi-database mining. Inf Syst 30 (1):71–88

Zaki MJ, Ogihara M (1998) Theoretical foundations of association rules. In: Proceedings of the DMKD workshop on research issues in data mining and knowledge discovery, pp 71–78

Zhang S (2002) Knowledge discovery in multi-databases by analyzing local instances, Ph.D. thesis, Deakin University

Zhang S, Zaki MJ (2002) Mining multiple data sources: local pattern analysis. Data Mining and Knowledge Discovery, Springer, pp 121–125

Zhang C, Liu M, Nie W, Zhang S (2004) Identifying global exceptional patterns in multi-database mining. IEEE Comput Intell Bull 3(1):19–24

Chapter 17
Concluding Remarks

With the advancement of technologies, mass storage devices are now capable of storing more data. Also, they have become cheaper. Moreover varieties of data collection channels are now available in the market. Data mining is an emerging field of study, and has been applied to various domains. Some new patterns such as conditional pattern, arbitrary Boolean expression induced by itemset, type I global exceptional itemset and type II global exceptional itemset are discussed in this book. Also, some association measures viz., A_1, A_2, association rules induced by item and quantity, overall association between items, heavy association rule, exceptional association rule, $simi_1$ and $simi_2$ and influence of an item on another item, are reported in different chapters.

17.1 Mass Storage Systems

In these days technologies for mass storage have become cheaper, as well as many devices are available for data collection. Also, a system can hold many media units, disks or tapes, and use a robotic mechanism to load and unload drives. It offers automated access to data. Library capacities are huge and can be used in a networked computer environment. Newer storage technologies include helical scan magnetic tapes, and optical tapes. The storage densities, access speeds and costs for these new technologies are practical. Within a short time a terabyte size archive will be possible. To efficiently feed high-performance fabricated CPUs, high capacity and high-speed I/O buses have been developed for mass storage system. High Performance Parallel Interface (HIPPI) standard allows data transfer speeds of up to one hundred megabytes per second. This leads to high storage devices being directly attached to networks. Now innovative software can control the movement of data between storage devices, robotics, and networks. It also makes a large, fast, and flexible file system available. All these recent innovations results in acquiring, storing and accessing big data easily and efficiently[1] (Ranade 1991; Buyya et al. 2001). There is

---

[1] IEEE MSST: http://storageconference.org.

© Springer International Publishing Switzerland 2015
A. Adhikari and J. Adhikari, *Advances in Knowledge Discovery in Databases*,
Intelligent Systems Reference Library 79, DOI 10.1007/978-3-319-13212-9_17

no doubt that data analysis and pattern recognition in big data will be next wave of computing in the area of data mining. Let us learn more about some of the emerging areas of data mining.

17.2 Some Emerging Domains in KDD

Knowledge discovery in databases (KDD) is a well known area of research in data mining. It was introduced by Gregory Piatetsky-Shapiro in 1989, and since then it continues to grow. The application of different data mining and knowledge discovery techniques is found on different types of data such as uncertain data, social network data, sensor data, biological data, high dimensional data, imbalance data, big data, big data on small devices, privacy preserving data, text data, heterogeneous data, noisy data, outliers, and graph data. Aggarwal (2009) presents surveys by well known researchers in the field of uncertain databases. It contains the most recent models, algorithms, and applications in the field of uncertain data. Social networks provide an intuitive model of the relations between individuals in a social group. Mining and data analysis of social network data is gaining popularity. Memon et al. (2010) have edited a book on analysis of social network data and network related data analyses. With advancements of hardware technology we can have now cost effective devices such as miniaturized sensors, GPS enabled devices and pedometers to collect data. Sensor data are generated from varieties of applications. Aggarwal (2013) has edited a book on sensor data analytics. It presents different sensor data processing, challenges in analyzing sensor data and different research areas. Biologists are stepping up their efforts in understanding the biological processes that underlie disease pathways in the clinical contexts. This has resulted in a flood of biological and clinical data from genomic and protein sequences, DNA microarrays, protein interactions, biomedical images, to disease pathways and electronic health records. Li et al. (2014) have edited a book on handling such challenging data analysis problems, and demonstrate with real applications how biologists and clinical scientists can employ data mining to enable them to make meaningful observations and discoveries from a wide array of heterogeneous data from molecular biology to pharmaceutical and clinical domains. With the rapid development of computational biology and e-commerce applications, high-dimensional data becomes more and more powerful. Thus, it is an urgent problem of great importance when mining is performed on high-dimensional data. There are some challenges for mining data of high dimensions, the first one is the curse of dimensionality and the second one is the meaningfulness of the similarity measure in the high dimension space (Takashi and Jun 2012). Solving imbalanced learning problems is critical in numerous data-intensive networked systems, including surveillance, security, Internet, finance, biomedical, defense, and more. Due to the inherent complex characteristics of imbalanced data sets, learning from such data requires new understandings, principles, algorithms, and tools to transform vast amounts of raw data efficiently into information and knowledge representation. On these as aspects, He and Ma (2013) have edited recent research articles on various issues of imbalanced data. The field of data mining has made significant and far-reaching advances over the past three decades. Because of its potential power for solving complex problems, data mining has been successfully applied to diverse areas such as business, engineering, social media, and biological science (see Chu 2013). Owing to continuous advances in the computational power of handheld devices like smartphones and tablet computers, it has become possible to perform big data operations including modern data mining processes onboard on these small devices. A decade of research has proved the feasibility of what has been termed as mobile data mining, with a focus on one mobile device running data mining processes. For more details one could refer to the book authored by Gaber et al. (2014). Privacy-preserving data mining is an interesting area, especially applicable to multi-party data. Aggarwal and Yu (2008) present summary of the state-of-the-art accomplishments in the area of privacy-preserving data mining, discussing the most important algorithms, models, and applications in each direction. The computational linguistics community has viewed large text collections as a resource to be tapped in order to produce better text analysis algorithms. Hearst (1999) has outline recent ideas about how to pursue exploratory data analysis over text. Aggarwal and Zhai (2012) have edited a book on text mining focusing on social networks and data mining. This volume is contributed by leading international researchers and practitioners. With the rapid growth of computer networks and data storage technologies, the volume and diversity of data has become a major challenge for data exploitation. These large volumes of heterogeneous data types need to be accessed and treated in a uniform manner for richer information extraction. However, mining heterogeneous data types is complex because of the unique nature of the heterogeneous data types, different ways of representing heterogeneous data types, and particular complexity of combining heterogeneous data mining results. Bourennani (2011) has presented a unified vectorization method proposed for simultaneous processing, classification and clustering of heterogeneous data types, which allows a unified visualization of the data. Machine learning and data mining are inseparably connected with uncertainty. The observable data for learning is usually imprecise, incomplete or noisy. Qin and Tang (2014) introduce 'label semantics', a fuzzy-logic-based theory for modeling uncertainty. Several new data mining algorithms based on label semantics are proposed and tested on real-world datasets. We have reached in a state where it is impossible to include all the recent advancements of KDD. This book has attempted to highlight the recent contributions made to the following topics of KDD: data mining in a market basket database, data mining in time-stamped databases and data mining in multiple databases.

17.3 New Patterns and Their Recognitions

Discovery of new patterns remains an active area of research. As we encounter different data new patterns are likely to emerge. Thus, the characteristics of patterns are very much depended on the data. As a result, new techniques of data mining have emerged to deal with new and complex type of datasets, multiple large databases; classical patterns have appeared in specialized forms; new patterns and associations have also been reported subsequently. Let us revisit the chapters to find out new patterns that have been identified in this book.

The study of items in an itemset remains incomplete if we know only the supports and the association rules. These information are not sufficient to answer all types of queries about the itemset. Let us consider an itemset $\{a, b, c\}$, and a few queries based on this itemset as given below.

(i) Given the itemset $\{a, b, c\}$, find the support that a transaction contains item a but not items b and c. (ii) Given the itemset $\{a, b, c\}$, find the support that a transaction contains items a and b but not item c.

The answers to those queries are not immediately available. In general, given the itemset X if we wish to study the association among the items in Y with negation of items in X-Y, then such analysis is not immediately available from frequent itemsets and positive association rules, for $Y \subseteq X$. Such association analyses could be interesting, since the corresponding Boolean expressions could have high supports. Therefore, we need to mine such patterns for effective analyses of items in a frequent itemset. We introduce the concept of conditional pattern and present a framework for capturing such associations, see Chap. 2. (Adhikari and Rao 2008c).

In market basket data analysis, frequent itemsets can be considered as the most popular pattern. Thus, a detailed analysis of such patterns is always desirable. Finding a generalized type of pattern, called an arbitrary Boolean expression induced by a frequent itemset, based on frequent itemset remains an interesting issue. In Chap. 3 a framework has been designed for mining well formed Boolean expressions induced by frequent itemsets (Adhikari and Rao 2007a).

Analysis of sales series of items is an interesting topic. In view of performing this task, one could analyze the sales series for each item. In analyzing a sales series in-depth, it is evident that an existence of a notch might be an indication of a bigger notch. This represents an exceptionality of sales of an item. Based on such an exception, we define iceberg in time databases (see Chap. 7) (Adhikari et al. 2011b). An iceberg possesses two interesting characteristics given as follows: (i) It is a generalized notch having long height, and (ii) It is a generalized notch with large width. Although the concept of long height and large width are relative measures, a threshold based algorithm is presented to report icebergs in multiple time-stamped databases.

With the passage of time multiple databases are being encountered in different domains. Obviously data analysis with multiple related databases becomes an urgent need. Some patterns such as association rule (Agrawal et al. 1993) and conditional pattern (Adhikari and Rao 2008c) have already been reported. These types of patterns are also applicable to multiple related databases. But there are patterns that are specific to multiple databases. Such patterns are discussed in Chap. 8. There are two types of exceptional itemsets: type I global itemsets and type II global itemsets. Type I global exceptional itemset in multiple databases has high frequency but low support, where as type II global exceptional itemset in multiple databases has low frequency but high support. An algorithm is designed to mine type II global exceptional frequent itemsets (Adhikari 2012).

17.4 New Measures

In market basket data we deal with various types of items. In view of making different types analyses one may need to measure associations among items. In this book, we have introduced two such measures: A_1 and A_2 (Adhikari and Rao 2008b). Measure A_1 is the proportion of the number of transactions containing all the items of X and the number of transactions containing at least one of the items of X. The association among items of an itemset and the number of association rules generated from the itemset are positively correlated, provided the support of the itemset is high. If the association among items of an itemset is more then it is expected to generate more association rules and vice versa. A_2 measure is defined based on the following observations. A transaction in a database D provides the following information regarding association among items of X: (i) A transaction that contains all the items of X contributes maximum value towards overall association among items of X. We attach weight 1.0 to each such transaction. (ii) A transaction that contains k items of X contributes some value towards overall association among the items of X, for $2 \leq k \leq |X|$. We attach weight $k/|X|$ to each such transaction. (iii) A transaction that does not contain any item of X contributes no information regarding association among the items of X. (iv) A transaction that contains only one item of X contributes maximum value towards overall dispersion among the items of X. At a given X, we attach a weight to each transaction that contributes some value towards overall association among the items of X. More details and various properties about these two measures are given in Chap. 4.

Association rule mining is undoubtedly one of the most important aspects of data mining. Several intelligent algorithms (Agrawal and Srikant 1994; Han et al. 2000; Antonie and Zaïane 2004; Savasere et al. 1995) are proposed since Agrawal et al. (1993) proposed association rules. In many real-world data contain item as well as the quantity purchased. Therefore, the classical support-confidence framework would not be sufficient. Let *TIMT* be the type of a database such that a Transaction in the database might contain an Item Multiple Times. We introduced transaction frequency and database frequency in Chap. 5. Based on these concepts, three categories of framework for mining association rule are proposed (Adhikari and Rao 2008d).

Analysis of market basket data is very interesting issue, and it has many applications to the industries. In this respect, the variation of sales of an item over time needs to be analysed, and therefore, we present the notion of stability of an item (see Chap. 9). Stable items are useful in making numerous strategic decisions for a company. An algorithm is designed for clustering items in multiple databases based on degree of stability. Afterwards we have introduced the notion of best class in a cluster. An algorithm is designed to find the best class in a cluster (Adhikari et al. 2009).

Association rule mining is an interesting issue. Significant amount of work has been taken place in last two decades. In Chap. 10, some extreme types of association rules are identified. Heavy association rule, high-frequency association rule, low-frequency association rule, and exceptional association rule are specific to multi-database mining environment. An algorithm is presented to detect such types of association rules (Adhikari and Rao 2008a).

Several data analyses are being performed in market basket data. In some applications the association between two items becomes an immediate requirement. With help of support measures different types of associations are calculated. Chapter 12 explains these issues and measures like positive association, negative association and overall association between items are discussed (Adhikari et al. 2011a).

Influence of items on some other items might not be the same as the association between these sets of items. The overall influence between two itemsets in a large database could be negative as well as positive (Adhikari and Rao 2010). Thus, the concepts of negative and positive influences are introduced in Chap. 14. A measure of overall influence of an itemset over another itemset is presented. For the purpose of calculating overall influence, it has been expressed by supports of the itemsets.

It may be required to compare two databases to find their similarity. One could make meaningful comparisons using knowledge in the similar databases. In view of clustering similar databases two measures, viz., $simi_1$ and $simi_2$, are presented in Chap. 15 (Adhikari and Rao 2008e). Various properties of these two measures are shown. Also, it has been observed that similarity measure $simi_2$ exhibits higher discriminatory power than that of the similarity measure $simi_1$.

17.5 Algorithms and Techniques

In many cases mining multiple databases follows a different approach to classical mining technique (Agrawal and Srikant 1994; Han et al. 2000; Savasere et al. 1995). We highlighted the main reasons for following a different approach in Adhikari et al. (2010). In Chap. 6, we have presented a new technique, called pipelined feedback model, for mining multiple large databases (Adhikari et al. 2007). In this technique the databases are sorted by their size in increasing order. Let $D_1, D_2, \ldots, D_n$ be the sorted databases, and they are mined one at a time in this order. While mining D_i, all the patterns in D_{i-1} are mined irrespective of their values of interestingness measures, and some new patterns that satisfy user-defined threshold values of interestingness measures, $i = 2, 3, \ldots, n$. It has been observed that the synthesized patterns are more accurate than the synthesized patterns obtained by another technique.

The issue of clustering objects is a well known problem. Data analysis using multiple databases is a recent field of research (Adhikari et al. 2014). Chapter 11 presents a new algorithm for clustering local frequency items in multiple databases (Adhikari et al. 2008e). It has been shown that A_2 measure is effective in measuring association among items in a high frequency itemset in multiple databases. The algorithm uses A_2 measure for clustering items.

Multi-database mining is normally approximate in nature. Such approximations come due to a large size of local databases. When we encounter a problem dealing with selective items in multiple large databases, it may be possible to optimize the techniques. One such problem is dealt with in Chap. 12. It is problem of finding global patterns of select items in multiple databases. In this case each local branch of the organizations filters the transactions and relevant transactions are retained. Such filtered databases are finally amalgamated and true patterns are synthesized with the help of local patterns (Adhikari et al. 2011a).

The problem of periodic pattern mining can be categorized into two types. One is full periodic pattern mining, where every point in time granularity (Bettini et al. 2000) contributes to a cyclic behavior of the pattern. The other and more general one is called partial periodic pattern mining, which specifies the behavior of the pattern at some but not all points of time granularity in the database. In Chap. 13 we have presented an effective algorithm by improving the limitations of the existing algorithm for mining calendar-based periodic patterns proposed by Mahanta et al. (2005). A hash-based data structure is used to improve the space efficiency of the proposed algorithm. Overall, it has improved the average complexity of the algorithm.

Clustering similar objects is an important activity. Effective decisions can be made by using inherent knowledge in multiple databases. For purpose grouping similar databases a clustering algorithm is given in Chap. 15 (Adhikari and Rao 2008e). Then each group (class) could be mined to gain desired knowledge.

Enhancement of knowledge extraction is a crucial activity in a knowledge-based system. Most of the multi-database mining tasks are approximate in nature. Thus, knowledge enhancement in this domain becomes always desirable. Chapter 16 describes a coding scheme of storing association rules in different data sources (Adhikari and Rao 2007b). It allows to store more association rules in a given storage space. Therefore, the extracted knowledge become more informative, and the decisions based on the enhanced knowledge becomes more valid. Similar coding can also be done for frequent itemsets. Such an idea is provided in Chap. 15 (Adhikari and Rao 2008e). Thus, one can conclude that storing of more patterns results in more valid decisions.

17.6 Challenges Ahead

We have witnessed smarter technologies for collecting, storing, transmitting and accessing big data. Thus, we are in the era of handling big data. Recently, we have read an article posted by Banafa (2014) in Linkedin. It gives an idea of the current state of data analytics and the type of data we need to handle in these days. Handling a big size of data could be a major challenge to all areas of data mining. However, there are specific challenges that need to be addressed for mining a certain type of data.

In case of uncertain data, the major challenges are given as follows (Aggarwal 2009):

- Modeling of uncertain data: Here the underlying complexities can be captured while keeping the data useful for database management applications.
- Uncertain data management: One may wish to adapt traditional database management techniques such as join processing, query processing, indexing, and database integration.
- Uncertain data mining: It is critical to design data mining techniques that can take such uncertainty into account during the computations.

While early research on social network data focused on structural questions, recent work has extended this to consider the social processes such as on-line social systems related to communication, community formation, information-seeking and collective problem-solving, marketing, the spread of news, and the dynamics of popularity. There are a number of fundamental issues, however, for which we have relatively little understanding, including the extent to which the outcomes of these types of social processes are predictable from their early stages, the differences between properties of individuals and properties of aggregate populations in these types of data, and the extent to which similar social phenomena in different domains have uniform underlying explanations (Kleinberg 2007).

Data collection is a big challenge in the context of sensor processing because of the natural errors and incompleteness in the collection process. Sensors are often designed for applications which require real-time processing. This requires the design of efficient methods for stream processing. In many cases, it is critical to perform in-network processing, wherein the data is processed within the network itself, rather than at a centralized service. This needs effective design of distributed processing algorithms, wherein queries and other mining algorithm can be processed within the network in real time (Aggarwal 2013).

Many researchers believe that mining biological data continues to be an extremely important problem, both for data mining research and for biomedical sciences. An example of a research issue is how to apply data mining to HIV vaccine design. In molecular biology, many complex data mining tasks exist, which cannot be handled by standard data mining algorithms. These problems involve many different aspects, such as DNA, chemical properties, 3D structures, and functional properties (Yang and Wu 2006).

There are many challenges in mining high dimensional data. The first one is the curse of dimensionality. The complexity of many existing data mining algorithms is exponential with respect to the number of dimensions. With increasing dimensionality, these algorithms soon become computationally intractable and therefore inapplicable to many real applications. Secondly, the specificity of similarities between points in a high dimensional space diminishes. It was proven that, for any point in a high dimensional space, the expected gap between the Euclidean distance to the closest neighbor and that to the farthest point shrinks as the dimensionality grows. This phenomenon may render many data mining tasks (e.g., clustering) ineffective and fragile because the model becomes vulnerable to the presence of noise (Wang and Yang 2010).

Imbalanced data sets lead to a special case for classification problem where the class distribution is not uniform among the classes. Typically, they are composed by two classes: the majority (negative) class and the minority (positive) class. These type of sets pose a new challenging problem for data mining, since standard classification algorithms usually consider a balanced training set and this supposes a bias towards the majority class (He et al. 2013).

The data characteristics of the problems have grown from static to dynamic and spatiotemporal, complete to incomplete, and centralized to distributed, and grow in their scope and size. The effective integration of big data for decision-making also requires privacy preservation (Chu 2014).

In collaborative data mining in mobile computing environments there are many research issues. Construction of different agents such as agent miner, mobile agent resource discoverer and mobile agent decision maker and number of such agents are important design issues. Also, the architecture of pocket data mining plays an important role. Moreover, the combination of data mining techniques to be used is essential to ensure that the built models are of optimum accuracy (Stahl et al. 2010).

The process of privacy-preservation leads to loss of information for data mining purposes. This loss of information can also be considered as a loss of *utility* for data mining purposes. Also, there are specific challenges to a specific method. For example, the randomization method has some weaknesses, since it treats all records equally irrespective of their local density (Aggarwal and Yu 2008).

The major challenging issues in text mining arise from the complexity of a natural language itself. The natural language is not free from the ambiguity problem. Ambiguity means the capability of being understood in two or more possible senses or ways. One phrase or sentence can be interpreted in various ways, thus various meanings can be obtained. Named entity recognition describes an identification of entities in free text. One of the major problems in biological text mining is ambiguous protein names; one protein name may refer to multiple gene products. Thus, ambiguity is still the major problem in text mining applications (Jusoh and Alfawareh 2012).

Success in various application domains including sensor networks, social networks, and multimedia, has ushered in a new era of information explosion. Despite the diversity of these domains, data acquired by applications in these domains are often voluminous, heterogeneous and containing much uncertainty. They share several common characteristics, which impose new challenges to storing, integrating, and processing these data, especially in the context of data outsourcing and cloud computing. Some challenges in managing and mining large, heterogeneous data include the following. First, autonomous data acquisition gives rise to privacy and security issues. Therefore, data management and mining must be elastic and privacy-conscious. Second, data is often dynamic and the trend in the data is often unpredictable. This calls for efficient incremental or cumulative algorithms for data management and mining. Load balancing and other real-time technologies are also indispensable for the task. Third, data repositories are distributed. Thus, gathering, coordinating, and integrating heterogeneous data in data management and mining will face unprecedented challenges (Hu et al. 2011).

Removal of noise and detecting outliers are of similar activities. The presence of noise is due to imperfections in the data collection process or noise that consists of irrelevant or weakly relevant data objects. Two critical challenges typically associated with mining data streams are concept drift and data contamination (Xiong et al. 2006; Chu et al. 2004a, b; Chu and Zaniolo 2004c).

References

Adhikari A (2012) Synthesizing global exceptional patterns in different data sources. J Intell Syst 21(3):293–323

Adhikari A, Rao PR (2007a) A framework for synthesizing arbitrary boolean expressions Induced by frequent itemsets. In: Proceedings of 3rd Indian international conference on artificial intelligence, pp 5–23

Adhikari A, Rao PR (2007b) Enhancing quality of knowledge synthesized from multi-database mining. Pattern Recogn Lett 28(16):2312–2324

Adhikari A, Rao PR (2008a) Synthesizing heavy association rules from different real data sources. Pattern Recogn Lett 29(1):59–71

Adhikari A, Rao PR (2008b) Capturing association among items in a database. Data Knowl Eng 67(3):430–443

Adhikari A, Rao PR (2008c) Mining conditional patterns in a database. Pattern Recogn Lett 29 (10):1515–1523

Adhikari A, Rao PR (2008d) Association rules induced by item and quantity purchased. In: Haritsa JR, Kotagiri R, and Pudi V (eds) Proceedings of international conference on database systems for advance applications. LNCS, vol 4947. pp 478–485

Adhikari A, Rao PR (2008e) Efficient clustering of databases induced by local patterns. Decis Support Syst 44(4):925–943

Adhikari J, Rao PR (2010) Measuring influence of an item in a database over time. Pattern Recogn Lett 31(3):179–187

Adhikari A, Rao PR, Adhikari J (2007) Mining multiple large databases. In: Proceedings of the 10th international conference on information technology, pp 80–84

Adhikari J, Rao PR, Adhikari A (2009) Clustering items in different data sources induced by stability. Int Arab J Inf Technol 6(4):394–402

Adhikari A, Ramachandrarao P, Pedrycz W (2010) Developing multi-database mining applications. Springer, London

Adhikari A, Ramachandrarao P, Pedrycz W (2011a) Study of select items in different data sources by grouping. Knowl Inf Syst 27(1):23–43

Adhikari J, Rao PR, Pedrycz W (2011b) Mining icebergs in time-stamped databases. In: Proceedings of Indian international conferences on artificial intelligence, pp 639–658

Adhikari A, Adhikari J, Pedrycz W (2014) Data analysis and pattern recognition in multiple data sources. Springer, Switzerland

Aggarwal CC (ed) (2009) Managing and mining uncertain data. Springer, Heidelberg

Aggarwal CC (ed) (2013) Managing and mining sensor data. Springer, Heidelberg

Aggarwal CC, Yu PS (2008) Privacy-preserving data mining: models and algorithms. Springer, New York

Aggarwal CC, Zhai CX (2012) Mining text data. Springer, Heidelberg

Agrawal R, Srikant R (1994) Fast algorithms for mining association rules. In: Proceedings of 20th international conference on very large databases (VLDB), pp 487–499

Agrawal R, Imielinski T, Swami A (1993) Mining association rules between sets of items in large databases. In: Proceedings of ACM SIGMOD conference management of data, pp 207–216

Antonie M-L, Zaïane OR (2004) Mining positive and negative association rules: an approach for confined rules. In: Proceedings of knowledge discovery in databases (PKDD), pp 27–38

Banafa A (2014) The future of big data and analytics. http://www.linkedin.com

Bettini C, Jajodia S, Wang SX (2000) Time granularities in databases, data mining and temporal reasoning. Springer, Heidelberg

Bourennani F (2011) Heterogeneous data mining. VDM, Saarbrücken

Buyya R, Cortes T, Jin H (2001) High performance mass storage and parallel I/O: technologies and applications. Wiley, New Jersey

Chu WW (2013) Data mining and knowledge discovery for big data: methodologies, challenge and opportunities. Springer, Heidelberg

Chu WW (ed) (2014) Data mining and knowledge discovery for big data. Springer, Heidelberg

Chu F, Zaniolo C (2004c) Fast and light boosting for adaptive mining of data streams. In: Proceedings on advances in knowledge discovery and data mining (PAKDD). Springer, Heidelberg, pp 282–292

Chu F, Wang Y, Zaniolo C (2004a) An adaptive learning approach for noisy data streams. In: IEEE international conference on data mining (ICDM), pp 351–354

Chu F, Wang Y, Zaniolo C (2004b) Mining noisy data streams via a discriminative model. In: Discovery science. Springer, Heidelberg, pp 47–59

Gaber MM, Stahl F, Gomes JB (2014) Pocket data mining: big data on small devices. Springer, Heidelberg

Han J, Pei J, Yiwen Y (2000) Mining frequent patterns without candidate generation. In: Proceedings of ACM SIGMOD conference management of data, pp 1–12

He H, Ma Y (eds) (2013) Imbalanced learning: foundations, algorithms, and applications. Wiley, New Jersey

Hearst MA (1999) Untangling text data mining. In: Proceedings of the 37th annual meeting of the association for computational linguistics (ACL)

Hu H, Wang H, Zheng B (2011) Challenges in managing and mining large, heterogeneous data. In: Database systems for advanced applications (DASFAA) (2). Springer, Heidelberg, p 462

Jusoh S, Alfawareh HM (2012) Techniques, applications and challenging issue in text mining. Int J Comput Sci Issues 9(6):1–6 ISSN: 1694-0814

Kleinberg JM (2007) Challenges in mining social network data: processes, privacy, and paradoxes. In: Proceedings of international conference on Knowledge discovery and data mining (KDD), pp 4–5

Li X, Ng A-K, Wang JTL (eds) (2014) Biological data mining and its applications in healthcare. World Scientific Publishing, Singapore

Mahanta AK, Mazarbhuiya FA, Baruah HK (2005) Finding locally and periodically frequent sets and periodic association rules. In: Pattern recognition and machine intelligence LNCS, vol 3776. Springer, Heidelberg, pp 576–582

Memon N, Xu JJ, Hicks DL, Chen H. (eds) (2010) Data Mining for social network data. Springer, New York

Qin Z, Tang Y (2014) Uncertainty modeling for data mining

Ranade S (1991) Mass storage technologies. Information Today

Savasere A, Omiecinski E, Navathe S (1995) An efficient algorithm for mining association rules in large databases. In: Proceedings of the 21st international conference on very large data bases, pp 432–443

Stahl FT, Gaber MM, Bramer M, Yu PS (2010) Pocket Data Mining: Towards collaborative data mining in mobile computing environments. In: IEEE international conference on tools with artificial intelligence (ICTAI), vol 2 pp 323–330

Takashi W, Jun L (eds) (2012) Emerging trends in knowledge discovery and data mining. LNCS, vol 7769. Springer, Heidelberg

Wang W, Yang J (2010) Mining high-dimensional data. In: Maimon O, Rokach L (eds) Data mining and knowledge discovery handbook, Springer, New York

Xiong H, Pandey G, Steinbach M, Kumar V (2006) Enhancing data analysis with noise removal. IEEE Trans Knowl Data Eng 18(2):304–319

Yang Q, Wu X (2006) 10 challenging problems in data mining research. Int J Inf Technol Decis Mak 5(4):597–604

Index

A
ACP coding, 341
A_1 measure, 49, 52–54
A_2 measure, 53–63
Association, 52, 74, 175–177, 237–240
Association rule mining in multiple databases, 74–77, 175–177, 181–183

C
Calendar-based periodic patterns, 259
Certainty factor, 260–265
Clustering databases, 309
Clustering items, 161, 208
Clustering local frequency items, 208
Clustering technique, 161, 203, 208, 309
Coding association rules, 338, 341
Coding frequent itemsets, 325
Conditional patterns, 11

D
Database frequency, 74

E
Exceptional association rule, 135
Exceptional pattern, 129
Extended model of local pattern analysis, 178
Extreme types of association rule, 175

F
Future directions of KDD, 6

G
Generalized multi-database mining techniques, 85
Generalized notch, 103
Generator of expression, 37
Global pattern, 83, 129, 238
Grouping, 240, 244
Grouping frequent items, 244

H
Heavy association rule, 134
High-frequency association rule, 176

I
Iceberg notch, 104
Influential of an item on a set of specific items, 289
Incremental approach to mining icebergs, 109
Influence of an item, 286
Influence of an itemset on another itemset, 287
IS coding, 325
Itemset frequency, 74

L
Local pattern, 178, 305
Local pattern analysis, 84
Low-frequency rule, 176

M
Measure of association, 52, 199, 237
Measuring association among items, 52

© Springer International Publishing Switzerland 2015
A. Adhikari and J. Adhikari, *Advances in Knowledge Discovery in Databases*,
Intelligent Systems Reference Library 79, DOI 10.1007/978-3-319-13212-9

Printed in the United States
By Bookmasters